Progress in SOI Structures and Devices Operating at Extreme Conditions

NATO Science Series

A Series presenting the results of scientific meetings supported under the NATO Science Programme.

The Series is published by IOS Press, Amsterdam, and Kluwer Academic Publishers in conjunction with the NATO Scientific Affairs Division

Sub-Series

I. Life and Behavioural Sciences	IOS Press
II. Mathematics, Physics and Chemistry	Kluwer Academic Publishers
III. Computer and Systems Science	IOS Press
IV. Earth and Environmental Sciences	Kluwer Academic Publishers
V. Science and Technology Policy	IOS Press

The NATO Science Series continues the series of books published formerly as the NATO ASI Series.

The NATO Science Programme offers support for collaboration in civil science between scientists of countries of the Euro-Atlantic Partnership Council. The types of scientific meeting generally supported are "Advanced Study Institutes" and "Advanced Research Workshops", although other types of meeting are supported from time to time. The NATO Science Series collects together the results of these meetings. The meetings are co-organized bij scientists from NATO countries and scientists from NATO's Partner countries – countries of the CIS and Central and Eastern Europe.

Advanced Study Institutes are high-level tutorial courses offering in-depth study of latest advances in a field.
Advanced Research Workshops are expert meetings aimed at critical assessment of a field, and identification of directions for future action.

As a consequence of the restructuring of the NATO Science Programme in 1999, the NATO Science Series has been re-organised and there are currently Five Sub-series as noted above. Please consult the following web sites for information on previous volumes published in the Series, as well as details of earlier Sub-series.

http://www.nato.int/science
http://www.wkap.nl
http://www.iospress.nl
http://www.wtv-books.de/nato-pco.htm

Series II: Mathematics, Physics and Chemistry – Vol. 58

Progress in SOI Structures and Devices Operating at Extreme Conditions

edited by

F. Balestra

Institut de Microélectronique,
Electromagnétisme et Photonique, INPG-CNRS,
Grenoble, France

A. Nazarov

Institute of Semiconductor Physics,
National Academy of Sciences,
Kyiv, Ukraine

and

V.S. Lysenko

Institute of Semiconductor Physics,
National Academy of Sciences,
Kyiv, Ukraine

Kluwer Academic Publishers

Dordrecht / Boston / London

Published in cooperation with NATO Scientific Affairs Division

Proceedings of the NATO Advanced Research Workshop on
Progress in SOI Structures and Devices Operating at Extreme Conditions
Kyiv, Ukraine
15–20 October 2000

A C.I.P. Catalogue record for this book is available from the Library of Congress.

ISBN 1-4020-0575-X

Published by Kluwer Academic Publishers,
P.O. Box 17, 3300 AA Dordrecht, The Netherlands.

Sold and distributed in North, Central and South America
by Kluwer Academic Publishers,
101 Philip Drive, Norwell, MA 02061, U.S.A.

In all other countries, sold and distributed
by Kluwer Academic Publishers,
P.O. Box 322, 3300 AH Dordrecht, The Netherlands.

Printed on acid-free paper

Printed in the Netherlands.

TABLE of CONTENTS

PREFACE

This book contains the selected contributions of the speakers who attended the NATO Advanced Research Workshop on "Progress in Semiconductor-On-Insulator Structures and Devices Operating at Extreme Conditions", which was held in The Cultural Center "Jerelo" in Kyiv, Ukraine, on October 15-20, 2000.

One of the goal of this Workshop was to cement existing links established during the two previous NATO Silicon On Insulator Workshops held in Ukraine, during 1994 and 1998, and to develop new contacts with researchers in those countries. Another objective of the Workshop was of course to provide an international forum for discussing today's status and progress as well as tomorrow's challenges and trends in Silicon-On-Insulator technology and physics of new SOI devices operating at extreme conditions, due to the unique advantages of SOI in this important area (high and low temperatures, high-dose irradiation, low voltage, etc.).

All major SOI materials and devices were presented during the Workshop by many high-qualified scientists from the former Soviet Union and western countries. A particular attention has been paid on the innovation in material technology, the reliability of SOI operating in harsh conditions, the characterization of SOI materials and devices and the perspectives for SOI structures and circuits. In the first part of this book, dealing with material technology, the SIMOX and ELTRAN technologies, the Smart-Cut technique, the SiCOI structures and the MBE growth for the realization of advanced devices are described. The second part covers the reliability of SOI devices operating at extreme conditions. The low and high temperature operation of deep submicron MOSFETs as well as novel SOI technologies and circuits, the interest of SOI in a harsh environment and the properties of the buried oxide are examined. The third part treats the characterization of advanced SOI materials and devices. The laser-recrystallized SOI layers, ultra-short SOI MOSFETs and nanostructures, gated-diodes and SOI devices realized by various techniques are investigated. The last part of this book overviews the perspectives of SOI structures. The wafer bonding technique, the applications of oxidized porous silicon, semi-insulating silicon materials for HF circuits, the self-organization of silicon dots and wires on SOI and some new physical phenomena are analysed.

All the participants to the Workshop want to express their gratitude to the NATO ARW – International Scientific Exchange Programme, which has made the meeting possible, as well as to Ministry of Science and Education of Ukraine, National Academy of Sciences of Ukraine and leadership of Institute of Semiconductor Physics, which supported this workshop.

We also would like to thank Dr. Valeria Kilchytska for assistance in local technical and social arrangements, Dr. G. Rudko, Ja. Vovk, I. Barchuk, Dr. Tjagulskii and Dr. I. Osijuk, from Institute of Semiconductor Physics - Kyiv, for clerical and technical assistance, and Sylvie Mercier, from LPCS/IMEP-INP Grenoble, for her help in organizing this book.

Francis Balestra
Grenoble, France

Alexei N. Nazarov
Vladimir S. Lysenko
Kyiv, Ukraine

PERSPECTIVES OF SIMOX TECHNOLOGY

MARIA J. ANC
Ibis Technology Corporation
Danvers, MA 01923, USA

1. Introduction

Silicon-on-Insulator (SOI) technology has experienced significant growth during the past few years. Among the SOI fabrication methods, Separation by Implantation of Oxygen (SIMOX) technology producing thin layer SOI material was first to reach the maturity required for main stream VLSI CMOS applications [1,2]. At its early years SIMOX was believed to be expensive and inaccessible due to the necessity of utilizing specialized equipment to implant high oxygen doses and non-standard post-processing. Progress in equipment and process development [3,4,5] led to the improvements in material quality and manufacturability [6]. Successful manufacture of CMOS circuits in SIMOX stimulated growth of interest in using all SOI materials. At present, at its maturity, SIMOX is perceived to offer a relatively simple manufacturing process of thin layer SOI substrates of extremely competitive quality and cost.

2. SIMOX process

SIMOX structure forms with the range of doses at wide range of energy. Basic SIMOX process involves implantation of the oxygen doses on the order 10^{18} O^+/cm^2 and annealing at temperatures in excess of 1300 °C. Implantation of oxygen is performed at elevated temperatures to prevent amorphization of substrate during implantation, and needs to be followed by high temperature annealing in inert ambient to restore crystalline quality of the silicon layer and form high integrity buried oxide. High temperature oxidation processes may be used for material quality improvement and layer thickness adjustment [6-9]. SIMOX process is shown in a schematic view in Figure 1.

In SIMOX, layer thickness is primarily determined by the implantation energy and dose, and moderately adjusted by the annealing ambient and process schemes. SIMOX formed with the oxygen doses larger than the critical dose required to reach stoichiometric concentration of oxygen in the buried oxide layer during the implantation is called full dose or standard dose SIMOX. Typically, standard dose SIMOX process requires implantation of 1.8×10^{18} O^+/cm^2 at 180 keV - 200 keV. The stoichiometric buried oxide forms already during the implantation, as shown in Figure 2 (a). After ~5 hour anneal at temperatures >1300 °C in argon with trace of oxygen a multi layer structure consisting of approximately 180 nm of crystalline silicon on top of 370 nm thick buried oxide is formed. As shown in Figure 2 (b), the buried oxide of such SIMOX structure is composed of a thick stoichiometric SiO_2 and a characteristic silicon-rich transition layer at the bottom. SIMOX surface and interface exhibit microroughness in

F. Balestra et al. (eds.), Progress in SOI Structures and Devices Operating at Extreme Conditions, 1–10.

microscopic scale. Typical AFM image of the surface of standard dose SIMOX is shown in Figure 3.

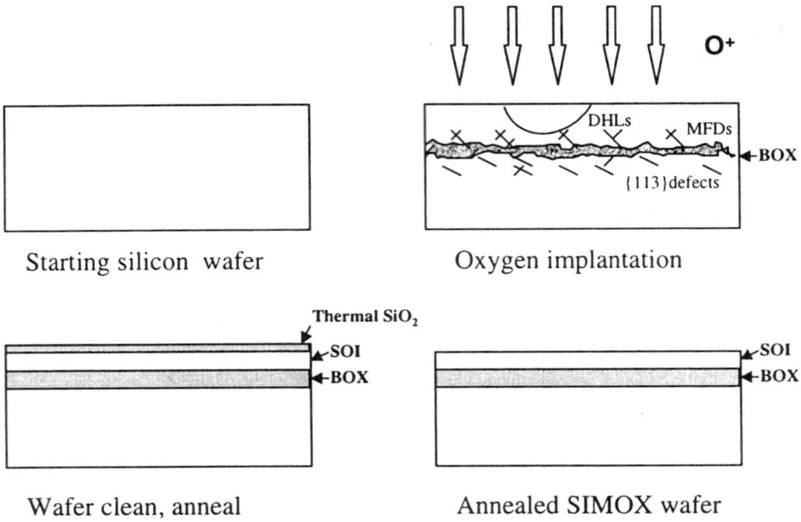

Figure 1. Schematic view of basic SIMOX process. In as-implanted material: DHLs – Dislocation Half-Loops, MFDs – Multiply-Faulted-Defects)

(a) as-implanted standard dose SIMOX

(b) annealed standard dose SIMOX

Figure 2. TEM cross section of as-implanted (a) and annealed full dose SIMOX with 370 nm thick buried oxide (b).

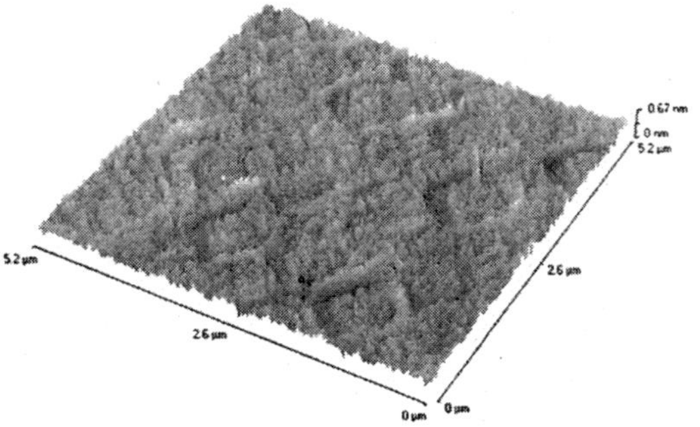

Figure 3. AFM image of the surface of standard dose SIMOX.

The SIMOX structure also forms with lower doses of oxygen. For example, a 100 nm thick buried oxide can be implanted with ~ 4 times lower dose at energy 180 keV – 200 keV [10], and the 60 nm thin buried oxide can be implanted at energy range 60 keV – 70 keV with 1/10 of the full dose [11,12]. As shown in Figure 4 (a) implantation of the sub-stoichiometric oxygen dose results in formation of the buried band of damage clusters intermixed with oxide precipitates. In this material the stoichiometry of the buried oxide is reached during the subsequent annealing processes. While in standard dose SIMOX small variations of the implanted dose do not have significant effects on material properties the deviation in the sub-stoichiometric oxygen dose implanted for low dose SIMOX induces significant changes. A discontinued buried oxide forms if the dose is too low while excess of the dose at the high end of the process window results in formation of a continuous albeit rich in silicon inclusions buried oxide. For the doses from the optimum range a high quality free of silicon inclusions buried oxide forms, as shown in Figure 4 (b).

Annealing schemes in low dose SIMOX process are more complex than in standard dose SIMOX. In general, the basic annealing process used in standard dose SIMOX does not give satisfactory electrical integrity of the thin BOX. It has been found that the annealing ambient has strong effects on the quality of the thin BOX, and annealing temperature needs to be higher (typically 1350°C) than in full dose SIMOX. In particular, high temperature oxidation process inducing Internal Thermal Oxidation (ITOX) has been found to have beneficial effects on the quality of the low dose SIMOX including integrity of the buried oxide, surface and interface microroughness [8.9]. Figure 5 shows the TEM cross section of high-energy low-dose SIMOX annealed in inert ambient (a), and in oxidizing ambient (b). Corresponding AFM images of surface and interface of these samples are shown in Figure 6.

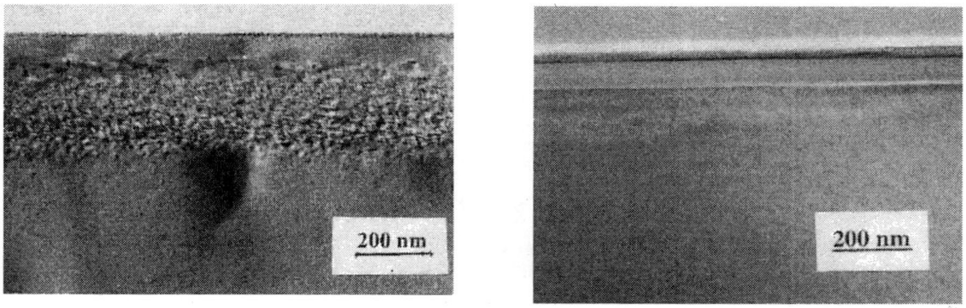

(a) as-implanted low dose SIMOX

(b) annealed low dose SIMOX

Figure 4. TEM cross section of as–implanted (a), and annealed high quality low dose SIMOX with 60 nm thick buried oxide (b).

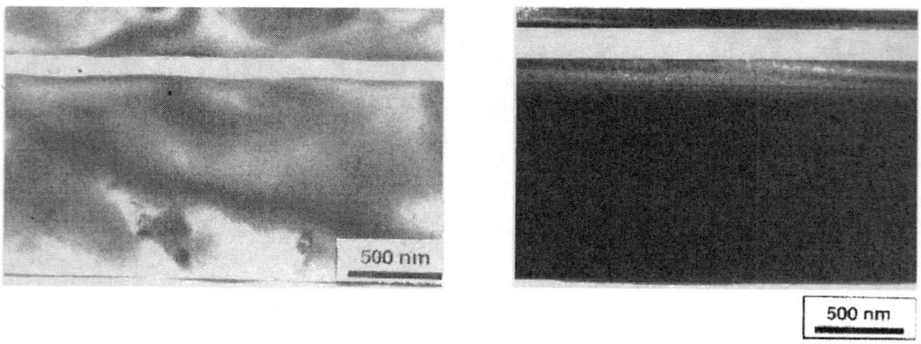

(a) annealed in inert ambient

(b) annealed with ITOX

Figure 5. TEM cross section of high-energy low-dose SIMOX annealed in inert ambient (a) and annealed with ITOX (b).

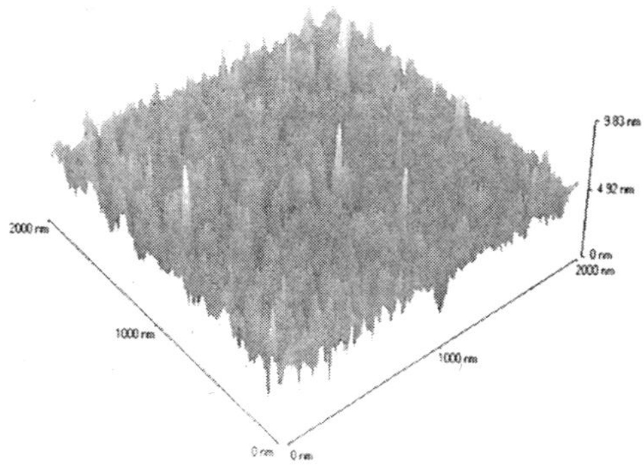

Silicon surface, RMS=0.72 nm
inert ambient anneal Ar(<1%O_2)

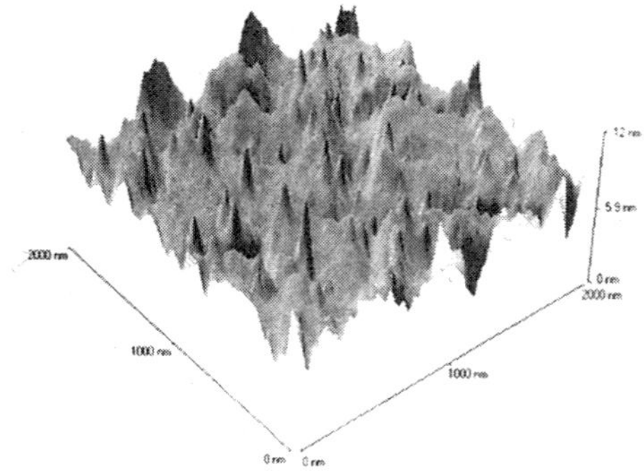

Silicon/BOX interface, RMS=1.2 nm
inert ambient anneal Ar(<1%O_2)

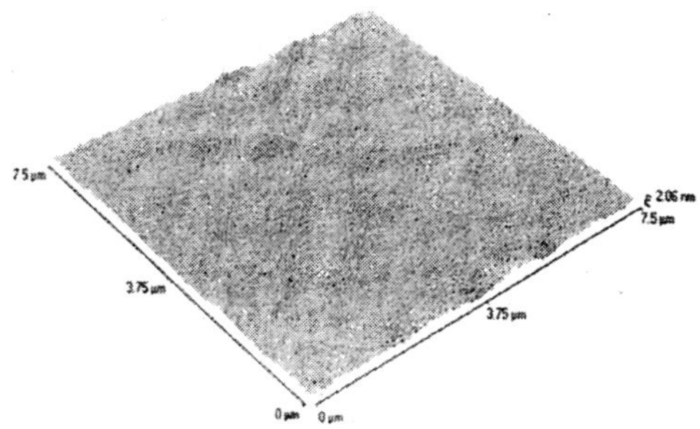

Silicon surface, RMS=0.25 nm
anneal with ITOX 30%O_2

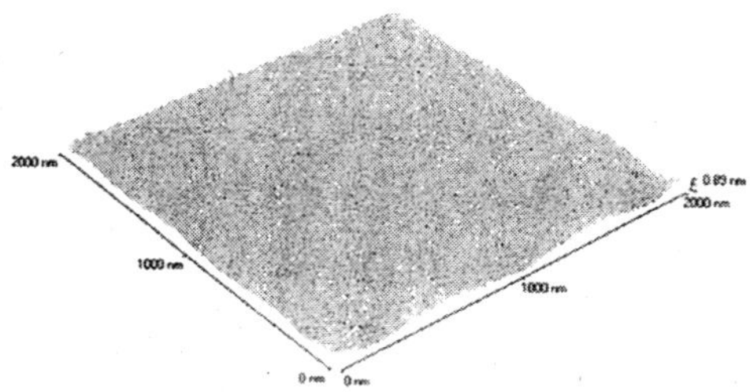

Silicon/BOX interface, RMS=0.1 nm
anneal with ITOX , 30%O_2

Figure 6. AFM images of surface and interface microroughness of high-energy low-dose SIMOX.

In addition to the above fundamental processes, several new process schemes have been developed for thin BOX fabrication. Among them damage induced buried oxide [13,14] and thin BOX SIMOX obtained by multiple energy implant in the energy regime below the one of standard high energy SIMOX level [15] have shown competitive characteristics and performance in circuit applications [2,16].

3. Characterization methods

SIMOX properties have been extensively characterized with variety of methods. Techniques such as Transmission Electron Microscopy (TEM), Spectroscopic Ellipsometry (SE), Optical Reflectometry, Atomic Force Microscopy (AFM), Secondary Ion Mass Spectroscopy (SIMS), Total Fluorescence X-ray Spectroscopy (TXRF), Inductively Coupled Plasma Mass Spectroscopy (ICPMS) and Surface Photovoltage (SPV) are routinely used to evaluate microstructure, layer uniformity and contamination level in SIMOX wafers. Defect density is examined with decorative defect etches including enhanced Secco etch [17] and HF immersion [18]. Buried oxide short density and leakage characteristics are routinely measured using electrochemical analysis method utilizing copper sulfate water solution [19]. This simple method found wide application in SIMOX characterization due to its accessibility, versatility, quick turn-around and adequacy in extracting details of the SIMOX BOX conduction. SOI structures can be thoroughly characterized with a wide range of device characterization methods. Specifically applicable to the SOI substrates electrical characterization methods are described in details in [20].

4. Defects in SIMOX

All silicon wafers contain certain level of defects whose nature and density depend upon crystal growth conditions and thermal history of the wafer in subsequent processing. SOI fabrication techniques introduce their own category of defects, some being common and some being specific to the method of fabrication [21,22].

4.1 SOI DEFECTS

In the TEM cross section view (method with the detection limit of $\sim 10^6/cm^2$) a contemporary SIMOX material exhibits perfectly crystalline top silicon layer, as shown in Figures 2 (b) and Figure 3(b). If larger area of the wafer is analyzed, both types of SIMOX, standard dose and low dose SIMOX, exhibit common defects, some - originating from the formation mechanisms and some – from the manufacturing issues. Silicon threading dislocations (through-layer dislocations or TD-defects) and pyramidal stacking faults evolve during the high temperature annealing cycle from the defects formed during oxygen implantation, and are found in various densities in SIMOX [23]. In early developments TD-defects constituted the dominating imperfection in SIMOX silicon layer and the observed density was found to increase exponentially with the increase of the implanted dose (increase of the buried oxide thickness). Over the decade of efforts the silicon defect density in SIMOX has been reduced by orders of magnitude due to the improvement in equipment design, process control and development of new processes. The dislocation density has been found to be dependent on substrate temperature during the implantation and also on process scheme. It has been proven that multiple implant/annealing process schemes result in reduced dislocation densities compared to the substrates with the same BOX thickness but fabricated in single step processes. Detailed characterization studies of low dose SIMOX have shown that the silicon defect density actually decreases with increasing thickness of the BOX within certain dose range [6,23].

Implementation of SIMOX substrate into the IC manufacturing has changed the understanding of defects and their role in SOI. In contemporary SIMOX products fabricated with optimized processes silicon dislocations occur in a density range from 10^2 /cm^2 to 10^6 /cm^2 depending upon the dose and process type. The upper limit of acceptable dislocation density in SIMOX is assumed to be somewhere at $5x10^6$ /cm^2 at presently fabricated circuits, although the exact correlation data has not been published. This limit may become more stringent in more demanding applications in the future. Of high importance is a fact that in present generation of SIMOX material, defects appearing in much lower density have been found to have a detrimental effect on SOI circuit yield. These defects may be related to the particles on the surface, crystal grown-in defects or metallic decoration of any defects (including dislocations) in top silicon layer. In particular, some of these defects can be delineated with concentrated HF and their density may not exceed 1/cm^2 to yield the circuit. In applications using thin buried oxide SIMOX wafers the density of silicon inclusions in the BOX may become a critical factor while in ultra thin layer applications (fully-depleted SOI) interfacial properties may affect the yield.

4.2 BURIED OXIDE DEFECTS

Properties of the buried oxide are strongly related to its microstructure and to the SIMOX process conditions, which include implant parameters and thermal history with all insides of these technological steps.

Pinholes and enhanced background conduction used to be the primary concerns in electrical performance of the buried oxide. Structural defects, distinct, highly conductive sites (1uA at < 0.5MV/cm) called pinholes or pipes have been shown to be virtual silicon bridges in the buried oxide or sites with thinned oxide around the silicon inclusions. The origin of pipes was found in large particles residing on the surface of the wafer during implantation and screening the ion beam. After significant efforts in particles reduction the pinhole density is typically below 0.2/cm^2 in contemporary SIMOX.

Enhanced background conduction called sometimes bulk-and-defect conduction is also related to the microstructure of the buried oxide, however in a different scale. The microstructure of the buried oxide has been found to be silicon enriched by presence of crystalline and non-crystalline silicon inclusions [24,25]. The silicon inclusions can be present in the buried oxide in a form of elongated islands with longer axis aligned with the {100} plane and the side walls bound by {111} crystallographic planes, or having a shape of truncated octahedrals. These inclusions are crystalline in nature with the orientation identical or close to the main crystal. In standard dose SIMOX they are typically found in the transition region near the buried oxide / substrate interface and occupy not more than 7% of the total BOX thickness in contemporary material. Their impact in commercial applications is insignificant. In contrast, in low dose SIMOX, silicon inclusions can be found across the entire thickness of the buried oxide in various densities, and in non-optimized processes may have inferior effect on the BOX integrity.

The photo-injection studies detected presence of silicon inclusions in the BOX in a form of amorphous silicon clusters that exhibit amphoteric character under UV illumination [24,26]. These defects are characterized by relatively large electron capture cross section compared to the defects not sensitive to the UV radiation, and are believed to be the main source of increased background conduction through the buried oxide and electron trapping. Their density strongly depends on process conditions. Embedded in the amorphous SiO_2 matrix they can not be revealed by the TEM.

Due to its confinement, similarly as other SOI buried oxides, the SIMOX BOX exhibits higher density of E' centers and higher hydrogen uptake than superficial thermal oxides. The specific network defects, highly strained Si-O bonds originating from the formation and high temperature processing of the confined oxide structure manifest themselves by hole trapping and oxide charge [24,27]. These effects are important in rad hard applications.

The interface between silicon and buried oxide has been shown to be atomically smooth in SIMOX. The microscopic roughness of the surface and interface varies upon the process conditions, especially in low dose SIMOX. The representative AFM images are shown in Figures 4 and 6. Since the interface becomes a working part of the devices in fully depleted circuits its characteristics are important.

As shown above, SIMOX buried oxide contains specific categories of defects. Studies of the nature of these defects allowed to improve understanding of their origin and reduce their density such that SIMOX substrates can be used in present generation of integrated circuits.

4.3 CURRENT SIMOX MATERIAL CHARACTERISTICS

For main stream ULSI applications SIMOX wafers need to meet the criteria set for substrates by the SIA technology roadmap [28]. In SOI, the SIA guidelines pertain to the uniformity of the silicon layer, silicon defect density, buried oxide short density, surface and interface microroughness. Typical parameters of present generation of commercial SIMOX wafers are summarized in Table 1.

Table 1. Typical parameters of SIMOX wafers.

Parameter	Units	Standard dose SIMOX	Medium dose SIMOX	Low dose SIMOX	Characterization method
Silicon layer thicknesses	nm	190	190	50	Spectroscopic Ellipsometry
Silicon layer uniformity	nm	+/-5	+/-5	+/-2	Spectroscopic Ellipsometry
Buried oxide layer thickness	nm	375	115 - 150	60	Spectroscopic Ellipsometry
Buried oxide layer uniformity	nm	+/-10	+/-5	+/-2	Spectroscopic Ellipsometry
Particles (Local Light Scatters)	cm^{-2}	0.18	0.03 - 0.14	0.03 - 0.14	Scatterometry
Crystalline Defects (Threading Dislocations or Stacking Faults)	cm^{-2}	$<10^7$	$<10^5$	$<10^5$	Enhanced Secco etch
Defects in the Si ("HF Defects")	cm^{-2}	0.55	0.22		HF immersion
Defects in the BOX (Shorts)	cm^{-2}	<0.1	<0.2	<2	Electrolytic Cu plating
Micro-Roughness Rms (Surface)	nm	0.7	0.5	0.2	Atomic Force Microscopy
Micro-Roughness Rms (Interface)	nm	<2.5	<2.5	0.5	Atomic Force Microscopy
Surface Metallic Contamination	at/cm^2	$<10^{11}$	$<10^{11}$	$<10^{11}$	Total X-Ray Fluorescence Spectrometry

5. Equipment and manufacture of SIMOX

Manufacture of SIMOX requires specialized high current oxygen implanters. The contemporary generation of oxygen implanters [3] is equipped with beamline design allowing to minimize variation of implant angle across the large diameter wafer, and independent beam current and substrate temperature control. High powered lamps heating substrate during the implantation enable implantation at lower energies, thus allowing choice of implant conditions, process optimization and dose reduction. Wafer holders and all materials used in the beamline must be compatible with silicon-silicon dioxide system to prevent particulate generation and metallic contamination of the wafers. Uniformity and reproducibility of the oxygen dose, automation in wafer handling and machine operation are primary yield factors in implantation processes in volume manufacture.

Annealing of SIMOX is performed in horizontal or vertical furnaces with high purity SiC and/or polysilicon hardware capable to sustain long term exposure to temperatures in excess of 1300 °C in inert and/or oxidizing ambient. With increase of wafer diameter consideration is needed to slip generation in thermal processes and appropriate hardware design. For quality assurance of SIMOX wafers contamination control in both, ion implantation and annealing processes, is equally important.

6. Perspectives for the future

Major technical limitation in SIMOX technology is related to the restricted range of implant energy and doses producing high quality SOI material with certain range of layer thickness (maximum 400 nm thick BOX). The strength of this technology, however, is in high uniformity of thin film layers attainable in large diameter wafers and feasible in 300 mm substrates. SIMOX process flow involves only few operations adding an advantage of simplicity and higher yield in high volume manufacturing.

For main stream ULSI applications SIMOX materials need to meet the criteria set for substrates by the SIA technology roadmap. In SOI, the SIA guidelines pertain to the uniformity of the silicon layer, silicon defect density, buried oxide short density, surface and interface microroughness, and provide major driving force for material quality improvement.

For large volume supply, the cost of fabrication needs to complement high quality material performance. Cost reduction in SIMOX technology is presently perceived as a reduction of implant time. Increasing throughput of the implanters can be achieved by increasing the beam current, reducing the implanted doses (low dose processes) and process automation. Development of manufacturable low dose processes, optimization of high temperature annealing process and equipment for large diameter wafers, and implementation of in-situ process characterization methods may provide a significant contribution to yield enhancement and cost reduction in a long term.

In summary, SIMOX technology has been successfully developed to meet the requirements of present generation of CMOS circuits. Future and long-term success can be envisaged by continuous efforts toward proficiency in equipment design, material development, effective cost reduction and consistency in manufacture of high quality and low cost SIMOX.

7. References

1. G. Shahidi, A. Ajmera, F. Assederaghi, R. Bolam, A. Bryant, M. Coffey, H. Hovel, J. Lasky, E. Leobandung, H.S. Lo, M. Maloney, D. Moy, W. Raush, D. Sadana, D. Schepis, M. Sherony, J. Sleight, L.F. Wagner, K. Wu, B. Davari, T.C. Chen (1999) Mainstreaming of the SOI Technology. *Proc. 1999 Int. IEEE/SOI Conf.* ISBN 0-7803-5456-7, 1-4.
2. E. Leobandung, E. Barth, M. Sherony, S.H. Lo, R. Schultz, W. Chu, M. Khare, D. Sadana, D. Shepis, R. Bolam, J. Sleight, F. White, F. Assaderaghi, D. My, G. Biery, R. Golfblatt, T.C. Chen, B. Davari, G. Shahidi (1999) High Performance 0.18 um SOI CMOS Technology. in *IEDM Tech, Digest*, ISBN 0-7803-5410-9, IEDM99-679-682.
3. G. Ryding, T.H. Smick, M. Farley, B.F. Cordts, R.P. Dolan, L.P. Allen, B. Mathews, W. Wray, B. Amundsen, M.J. Anc (1996) The Ibis 1000 SIMOX Production Implanter. *Proc. Int Conf. on Ion Implantation Technology, Austin, TX.*
4. *The Electrochem. Soc. Proc. Series* vol PV99-3, ISBN 1-56677-225-7, (1999)
5. *Proc. 2000 Int. IEEE/SOI Conf.* ISBN 0-7803-6389-2, (2000)
6. For review: M.J. Anc, D.K. Sadana (1999) Technology and Properties of SIMOX. in Properties of Crystalline Silicon. *EMIS Datareviews Series, Eds. R. Hull, S.S. Iyer, Chapter 18.9.*
7. S. Nakashima, N. Ohwada, T. Katayama, Y. Miyamura, A. Matsuzaki, M. Imai, K. Izumi (1994) Thickness increment of buried oxide in a SIMOX wafer by high temperature oxidation. *Proc. Int. IEEE/SOI Conf.*, ISBN 0-7803-2406-4,71-72
8. T. Katayama, Y. Miyamura, M.Kataoka, M. Danbata, M. Imai, S. Nakashima N. Ohwada (1994) *ibid.*, 75-76.
9. A. Matsumura, K. Kawamura, T. Mizutani, S. Takayama, I. Hamaguchi, Y. Nagatake (1999) Recent progress in low-dose SIMOX wafers fabricated with Internal Thermal Oxidation (ITOX) process. *The Electrochem. Soc. Proc. Series* vol PV99-3, ISBN 1-56677-225-7, 79-92.
10. S. Nakashima, K. Izumi *J. Mat. Res.* Vol. 5, No.9, Sept.1990, 1918-1922.
11. Y. Li, J.A. Kilner, R.J. Chater, P.L.F. Hemment, A. Nejim, A.K. Robinson, K.J. Reeson, C.D. Marsh, G.R. Booker (1992) The Effects of Dose and Target Temperature on the Low Energy (70 keV) SIMOX layers. in *Silicon-on-Insulator Technology and Devices /1992*, K. Izumi Editor, PV92-13, 368-387.
12. M.J. Anc, R.P. Dolan, J. Jiao, T. Nakai (1999) Thin-layer SIMOX for Future Applications. *Proc.1999 Int. IEEE/SOI Conf.*, ISBN 0-7803-5456-7, 106-107.
13. O.W. Holland et al. (1996) *Appl. Phys. Lett.* **69**, 674-676.
14. L. Chen, S. Bagchi, S.J. Krause, P. Roitman (1999) Effect of Single-Step, High-Oxygen-Concentration Annealing on Buried Oxide Layer Microstructure in Post-Implant-Amorphized, Low-Dose SIMOX Material. *Proc. 1999 Int. IEEE/SOI Conf.*, ISBN 0-7803-5456-7, 123-124.
15. R.P. Dolan, M. Alles, M. Anc, B. Cordts, J. Dunne, M. Gindlesperger, W. Hornblower, T.Y. Yang, M. Powell, J. Blake, T. Nakai (2000) Medium-Dose SIMOX Quality Improvement for Advanced CMOS Applications. in *Proc. Int. 2000 IEEE/SOI Conference*, ISBN 0-7803-6389-2, 46-47.
16. S.T. Liu, L.P. Allen, M.J. Anc, W.C. Jenkins, H.L.Huges, M.E. Twigg, R.K.Lawrence (1997) Reduction of Radiation-induced Back-channel Threshold-voltage Shifts in Partially-depleted SIMOX Devices by Using ADVANTOX Substrates. *IEEE Trans. on Nucl. Science*, **44**, (6), 2101-2105.
17. J. Margail, J.M. Lamure, J. Stoemenos, A.M. Papon (1992) Defects in SIMOX structures: Characterization and Some Formation Mechanisms. in *The Electrochem. Soc. Proc. Series* PV 92-13, 207-220.
18. D.K. Sadana (1996) Invisible defects in SOI, their process dependence and characterization. in *The Electrochem. Soc. Series* PV 96-3, 3-17.
19. L.P. Allen, M.J. Anc, M. Duffy, J.H. Parechanian, J.H. Yap (1996) Electrochemical Analysis of SIMOX buried oxides. *The Electrochem. Soc. Series* PV 96-3, p.18-27.
20. S. Cristoloveanu, S.S Li (1995) *Electrical Characterization of Silicon-on-Insulator Materials and Devices.* Kluwer Academic Publishers, Boston.
21. W.P. Maszara, R. Dockerty, C. Gondran, P.K. Vasudev (1997) SOI materials for mainstream SOI technology. in *The Electrochem. Soc. Proc. Series* vol PV97-23, 15-26.
22. M.A. Mendicino, (1999) Comparison of properties of available SOI materials. Eds. R. Hull, S.S. Iyer, *Properties of Crystalline Silicon* EMIS Datareviews Series, Chapter 18.10.
23. For review: S. Krause. M. Anc, P. Roitman (1998) Evolution and Future Trends of SIMOX material. *MRS Bulletin*, **23**, No.12, 25-29 and references therein.
24. A.G. Revesz, H.L. Hughes (1997) Properties of buried oxide layer in SIMOX. *Microelectr. Eng.* **36**, 343
25. V.V. Afanas'ev, A. Stesmans, A.G. Revesz, H.L. Hughes (1997) Structural inhomogenity and silicon enrichment of buried SiO_2 layers formed by oxygen implantation in silicon. *J. Appl. Phys* **82**(5), 2184 (1997)
26. V.V. Afanas'ev, A.G. Revesz, G.A. Brown, H.L. Hughes (1994) Deep and Shallow Electron Trapping in the Buried Oxide Layer of SIMOX Structures. *J. Electrochem. Soc.* **141** (10), p.2801.
27. V.V. Afanas'ev, A.G. Revesz, H.L. Hughes (1996) Confinement Phenomena in Buried Oxides of SIMOX Structures as Affected by Porcessing. *J. Electrochem. Soc.* **143** (2) , 695
28. Semiconductor Industry Association International Technology Roadmap for Semiconductors, 1999.

MBE GROWTH OF THE TOP LAYER IN Si/YSZ/Si STRUCTURE

V.G. **BESHENKOV**, V.A. **MARCHENKO**, A.G. **ZNAMENSKII**,
A.F. **VYATKIN**, V.S. **AVRUTIN**, N.F. **IZYUMSKAYA**,
O.A. **SOLTANOVICH**, +A.G. **AKIMOV**, and +L.P. **KAZANSKII**
Institute of Microelectronics Technology, Russian Academy of Sciences, Chernogolovka, 142432 Russia
+Institute of Physical Chemistry, Russian Academy of Sciences, Moscow, 117915 Russia

Yttria-stabilized zirconia (YSZ) films were used as insulator layers in silicon-on-insulator structures. YSZ films on Si substrates were grown epitaxially by "off-axis" rf magnetron sputtering of a single-crystal YSZ target. Then Si films were grown on YSZ by molecular beam epitaxy. Current–voltage characteristics of the Si/YSZ/Si(100) and YSZ/Si(100) structures were measured, and breakdown voltages and electric fields were determined. It was found that a thin SiO_2 layer between the YSZ film and the Si substrate prepared by anodic-plasma oxidation of the Si substrate through the grown YSZ film improves crucially the breakdown voltage of the structures.

1. Introduction

Interest in silicon-on-insulator (SOI) technology was aroused due to the need for developing three-dimensional large scale integrated circuits, high-speed devices, and devices operating under extreme conditions. The use of yttria-stabilized zirconia (YSZ) as an insulator layer for SOI structures shows a promise, since this approach has such advantages as its compatibility to the standard silicon technology and a possibility to grow epitaxial Si/YSZ/Si structures at rather low temperatures. In this work, we focus on optimization of growth conditions of Si/YSZ/Si(100) heterostructures and examination of dielectric properties of YSZ layers.

2. Experimental

Epitaxial YSZ films on Si(100) structures were prepared by "off-axis" RF magnetron sputtering of a single-crystal YSZ target (12 mol % Y_2O_3). As was found previously [1], thin SiO_2 layer on Si substrate is required for epitaxial grown of YSZ films on Si. In this study, substrates of two types were used: with natural and artificial chemical silicon oxides. The thicknesses of the oxide layers were 2–3 and 1–2 nm, respectively. The growth was carried out in two steps. At the first stage, thin (0.6 nm) layer of the YSZ

F. Balestra et al. (eds.), Progress in SOI Structures and Devices Operating at Extreme Conditions, 11–15.

alloy was deposited on the substrate in an oxygen-free (Ar) atmosphere at about 800°C. At this stage, metal atoms interacted with SiO_2 layer according to the reaction $Zr + SiO_2 \rightarrow ZrO_2 + ZrSi_2$. At the second stage, YSZ films were grown by sputtering of the YSZ target in an $Ar+O_2$ atmosphere at a substrate temperature of 780–800°C, Ar pressure of 5 Pa, and a partial oxygen pressure of $3x10^{-2}$ Pa [2]. Oxidation of the silicide presumably occurs at the beginning of this stage. However, further investigations are required to elucidate the mechanism of YSZ epitaxial growth on Si. The grown structures were examined by Auger spectroscopy, reflection high-energy electron diffraction (RHEED), and scanning electron microscopy (SEM). We found that the YSZ films grown on Si substrate with natural oxide contained inhomogeneities with characteristic sizes of several microns. We presumed that these defects formed as a result of interaction of Zr or Y atoms with Si substrate because the natural oxide is nonuniform in thickness. To the contrary, the use of Si substrates covered with chemically-grown oxide drastically decreased the concentration of these defects, and the YSZ films showed high structural perfection (Figure 1a). Therefore, the YSZ films (40 and 80 nm thick) grown on Si substrate with chemically-grown oxide were used in further experiments.

The YSZ/Si(100) structures thus prepared were transferred into a growth chamber of a molecular beam epitaxy (MBE) machine. Prior to growth, the YSZ/Si(100) substrates were outgassed in ultra high vacuum ($\sim10^{-10}$ Torr) at a temperature of 550–580°C for 2–3 h, and then heated to about 1000°C for 2 min. Note that longer heating time led to the YSZ film evaporation. Si films 100–300 nm thick were grown at a rate of 1.5–2 nm/min. The pressure of residual gases during the growth was $<10^{-8}$ Torr. To optimize Si-film crystal perfection and film-surface smoothness, we varied the substrate temperature in the range from 530 to 760°C. RHEED studies revealed that all the Si/YSZ/Si(100) films were single-crystal (Figure 1b). The film smoothness improved with decreasing the growth temperature.

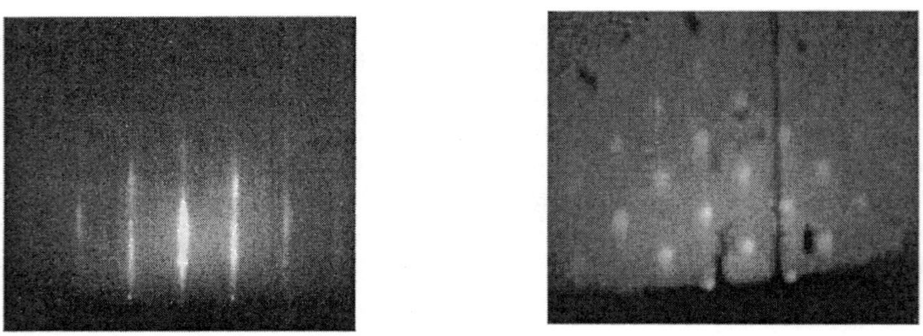

a b

Figure 1. RHEED patterns of (a) YSZ/Si(100) and (b) Si/YSZ/Si(100) structures measured at beam energies of 50 and 75 keV, respectively.

To examine dielectric properties of the YSZ films, current–voltage characteristics of the YSZ/Si(100) and Si/YSZ/Si(100) structures were measured. Contacts to the Si substrates were made from In-Ga eutectic alloy applied to the backside. Contacts to the top layers of the structures were of two types: a thin Au wire applied to the surface and 1-mm^2 Al dot evaporated on the surface through a Ta mask.

3. Results and Discussion

We found no difference in breakdown voltages between YSZ/Si(100) and Si/YSZ/Si(100) structures. As an example, Figure 2 shows the current–voltage characteristic of the YSZ/Si(100) structure. As seen from the figure, breakdown voltage is not high enough for device application. To improve the breakdown voltage, we prepared structures containing SiO_2 layer located between the YSZ film and the Si substrate. Our previous investigations revealed that high ionic conductivity of YSZ makes it possible to perform anodic-plasma oxidation of Si substrate through the grown YSZ film by applying positive voltage to the substrate (about 3 V), which provides a transport of negative oxygen ions through the YSZ film to the Si surface [1]. Thus prepared YSZ/SiO_2/Si(100) structures were compared with those prepared by heat treatment of the YSZ/Si(100) structures structures in air at 1000°C for 20 min. With both methods, the upper YSZ films remained perfect with smooth surfaces. The SiO_2 layers were 80 nm thick. Figure 3 shows the current–voltage characteristic of the YSZ/SiO_2/Si(100) structure with a SiO_2 layer prepared by anodic-plasma oxidation. One can see that the SiO_2 layer formed between the YSZ film and the Si substrate markedly improve dielectric properties of the structure. The table summarizes the results of the analysis of the current–voltage characteristics. Note that the dielectric characteristics measured with the use of 1-mm^2 Al-film contacts are worse than those measured using point Au-wire contacts on the same structure (see Table). This is explained by the presence of defects in the YSZ films, which offer channels for electrical current.

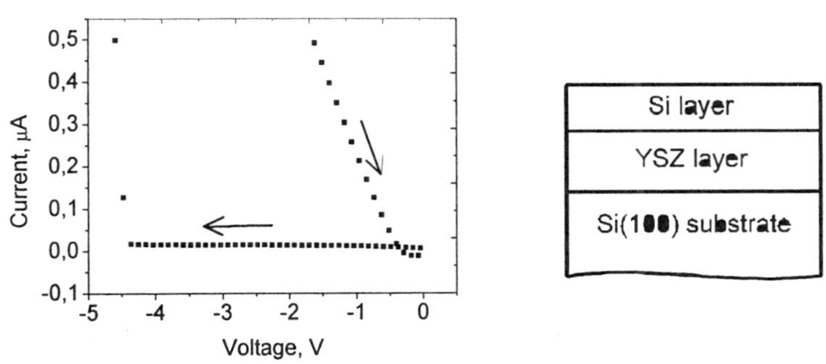

a b

Figure 2. (a) Current–voltage characteristic and (b) schematic sketch of a Si/YSZ/Si(100) structure. Evaporated Al dot was used as a contact to the top surface of the structure.

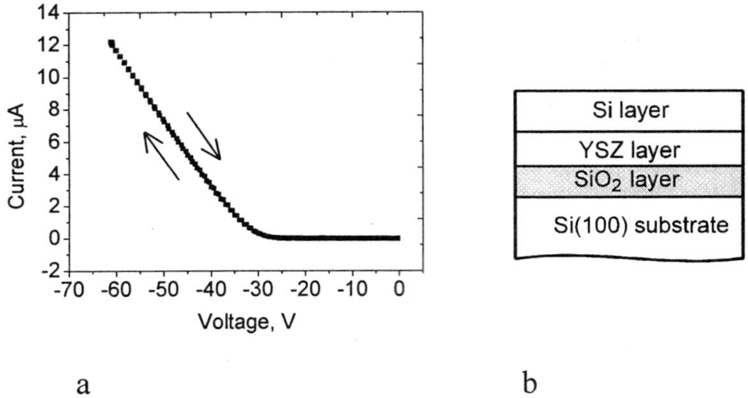

Figure 3. (a) Current–voltage characteristic and (b) schematic sketch of a $Si/YSZ/SiO_2/Si(100)$ structure. Evaporated Al dot was used as a contact to the top surface of the structure.

Table. Electrical characteristics of YSZ/Si(100) and $YSZ/SiO_2/Si(100)$ structures

Structure	**Electrode**	**Total thickness, nm**	**Breakdown voltage, V**	**Breakdown electric field,V/cm**	**Leakage resistance determined from flat portions of I–V characteristics, GΩ**
YSZ/Si	Au wire	40	16	$4.0x10^6$	30
YSZ/Si	Al film	40	5	$1.25x10^6$	2
$YSZ/SiO_2/Si$	Au wire	120	80	$6.67x10^6$	30
$YSZ/SiO_2/Si$	Al film	120	30	$2.5x10^6$	10

Figure 4 presents a SEM micrograph of the YSZ film containing such defects. To our opinion, these defects form due to surface contamination of the Si substrate or/and as a result of formation of silicide grains owing to the above-mentioned interaction of Zr or Y atoms with Si substrate at the initial stage of the YSZ deposition. The formation of the SiO_2 layer between the YSZ film and the Si substrate presumably isolates the defect YSZ layer from the substrate, improving the dielectric properties of the structures. The nature and origin of these defects are under investigation.

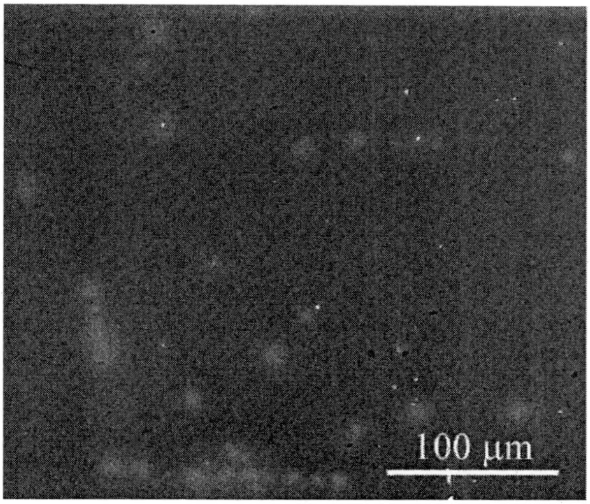

Figure 4. SEM micrograph of the surface of a YSZ/Si(100) structure. Defects are clearly seen in the image.

4. Conclusion

In summary, epitaxial Si/YSZ/Si(100) structures were grown to produce SOI structures, and their dielectric properties were studied. It was found that a thin SiO_2 layer between the YSZ film and the Si substrate prepared by anodic-plasma oxidation of the Si substrate through the grown YSZ film crucially improves the breakdown voltage of the structures.

5. Acknowledgment

We thank E.B. Yakimov for the assistance in SEM studies.
This work was supported by Ministry of Science and Technology, project no. 204-3(00)-P

6. References

1. Beshenkov, V.G., Znamenskii, A.G., and Marchenko, V.A. (1998) Perfect Fianit Films (ZrO_2+Y_2O_3) on Silicon (100). *Surface Investigation*, **14,** 61-65.
2. Beshenkov, V.G., Marchenko, V.A., and Znamenskii, A.G. (1994) Silicon Anodic-Plasma Oxidation During Magnetron Sputtering of Yttria-Stabilized Zirconia. *Appl. Phys. Lett.* **65,** 156-158.

SiCOI STRUCTURES. TECHNOLOGY AND CHARACTERIZATION

C. Serre,[1] A. Peréz-Rodríguez,[1] A. Romano-Rodríguez,[1] J.R. Morante,[1] J. Esteve,[2] M.C. Acero,[2] R. Kögler[3] and W. Skorupa[3].
[1]EME, Dept. Electrònica, Universitat de Barcelona, Martí i Franquès 1, 08028 Barcelona, Spain.
[2]Centre Nacional de Microelectrònica CNM-CSIC, Campus UAB, 08193 Bellaterra, Spain.
[3]Forschungszentrum Rossendorf, PF 510119, D-01314 Dresden, Germany.

Abstract

In this work, we present an alternative approach for the fabrication of SiCOI material based on the ion beam synthesis technique. In this process, β-SiC layers are synthesized by a multiple high dose C^+ implantation designed to form a buried stoichiometric flat carbon profile. The structural characterization confirms the formation of β-SiC, while the micromechanical assessment confirms the ability of these layers for MEMS applications. Combining IBS with wafer bonding, high crystalline quality ion beam synthesized β-SiC layers with low residual strain, have been successfully transferred onto oxidized Si wafers, obtaining SiCOI structure with abrupt SiC/SiO_2 interfaces and low surface roughness

On the other hand, the high flexibility of IBS has allowed the fabrication of poly-crystalline SiC on insulator by direct conversion of amorphous and polycrystalline Si on SiO_2 layers. The structural characterization indicates that in this case the structure of the poly-SiC layers is mainly determined by the implant temperature rather than by the topotactic transformation usually invoked in the case of crystalline materials. It is also found that when implanting pre-doped layers, the presence of interstitial P atoms leads to a significant deformation of the lattice, whereas activated P atoms in substitutional sites during the carbon implantation does not cause significant additional stress in the final poly-SiC layers, although neither does it seem to have any stress reduction effect. The micromechanical structures made from these samples suggest the viability of the direct IBS of poly-SiCOI structures for MEMS applications.

1. Introduction. State of the art.

SiC constitutes nowadays the most interesting semiconductor for the fabrication of electronic and sensor devices for harsh environment conditions, such as high temperature, mechanical stress, high radiation ambient, high electric field or corrosive ambient. This is related to the unique combination of electronic (wide band-gap, high

F. Balestra et al. (eds.), Progress in SOI Structures and Devices Operating at Extreme Conditions, 17–29.

breakdown field, high saturation carrier velocity and high mobility), chemical (extreme inertness) and mechanical (high stiffness and hardness) properties of SiC. The strong capabilities of this material for applications under harsh environment conditions are enhanced when using SiC on Insulator (SiCOI) structures. In this case, the buried dielectric layer in the SiCOI structure provides electrical isolation from the substrate, as well as a convenient sacrificial layer for the design and fabrication of sensor devices and micro-electro-mechanical systems (MEMS), combining operation under harsh environment capabilities with easy micromachining processing and possible use of low cost and large size available Silicon on Insulator (SOI) or Si wafers.

Various technologies are proposed for the fabrication of SiCOI structures. One of them is an extension of the well-known SIMOX technique to SiC [1]. A buried oxide layer (BOX) is synthesized by O^+ ion implantation at 650°C, so that the SiC overlayer contains minimal damage owing to in-situ annealing during the high temperature implantation. However, high defect concentrations are observed in regions adjacent to the BOX, which is an important drawback, taking into account the very high stability of ion implantation induced damage in the SiC lattice. Also at high doses, excess C atoms are expelled on both sides of the BOX, leading to compositional fluctuations, with accumulations of sp^2-self-bonded C atoms in the BOX close to the interfaces SiC/BOX. Another approach consists in growing SiC onto SOI [2]. In this case, the ideal situation would be to convert completely the Si overlayer (SOL) into SiC, which requires ultra-thin SOL. Moreover, it has been stated that very thin films should, in theory, be able to act as "compliant substrates" to seed heteroepitaxial deposition of materials with some lattice mismatch. In the case of SOI substrates, after thinning of the SOL, the low viscosity of the oxide at the SiC deposition temperatures is anticipated to provide isolation between the thin Si compliant film and the underlying Si substrate [3]. However, chemical vapor deposition (CVD) growth of SiC at the usual temperatures of about 1300°C on SOI results in the formation of very large cavities in the silicon overlayer (SOL) which are extended in the buried oxide layer (BOX). The cavities are formed due to Si migration from the SOL towards the surface in order to form SiC. The use of a slow heating ramp in the pre-carbonization step was found to totally avoid cavities, although, the SOL thickness remains almost unchanged, i.e. almost nothing of the SOL is converted into SiC [4]. Using reduced temperatures (1200°C) can preserve the integrity of the BOX. However, the SOL becomes the only Si supplier for the formation of SiC during the carbonization process, whereas the BOX acts as a barrier for Si migration from the substrate. Therefore the cavities are extended laterally through the Si-overlayer, and it appears that the SiC layer becomes more and more defective when the SOL thickness decreases and even delaminates if the SOL thickness reduces too much [5]. In addition, significant redistribution of Si occurs at the edges of the cavities due to the poor wetting of Si on SiO_2 at these high temperatures. In contrast, the improved wetting behavior of Si on silicon nitride decreases the tendency of Si to redistribute and to form cavities. According to this, the synthesis of a thin SiN layer at the Si/SiO_2 interface has been reported for the improvement of the structural quality of the SiC layer and reduction of voids at the bottom SiC interface [6].

Most of these problems can be solved by wafer-bonding and etch-back based technologies. Hetero-epitaxial β-SiC layers grown by CVD processes on Si substrates have been successfully transferred to insulating substrates such as oxidized silicon or sapphire wafers [7,8]. Since surface roughness is a critical factor for successful bonding, a previous chemical-mechanical polishing is necessary. After etching of the Si substrate, the final SiC exposed surface is the original SiC/Si interface, which is in fact the highly defective carbonized region. Therefore, an additional surface treatment is required. Bonding is also used in the Smart Cut process, which has been successfully applied to SiC wafers to form SiCOI [9]. This process combines hydrogen implantation into SiC wafers (to induce blistering), wafer bonding of the implanted wafer to a pre-oxidized Si wafer and thermal processing to induce splitting of the SiCOI structure. The resulting SiC layer in the structure is characterized by a crystalline quality similar to that of the original SiC wafer. Moreover, the high flexibility of the bonding processes allows this technology to be used with different kinds of handling wafers, such as low cost polycrystalline SiC ones [10] which high thermal conductivity would be more suitable than Si for high temperature applications or for further epitaxy.

In this work, we will present an alternative approach for the fabrication of SiCOI material based on the ion beam synthesis (IBS) combined with a bond and etch back technique. High temperature carbon ion implantation into Silicon has been proved as a suitable technique to produce high crystalline quality β-SiC buried layers with low residual stress [11], and low surface roughness. After selective oxidation of the SOL, bonding onto an oxidized Si wafer can take place without the need of further polishing. A room temperature, mechanically assisted, weak bonding is performed, followed by a high temperature strengthening step. After etching of the original Si substrate, using the SiC buried layer itself as an etch-stop layer, SiCOI structures with a sharp SiC/insulator interface free of voids or cavities have been obtained, with an average surface roughness of the SiC overlayer as low as 45 to 50 Å. As in the case of Smart Cut, the great flexibility of this process can be easily extended to the use of a wide range of handling substrates more suitable than Si for harsh environments, such as polycrystalline SiC wafers [10].

On the other hand, for many MEMS applications, polycrystalline SiC (poly-SiC) may constitute a more interesting choice, due to its higher technological versatility. However, poly-SiC films deposited on SiO_2 are discontinuous and present a poor adhesion due to a low-density interface region that contains voids and etch pits, whereas those deposited on Si_3N_4 have a dense microstructure, and continuous films can be formed with a good adhesion [12]. In comparison, the flexibility of IBS has allowed us to fabricate poly-SiC on insulator (poly-SiCOI) structures by converting SOI layers into poly-SiCOI, with the same IBS process as described above, but performed into polycrystalline and amorphous Si layers grown on SiO_2. The possibility to easily grow poly-Si layers on different substrates such as glass or alumina, gives further interest to this approach, allowing the direct synthesis of a wide range of different SiC multilayer structures without the need of further oxidation or wafer bonding processes. X-ray diffraction (XRD) and Auger electron spectroscopy (AES) measurements have been used to confirm the formation of a stoichiometric poly-SiC layer above the SiO_2 one. The use of initially amorphous and poly-Si layers, both undoped and P^+ doped, has been compared from the microstructural and residual stress points of view. Finally, these IBS poly-SiCOI layers have been used for the fabrication of micromechanical test structures, with a process directly derived from standard bulk Si micromachining technology [13].

2. Fabrication of SiCOI structures

2.1. ION BEAM SYNTHESIS OF SiC LAYERS

(001) B-doped Si wafers (16-24 Ωcm) were implanted at 500°C, with a multiple implantation process, at energies of 100, 120, 150 and 195 keV, and doses of 2.6×10^{17}, 3.3×10^{17}, 4.7×10^{17} and 10^{18} cm^{-2} respectively. These parameters were determined by TRIM simulation [14] in order to obtain a flat stoichiometric carbon profile. These implantations were performed at 500°C, in order to avoid amorphization of the target. The samples were subsequently annealed at 1150°C for 6 hours under nitrogen atmosphere. Similarly to the results reported in [11] for single implantation at 500°C, the structural analysis revealed the direct formation of a β-SiC layer with abrupt SiC/Si interfaces. In agreement with the topotactic transformation from the matrix Si lattice to the SiC one invoked in the literature, this layer is formed by SiC crystals with the same crystalline structure and orientation as the Si substrate. XPS measurements (fig.1) show a thick flat carbon profile, stable under annealing, with a SiC stoichiometric concentration and sharp interfaces with the substrate. The thickness of the SiC layer is about 3000 Å, while that of the top Si layer is about 2500 Å. The high crystalline quality of the layer after annealing can be seen in figure 2, where the FTIR spectra from an annealed sample exhibits a lorentzian absorption peak centred at 796 cm^{-1}, with a FWHM of about 50 cm^{-1}, characteristic of crystalline β-SiC. From the position of the peaks in the 2θ/ω XRD measurements, a slightly compressive strain has been found in the SiC layers, which has been estimated to be η_{200}=-0.013%, η_{111}=-0.003% and η_{220}=-0.015%. Assuming a SiC Young's modulus of 470 GPa, this corresponds to a stress of about 108MPa, 13MPa and 71MPa respectively. Finally, the layers have proved excellent etch-stop properties, as can be seen from the XPS analysis (fig. 1) of a sample etched one hour in Tetra-Methyl-Ammonium Hydroxide (TMAH) 25% wt at 80°C

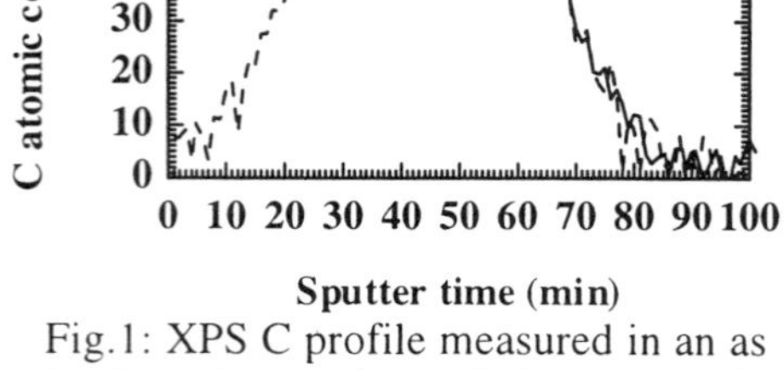

Fig.1: XPS C profile measured in an as implanted sample and in a sample etched 1hour in TMAH.

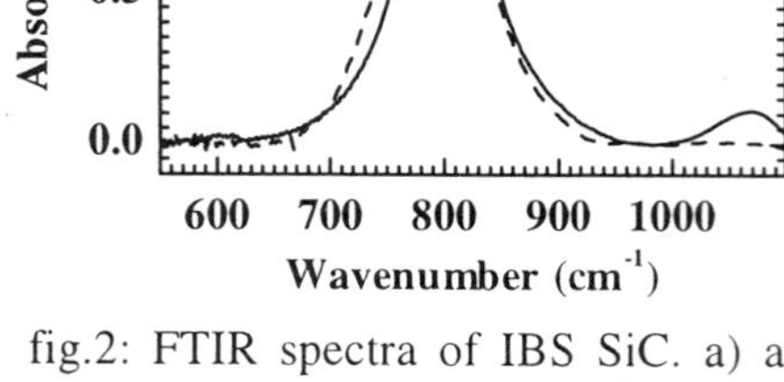

fig.2: FTIR spectra of IBS SiC. a) as implanted, FWHM ≈ 80 cm^{-1}, and b) annealed, FWHM ≈ 50 cm^{-1}.

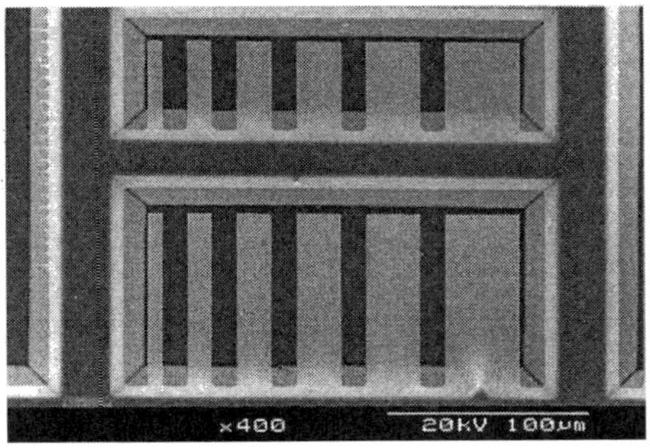

fig.3: SiC microstructures for mechanical properties assessment by beam bending.

(standard CMOS compatible etchant). This etch-stop behavior has been interpreted in terms of a percolation mechanism [11], assuming the layer to be formed by a binary SiC/Si system in a three-dimensional site-percolation model, and taking into account the inertness of SiC to the TMAH etchant.

The very low level of residual stress in the layers together with their excellent etch-stop properties has allowed the fabrication of high stiffness test microstructures suitable for the measurement of their mechanical properties (fig. 3). The fabrication process of the SiC microstructures was directly derived from the standard silicon bulk micromachining technology: photolithography of an aluminum mask through a positive photoresist, reactive ionic etching (RIE) patterning of the SiC layer (SF_6:O_2 gas mixture) and selective anisotropic etching of the top and underlying Si layers in TMAH 25 % wt at 80°C. The RIE conditions were finally chosen to be the same as usually used for Si, in order to keep a good control of the lateral overetching and of the etching rate of the underlying Si. However, since the etch rate of SiC is much lower than that of Si (typically about 15μm/min for Si and 20-30 nm/min for SiC in our equipment), the use of the Aluminum mask was necessary to stand the longer etching time needed. The mechanical properties of the IBS SiC layer were measured by beam bending with an atomic force microscope (AFM) [15]. This technique, which principle is schematically represented in figure 4, combines a very high load resolution with a nanometric precision in the measurement of the cantilever deflection. Owing to the nanometric deformations that can be applied by the AFM, the behavior of both the cantilever beam and the probe remain within the elastic domain, and can be described by Hooke's law,

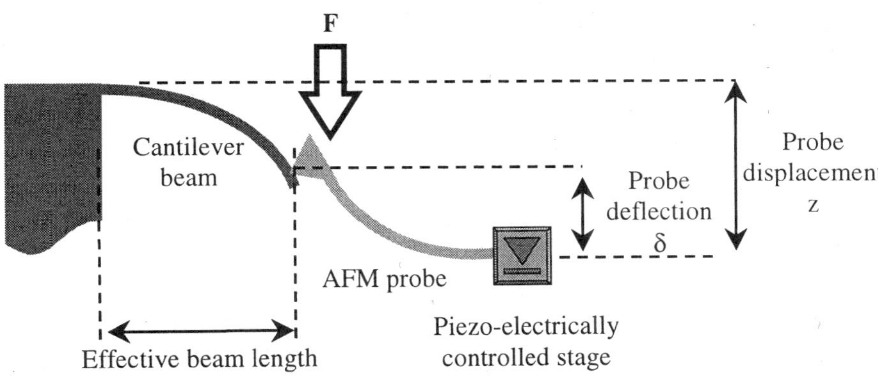

fig.4: measurement of micromechanical properties by beam bending with an AFM

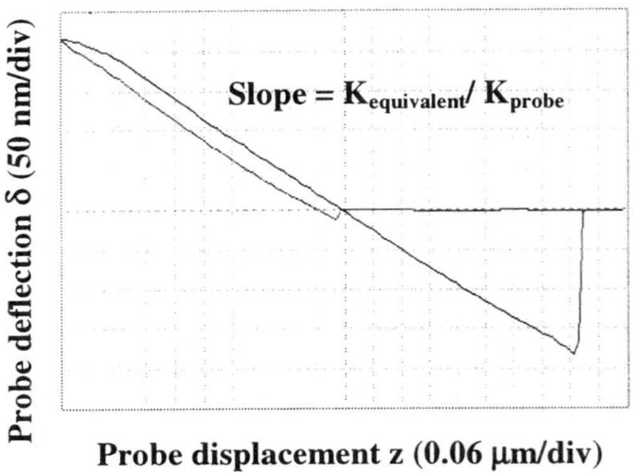

fig.5: typical AFM approach curve on SiC cantilever beams.

where the system beam-probe is simply the combination of two springs in series. Therefore, the Young's modulus can be straightforwardly extracted from the approach curve (deflection vs displacement of the probe, fig. 5) which is obtained when the probe is cyclically approached to and retracted from the surface of the beam. A value of 470 GPa equivalent to that of bulk SiC was obtained [15]. These samples will further be referenced to as "SiC samples".

Finally, an electrical analysis is now being carried out in order to evaluate the viability of these layers for microelectronic devices. For this, the effects of residual lattice strain, structural defects, and/or presence of SiC precipitates along the implantation tail on the electrical properties of the SiC active layer have to be analyzed.

2.2. BONDING AND ETCH-BACK.

Prior to the bonding itself, a thorough surface preparation needs to be carried out. First of all, in order to obtain a SiCOI structure, the top Si layer (SOL) of the SiC samples must disappear. The easiest approach would be to remove it by selective etching. However, we found in a previous etch-stop study [11] that the resulting RMS roughness of the SiC surface left by etching in TMAH is of the order of 50 to 60 Å as measured by AFM (fig 6). This is much higher than that of a bare Si wafer, which is usually about 1.5 to 2 Å, and would seriously decrease the efficiency of subsequent bonding. In view

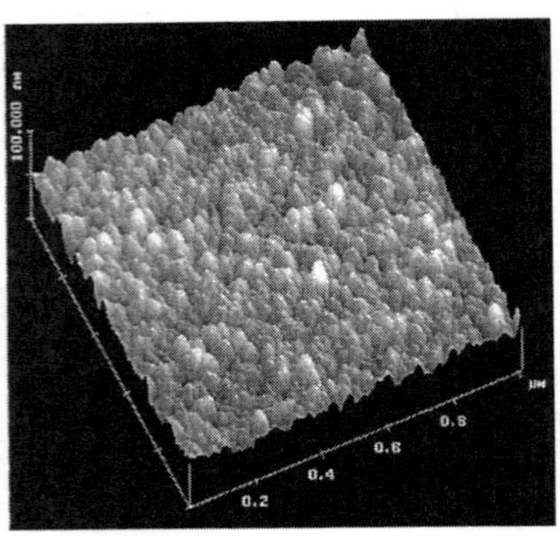

fig.6: AFM image of the surface of an IBS SiC layer after removing of the SOL.

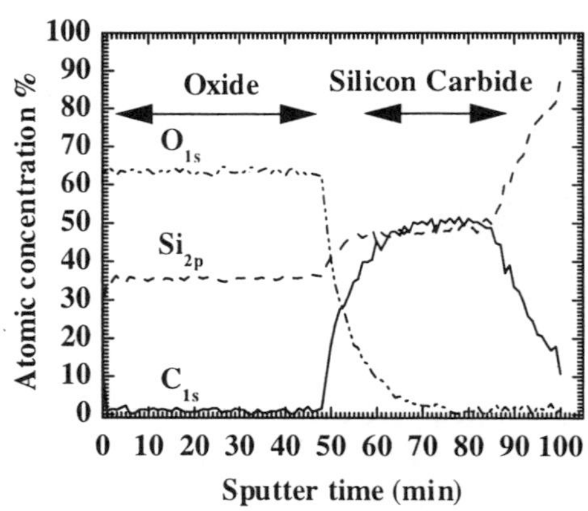

fig.7: XPS C and O profiles of SiC wet oxidation.

of this, we finally chose to selectively oxidize this top Si layer, taking advantage of the lower oxidation rate of SiC with respect to Si. The surface roughness of the SiC samples after wet oxidation was found to be around 15 to 16 Å. This is lower than that of 29 Å for the CVD SiC layers that were successfully used for bonding in ref. [8]. On the other hand, emphasis should be put on the importance of a good adjustment of the oxidation depth. Any residual SOL between the SiO_2 and the SiC in the final SiCOI structure has to be eliminated. At the same time the penetration of oxidation into SiC has to be minimized in order to limit the loss of SiC. Figure 7 shows the oxygen and carbon profiles from XPS analysis for a set of wet oxidation conditions that would lead to 8000 Å of SiO_2 in Si: in this case the whole 2500 Å top Si and about 1200 Å of SiC have been oxidized. The finally retained set of conditions corresponds to an equivalent SiO_2 thickness of 6000 Å in Si, which consumed less than 100 Å of SiC. The next step in the preparation of the samples consists in a careful cleaning of the oxidized Si wafers and SiC samples. This is a standard RCA cleaning, which in addition of removing any contamination, inclusions or particles of dust, leaves an hydrophilic surface. Hydrophilicity of the surfaces has been found to have a crucial influence on the efficiency of the bonding [8].

The bonding involves a room temperature weak bonding followed by a high temperature strengthening step. In contrast with the case of Si wafers, which spontaneously adhere to each other when brought into contact at room temperature, an external force is necessary to bond the SiC samples onto the Si oxidized wafers [8]. The strengthening process is done with commonly used conditions for standard Si wafer bonding: 2 hours dry oxidation followed by annealing at 1100°C during 8 hours in N_2. The subsequent removal of the original Si substrate is done by chemical etching in TMAH. This etch-stop process is controlled by the IBS buried SiC layer itself, owing to its excellent etch-stop properties as stated above.

According to the overall bonding process schematically represented in figure 8, the original SiC buried layer has been successfully transferred onto the surface of the oxidized Si wafer, leading to an $SiC/SiO_2/Si$ structure as shown in figure 9. The interface between the two bonded oxides cannot be distinguished, and neither voids nor bubbles are observable. At the interface SiC/SiO_2, a 20-25 nm thick layer with small

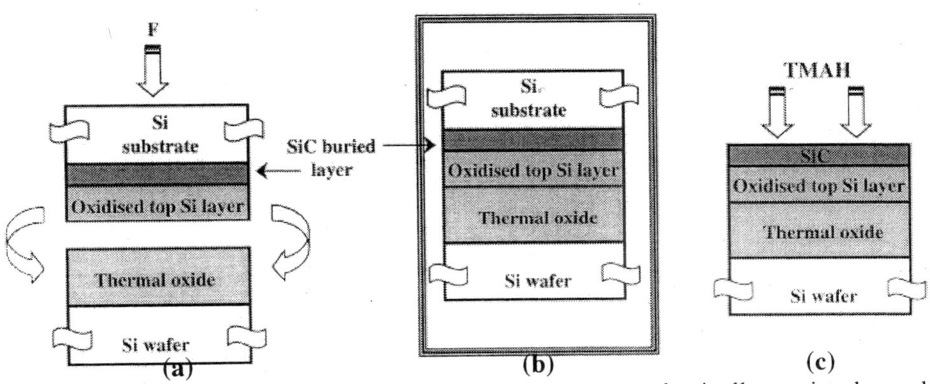

fig.8: Fabrication of SiCOI. a) Room temperature mechanically assisted weak bonding b) Strengthening annealing at 1150°C/8h in N_2 c) Etch-back of implanted wafer, controlled by the etch-stop IBS buried SiC layer.

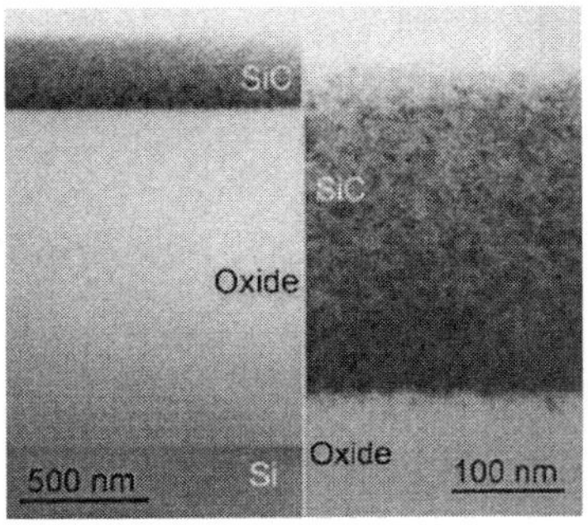

fig.9: Cross section TEM image of the SiCOI structure.

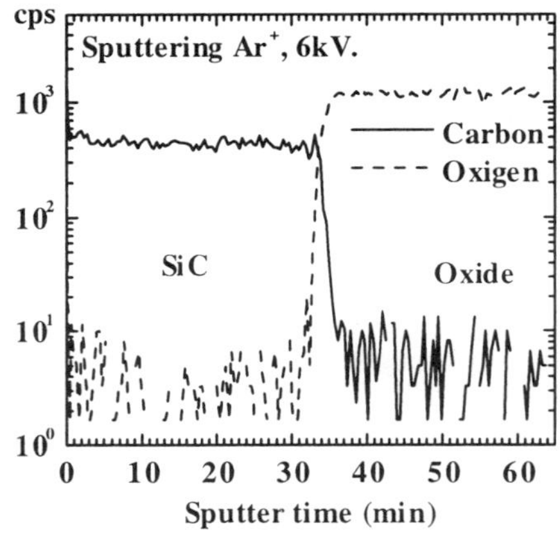

Fig.10: SIMS analysis of the SiCOI structure.

SiC precipitates in an oxide matrix can be seen, which likely corresponds to the tail of the implantation profile. The resulting SiCOI structure presents an average RMS surface roughness of about 50 Å, and a sharp interface between the SiC and the oxide with a remarkably flat stoichiometric C profile as shown by SIMS analysis (fig. 10). In this process, the crystalline quality of the final SiC layer is similar to that of the original IBS one.

Bonding of about 75 % of the initial area has been accomplished. This coverage ratio might be increased mainly by improving the efficiency of the room temperature bonding step. Two critical issues related to this are hydrophilicity enhancement treatments and a better homogeneity of the applied external force. On the other hand, although the previous mechanical assessment of the SiCOI structures has already demonstrated their viability for micromechanical applications., further experiments are also needed to characterize the electrical behavior of the SiC/SiO_2 interface, since it has been observed that fixed charges and interface traps densities in SiC/oxide interfaces strongly depend on the oxidation method as well as on the quality of the starting material [16]. These features, together with the electrical transport properties of the layers, are relevant for the performance of the electronic devices to be made in the structures.

3. Fabrication of poly-SiCOI structures

On the other hand, the flexibility of ion implantation has allowed the extension of the IBS technology to the direct synthesis of polycrystalline SiC on insulator for Micro-Electro-Mechanical Systems. In this process, amorphous or polycrystalline Si layers on SiO_2 are converted into polycrystalline SiC by the high dose carbon ion implantation. The starting material was (001) p-type boron doped 17.5 Ωcm device grade Si wafers, covered with an 1 μm thermal wet oxide grown at 1100°C. Amorphous Si layers were then deposited by low pressure CVD at 580°C with a thickness of about 5800 to 5900 Å (5500 Å aimed). In some samples, an additional residual stress reduction has been achieved by P^+ ion implantation, at a dose of 10^{15} cm^{-2}, and energy of 120 keV. The samples that will constitute the amorphous targets are left as-deposited and/or as implanted, while an annealing at 950°C/2 hours in N_2 ambient is necessary to form the final polySi layers [17]. The same four-step implant process as defined in section 2.1 has been used to obtain a flat carbon profile with stoichiometric SiC composition within the implanted layer just above the oxide. Polycrystalline substrates were implanted at 500°C with the purpose of maintaining the initial crystalline structure, whereas the as-deposited amorphous ones were implanted at room temperature since amorphization during the implantation is not a concern in this case. All the studied samples are finally annealed at 1150°C during 6 hours in N_2.

In all the samples, the structural and chemical characterization, carried out by FTIR, XRD and in-depth AES measurements, corroborate the direct synthesis by the ion implantation process of a stoichiometric poly-SiC layer above the sacrificial one, already before annealing. This has been observed even in the case of room temperature implantation into an initially amorphous Si substrate. This suggests the existence of a local beam heating of the samples at the implanted region, which could be enhanced by the presence of the SiO_2 thermal barrier. Between this layer and the SiO_2 substrate, a thin residual Si rich region is observed (fig. 11). This points out the need to carefully adjust the processing parameters to control the thickness of the different layers, so that the lower interface of the implanted layer can match the Si/SiO_2 interface. XRD measurements (fig. 12) confirm the polycrystalline nature of all samples, without any preferential orientation. These measurements show no effect from the P doping, i.e of the residual stress level of the starting poly-Si. On the other hand, three groups can clearly be distinguished from figure 12, according to the implantation conditions. All the layers made from polycrystalline Si and implanted at 500°C give essentially the same spectra in position and intensity (a). Conversely, amorphous substrates, which were implanted at room temperature, as well as a c-Si RT implanted wafer, all belong to the second group (b), and give a lower

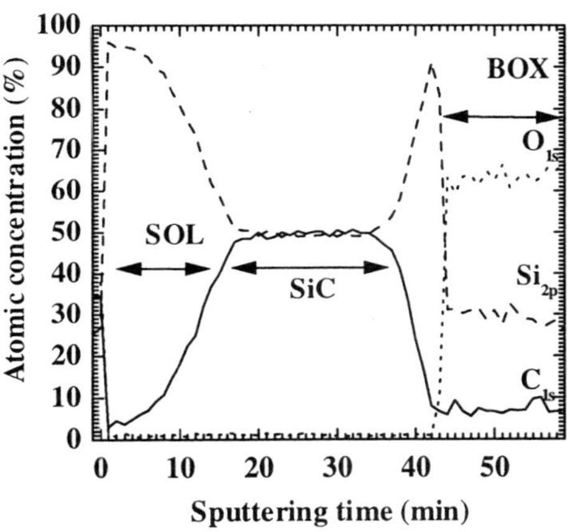

Fig.11: AES analysis of the poly-SiCOI structure.

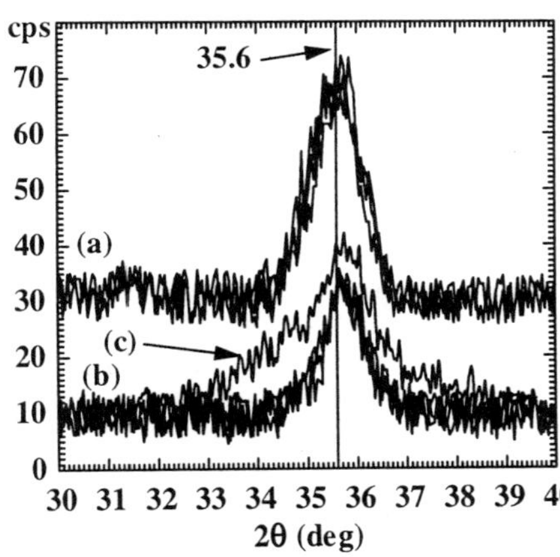

Fig.12: XRD analysis of the IBS poly-SiCOI structures (see text).

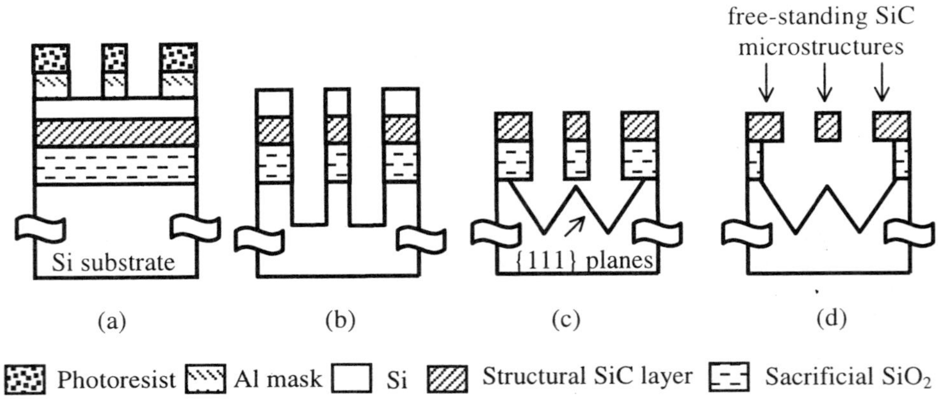

Fig.13: Fabrication process of the micromechanical structures in the IBS polySiCOI. a) Photolithography, b) RIE etching followed by photoresist and Al cleaning, c) Anisotropic TMAH etching, and d) HF etching

signal. The (1 1 1) peak of these samples is slightly shifted towards higher angles, which indicates some compressive strain of the lattice. Finally, the sample previously P^+ implanted, without activation annealing (c) exhibits a shoulder at lower angles, which seems to indicate the presence of a distribution of tensile residual stress and a significantly higher lattice deformation. From these results, it seems that the final microcrystalline structure of the ion beam synthesized layers essentially depends on the implantation temperature, more than on the initial structure of the starting wafer. This suggests that the microstructure of the polycrystalline IBS layers is not determined by a topotactic transformation, in contrast with the case of crystalline materials.

To evaluate the viability of this material for MEMS applications, micromechanical test structures have been fabricated, using techniques of photolithography, Reactive Ionic Etching (RIE) and wet etching directly derived from the standard bulk Si micromachining, in a process similar to that already described in section 2.1. Figure 13

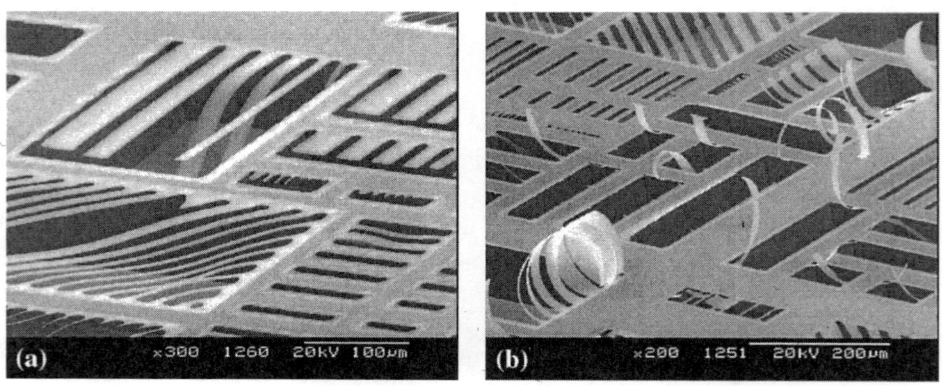

Fig.14: micromechanical structures made from IBS poly-SiCOI (see text)

shows a schematic description of this process, where a final dip in hydrofluoric acid is added in order to etch away the sacrificial BOX and release the self-standing structures. Quite similar results are found for the structures fabricated from an initially amorphous undoped Si layer and from initially polycrystalline ones (whether doped or not). The cantilever beams are almost flat, indicating a very low stress gradient within the polySiC layer, while the bridges are slightly buckled, which denotes some compressive residual stress (fig. 14a). Hence, it seems that the presence of P atoms in substitutional sites during the carbon implantation does not cause significant additional stress in the final poly-SiC layers, although neither does it seem to have any stress reduction effect. On the contrary, the sample made from an amorphous previously P^+ doped but without activating exhibits a different behavior: the structures are strongly bent upward, due to a very high stress gradient, while the stress is essentially tensile as can be deduced from the unbuckled bridges (fig. 14b). This corroborates the XRD measurements, which show a significant deformation of the lattice for this sample, as stated above. The presence of a significant concentration of interstitial P atoms (not activated P impurities corresponding to the gaussian-like pre-IBS implanted profile) during the C implantation might be responsible for such a high stress gradient in the final poly-SiC layer. Further experiments are required to determine the exact effect of implanting at 500°C as compared to room temperature implantation, in order to determine if stress effects from interstitial P atoms in this latter sample could be avoided. In any case, these data demonstrate the ability of the IBS process for the direct synthesis of complex SiCOI structures.

Conclusions

In this work, we have presented an alternative approach for the fabrication of SiCOI material based on the ion beam synthesis technique. In this process, the SiC layers are synthesized by a multiple high dose C^+ implantation designed to form a buried stoichiometric flat carbon profile. The structural characterization confirms the formation of β-SiC, while the micromechanical assessment confirms the ability of these layers for MEMS applications. Combining IBS with wafer bonding, high crystalline quality ion beam synthesized β-SiC layers with low residual strain, have been successfully transferred onto oxidized Si wafers, obtaining SiCOI structure with abrupt SiC/SiO_2 interfaces and low surface roughness. The effects of the presence of SiC islands close to SiC/SiO_2 interface is under investigation.

On the other hand, the high flexibility of IBS has allowed the fabrication of polycrystalline SiC on insulator by direct conversion of amorphous and polycrystalline Si on SiO_2 layers. The structural characterization has revealed that the structure of the poly-SiC layers is mainly determined by the implant temperature rather than by the topotactic transformation usually invoked in the case of crystalline materials. It has also been found that when implanting pre-doped layers, the presence of interstitial P atoms leads to a significant deformation of the lattice, whereas activated P atoms in substitutional sites during the carbon implantation does not cause significant additional stress in the final poly-SiC layers, although neither does it seem to have any stress reduction effect. In agreement with these results, it has been found that micromechanical structures made from these former samples exhibit strong deformations characteristic of a high stress gradient in the layers. In all the other cases, the results suggest the viability of the direct IBS of poly-SiCOI structures for micromachining applications. A further optimization to avoid the presence of the residual Si rich interfacial region between SiC/SiO_2, as well as a more complete structural and electrical characterization of these structures are presently under progress.

References

1. Ishsimaru, M., Dickerson, R.M. and Sickafus, K.E. (2000) Scanning transmission electron microscopy-Energy dispersive X-ray/Electron energy loss spectroscopy studies on SiC-on-insulator structures, *J. Electrochem,. Soc.* **147**, 1979-1981.
2. Reichert, W., Lossy, R., González Sirgo, J.M., Obermeier, E. and Stoemenos, J. (1996) β-SiC deposited on SIMOX substrates: characterization of the SiC/SOI system at elevated temperatures, *Inst. Phys. Conf. Ser.* **142**, 129-132.
3. Okhuysen, M.E., Mazzola M.S. and Lo, Y.-H. (1999) Low temperature growth of 3C-SiC on silicon for advanced substrate development, in C. H. Carter Jr, R. P. Devaty and G.S.Rohrer (eds). *Silicon Carbide and Related Materials/1999* , Trans Tech Publication, Switzerland, pp 305-308.
4. Wischmeyer, F., Wondrak, W., Leidich, D. and Niemann E. (1999) CVD growth of 3C-SiC on SOI (100) substrates with optimized interface structure, *Mat. Scie. Eng.* **B61-62**, 563-566.
5. Planes, N., Möller, H., Camassel, J., Stoemenos, Y., Falkovski, L., Eickhoff, M. and Krötz, G. (1999) SOL thinning effects on 3C-SiC on SOI, in C. H. Carter Jr, R. P. Devaty and G.S.Rohrer (eds). *Silicon Carbide and Related Materials/1999*, Trans Tech Publication, Switzerland, pp. 301-304.
6. Zappe, S., Obermeier, E., Stoemenos, J., Möller, H., Krötz, G., Wirth, H. and Skorupa, W. (1999) Stabilization of the 3C-SiC/SOI system by an intermediate silicon nitride layer, *Mater. Scien. Eng.* **B 61/62**, 522-525.
7. Vinod, K.N., Zorman, C.A., Yasseen, A.A. and Mehregany, M. (1998) Fabrication of low defect density 3C-SiC on SiO_2 structures using wafer bonding techniques, *J. Electron. Mater.* **27**, L17-L20.
8. Tong, Q.Y., Gösele, U., Yuan, C., Steckl, A.J. and Reiche, M. (1995) Silicon Carbide Wafer Bonding, *J. Electrochem. Soc.* **142**, 232-236.
9. Di Cioccio, L., Le Tiec, Y., Letertre, F., Jaussaud, C. and Bruel, M. (1996) Silicon carbide on insulator formation using the smart cut process, *Electronics Letters* **32**, 1144-1145.
10. Joly, J.P., Aspar, B., Bruel, M., Di Cioccio, L., Letertre, F. and Hugonnard-Bruyere, E. (2000) *NATO Advanced Research Workshop 2000 abstracts*, pp. 17-18.
11. Serre, C., Pérez-Rodríguez, A., Romano-Rodríguez, A., Calvo-Barrio, L., Morante, J.R., Esteve, J., Acero, M.C., Skorupa, W. and Kögler, R. (1997) Synthesis of SiC microstructures in Si technology by high dose carbon implantation: etch-stop properties *J. Electrochem. Soc.* **144**, 2211-2215.
12. Wu, C.-H., Zorman, C. A. and Mehregany, M. (1999) Characterization of polycrystalline SiC grown on SiO_2 and Si_3N_4 by APCVD for MEMS applications,

in C. H. Carter Jr, R. P. Devaty and G.S.Rohrer (eds). *Silicon Carbide and Related Materials/1999*, Trans Tech Publication, Switzerland, pp. 541-544.
13. Serre, C., Pérez-Rodríguez, A., Morante, J.R., Esteve, J., Acero, M.C., Kögler, R. and Skorupa, W. (2000) Ion beam synthesis of polycrystalline SiC on SiO_2 structures for MEMS applications, *J. Micromech. Microeng.* **10**, 152-156.
14. Ziegler, J.F., Biersack J.P. and Littmark, U. (1985) The stopping and range of ions in solids, Pergamon Press, New York, , Vol. 1.
15. Serre, C., Pérez-Rodríguez, A., Morante, J.R., Gorostiza, P. and Esteve, J. (1999) Determination of micromechanical properties of thin films by beam bending measurements with an atomic force microscope, *Sensors & Actuators*, **A74**, 134-138.
16. Chaudry, M.I. (1989) A study of native oxides of β-SiC using Auger electron spectroscopy , *J. Mater. Res.* **4** (2), 404-407.
17. Benítez, M.A., Fonseca, L., Esteve, J., Benrakkad, M.S., Morante, J.R., Samitier, J. and Schweitz, J-A. (1996) Stress-profile characterization and test-structure analysis of single and double ion-implanted LPCVD polycrystalline silicon, *Sensors and Actuators* **A 54**, 718-723.

NEW SIC ON INSULATOR WAFERS BASED ON THE SMART-CUT® APPROACH AND THEIR POTENTIAL APPLICATIONS

J-P JOLY, B.ASPAR, M.BRUEL, L.DI CIOCCIO, [A] F.LETERTRE, [B] E.HUGONNARD-BRUYÈRE

Leti CEA, Département de Microtechnologies, 17 rue des Martyrs F-38054 Grenoble France

[A] SOITEC, Parc Technologique des fontaines F38190 Bernin France

[B] LPM (UMR CNRS 5511), INSA, 20 Avenue Albert Einstein, F-69121 Villeurbanne France

Abstract

Important progress have been made in the fabrication of SiCOI (Silicon Carbide On Insulator) structures using the Smart-Cut® approach. The different structures which have been demonstrated both in terms of transferred layer polytypes (4H and 6H), of handle substrate (Silicon or Polycrystalline Silicon carbide) and of buried insulator layers (Silicon Dioxide and Silicon Nitride) will be described. Deep traps present in the SiC layer after transfer and annealing of the structure and which are generated by the ion implantation process has been studied using different techniques (Hall measurements, DLTS, Photoluminsecence, RPE). We will see that their density can be strongly minimised making the as transferred layer quality compatible with many applications. Considering both the improved layer quality and the different possible SiCOI structures now available the different possible applications and the perspectives will be reviewed.

1. Introduction

Large band gap semiconductors will find more and more applications in such important fields like Power electronics, High temperature electronics or Opto-electronics where traditional semiconductors are not adapted. Very important efforts have been made in the last decade on the development of wide band gap materials. It is crucial for any industrial development to get large size materials with good quality at a reasonable cost. Unfortunately crystal growth of these refractory materials is difficult. SiC for example can only be obtained using very high temperature sublimation techniques. III-V nitrides are only obtained using hetero-epitaxy on single crystalline substrates such as sapphire or Silicon Carbide but the poor lattice matched between GaN and these substrates make the growth of good quality material quite difficult.

F. Balestra et al. (eds.), Progress in SOI Structures and Devices Operating at Extreme Conditions, 31–38.

A new concept in the field of semiconductors has been introduced recently which consists in exfoliating a thin layer of a single crystalline substrate and transferring it to an other handle substrate. Amongst the different approaches which have been proposed the Smart-Cut® concept [1] is now recognised as a major breakthrough and it has been shown that it can be extended to a very large range of materials. It has been introduced by LETI and developed by LETI and SOITEC and is based on proton implantation and wafer bonding as depicted in figure 1. It is now used by SOITEC to produce large volume high quality SOI wafers. Demonstration has been obtained for many other materials including main semiconductors like InP, GaAs, Ge and SiC [2,3]. Both 4H- and 6H-SiC polytypes On Insulator structures (SiCOI) have been successfully obtained using the Smart-Cut® process.

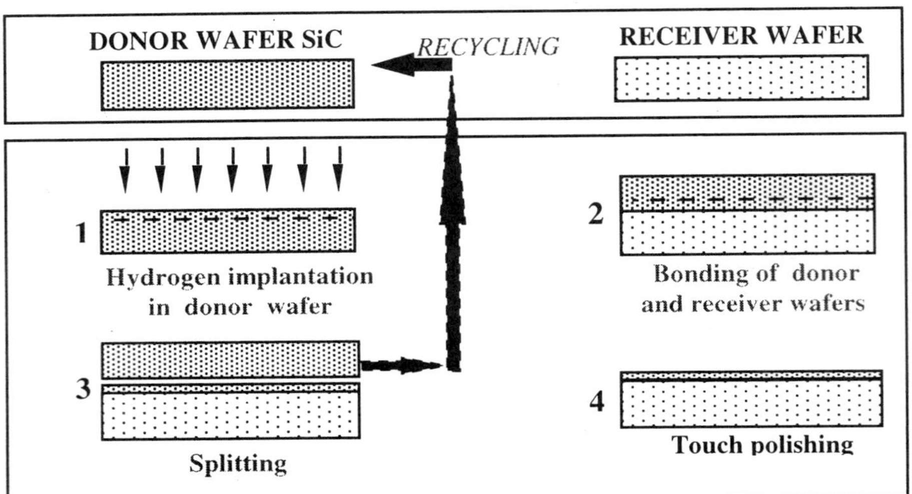

Figure 1: Schematic of the Smart-Cut® process

The Smart-cut® concept is particularly adapted to wide band gap semiconductors which as described earlier are expensive and difficult to grow. Using Smart-Cut we can indeed separate the requirements related the top layer in which the device will be fabricated from those of the handle substrates which will only play a mechanical role. Gain is obtained first in terms of material cost but also in terms of new functionalities. As SOI allows new device concepts or improve device characteristics, thin film SiC on insulator (SiCOI) or SiC on other structures make possible new devices structures. Association of a large band gap and refractory material like SiC with full dielectric isolation is especially promising for very high temperature and radiation hardened applications. If SOI and bulk SiC already much more adapted to harsh environment than bulk Silicon, SiCOI structures will represent an other step forward in the adaptation of semiconductors to more and more stringent conditions.

Furthermore having a membrane of SiC on a substrate separated from this substrate by a layer which can selectively etched allows the fabrication of new MEMS device working in harsh environments for which standard semiconductors such as Silicon are not adapted.

Aim of this paper will be in a first part to describe the latest developments and progress made on SiC Smart-Cut® structures and especially SiCOI structures. We will see that different structures in response of different possible applications have been developed. As previously published [4], initial SiCOI samples exhibited a high resistivity due to deep levels generated by implantation of large dose of protons required to induce thin film exfoliation. We will see in a second part that important progress has been made and that it is now possible to strongly decrease this compensation effect. Last part will describe the different applications which can be found to such SiCOI wafers. In the same part we will also see how the structures can be adapted to these applications in terms of substrate and buried layers.

2. Different structures investigated

SiC is subjected to polytypism. Mainly three polytypes are currently available. 3C polytype is epitaxially grown on Silicon by CVD at about 1350°C. The quality of these 3C layers is rather poor with a large density of extended defects which are due to the poor lattice matching between SiC and Silicon. Therefore these layers can only find applications in devices such as piezo-resistive pressure sensors. 6H and 4H single crystalline material are now commercially available up to 50 mm in diameter with extended defect densities much lower than in 3C. Both polytypes are grown by the so-called "Modified Lely" technique, a high temperature sublimation technique. It is more easy to grow 6H than 4H by this technique but unfortunately electron mobilities in the 6H polytypes are not adapted to the main Power device applications. 6H find therefore applications such as substrates for the hetero-epitaxial growth of III-V nitrides and for the fabrication of short wavelength Light emitting devices while 4H polytypes find applications in the field of Radio frequency and power device applications.

We first demonstrated the possibility to transfer a SiC and to obtain a SiCOI using the Smart-Cut® approach using a 6H polytype. We used for that purpose a Silicon handle substrate and a Silicon dioxide intermediate layer. We then investigated the possibility of making 3C and 4H layer transfer. For that purpose we first studied the so-called blistering effect in the different polytypes. Hydrogen ion implantation in materials at doses larger than strictly required for a Smart-Cut process creates blisters which are a good indication of the possibility of making a Smart-Cut transfer. As can be seen from figure 2 this blistering effect can be obtained in the three polytypes. 4H layer transfer has been indeed obtained under conditions rather similar to 6H transfer.

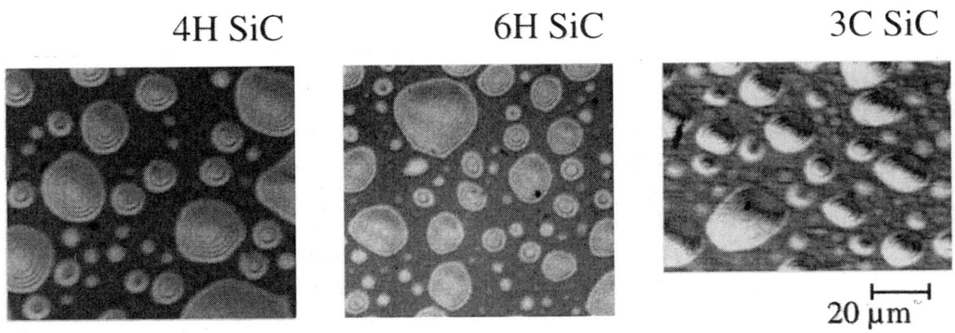

Figure 2: Blistering effects in three SiC polytypes after H ion implatation at the same dose and energy.

Quite large bonding energy can be obtained at room temperature using hydrophilic molecular bonding of Silicon dioxide terminated surface and Silicon wafers are available with excellent surface quality. But the large SiC stiffness and the large surface steps associated with CVD epitaxial growth of SiC are not favourable conditions for spontaneous bonding even when Hydrophilic bonding is used. Efforts have been therefore undertaken to improve SiC surface quality and to reduce low density macroscopic bonding defects as can be seen on the specimen shown on figure 3.

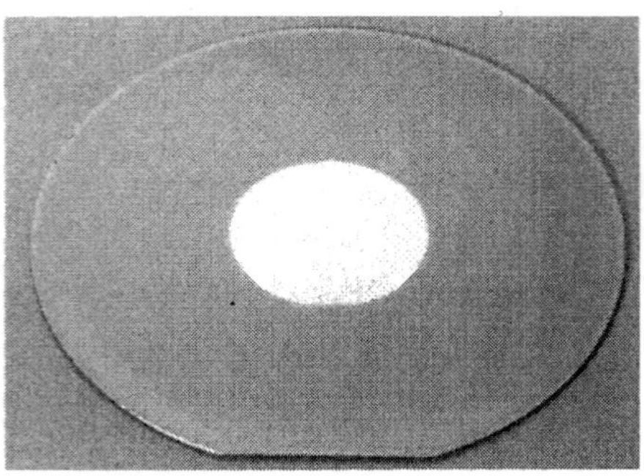

Figure 3: Top view of SiCOI wafer on Silicon substrate ($SiC/SiO_2/Si$)

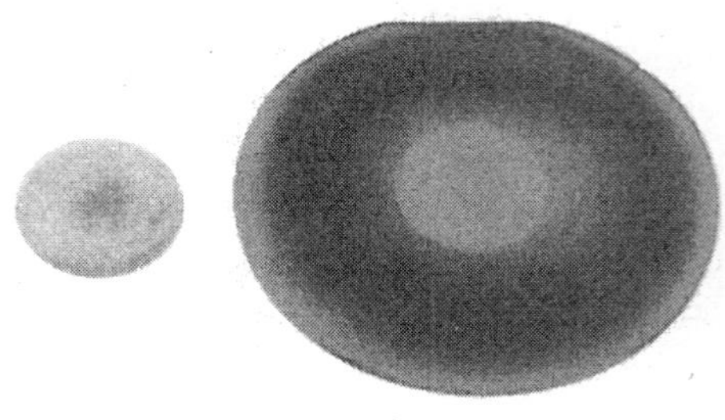

Figure 4: Top view of a SiCOI structure on a Polycristalline CVD substrate (SiC/SiN/SiC) and of the seed SiC wafer

It has been also demonstrated that several layers can be successively exfoliated from the same seed substrates. Specific wafer recycling techniques have been developed for that purpose. This was very important to validate the Smart-Cut approach on SiC from an economical point of view.

In order to better understand the splitting mechanism in SiC the splitting kinetics has been studied in. In case of pure thermally activated splitting we have shown [3] that, as in silicon, the splitting mechanism is thermally activated (figure 5). Activation energies have been related to Hydrogen diffusion behaviour in SiC . As Hydrogen diffuses slowly in SiC larger times or temperatures are needed when compared to Silicon.

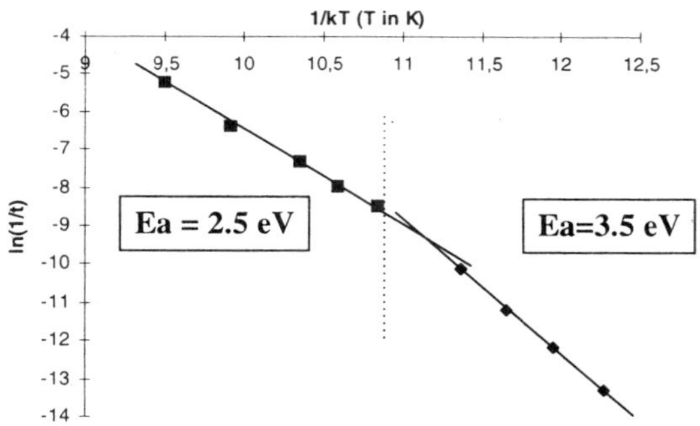

Figure 5 : Kinetics of splitting in SiC after H ion implatation versus temperature

Use of Silicon as handle wafer is not compatible with some applications. For those devices which can in principle sustain temperatures larger than 450°C the mechanical stability of Silicon substrates is not sufficient. Furthermore SiC device processing requires very high temperature (typically up to 1700°C) anneal for dopant activation. These temperatures are far above the Silicon melting point (1410°C). That is the reason why new structures based on Polycrystalline SiC substrates as handle wafer has been developed. Substrates from Morton Inc initially grown by CVD on graphite have been used For the applications requiring a very low thermal conductance of the substrate we have also developed structures where the silicon dioxide layer is replaced by Silicon nitride (figure 4). The thermal conductivity of Silicon nitride is typically of 30 $Wm^{-1}K^{-1}$ 20 times Larger than the one of Silicon dioxide. Taking into account the very large thermal conductivity of CVD Polycrystalline SiC substrates (up to 330 $Wm^{-1}K^{-1}$) the global heat dissipation properties of the SiCOI wafers can approach the SiC bulk properties and be compatible with typical power applications.

3. Properties of the SiC layer after Smart-Cut® transfer

Compared to usual 3C SiC structures on Insulator which are usually obtained by growing a 3C layer on a SOI substrates, SiCOI structures obtained using the Smart-Cut® approach are of much better crystalline quality. No extended defects can be found on TEM observations. Despite this lack of extended defect in the layer previous specimen fabricated using a non optimised process have shown a significant level of dopant compensation in the transferred SiC layer even after 1300°C annealing (see figure 6a)

[4]. This compensation is due to point defects introduced by the ion implantation process.

Combinations of process conditions have been therefore studied to reduce density of these defects [5]. With optimised conditions it is now possible to reduce compensation to a level acceptable for many applications. In layers separated from an initial epitaxial layer with doping in the range of 10^{17} cm^{-3} (n type) it is now possible to measure doping and room temperature mobilities almost identical with those measured on the initial epitaxial layers (see figure b).

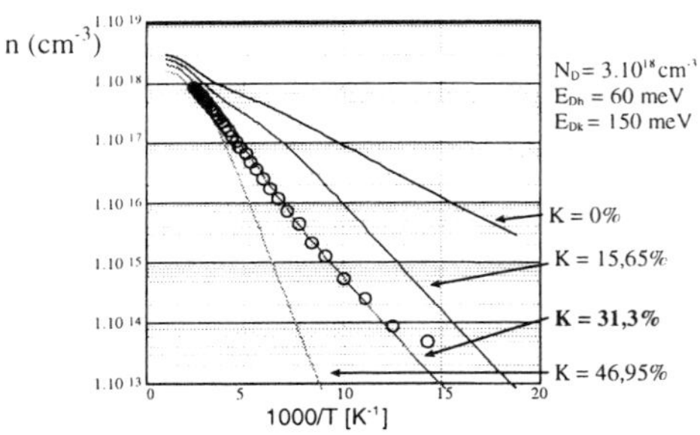

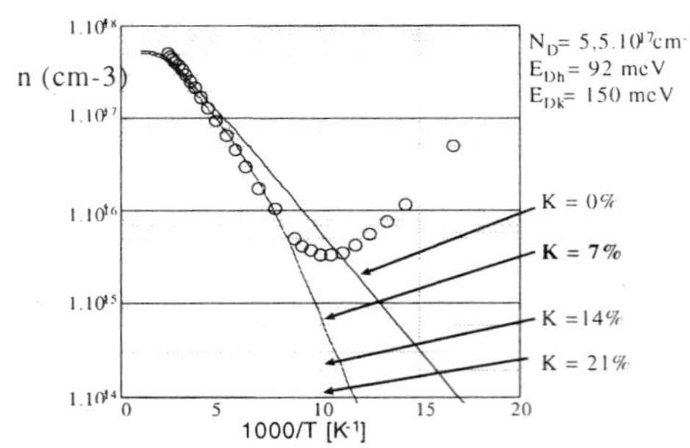

Figure 6 a : Carrier concentration in early 6H SiCOI layers (Hall measurements) fitted with calculated ones using compensation factors K and given trap parameters

Figure 6b : same after process improvement (lower initial doping)

It has been confirmed by DLTS (see figure 7) that unknown deep traps that were responsible of deep compensation in the previous layers are not any more measured in the new specimen. If standard Z_1/Z_2 centers usually attributed to irradiation in SiC are still seen, their density is rather low in the range of 10^{16} cm^{-3} and well correlated to the level of compensation as obtained using Hall measurements.

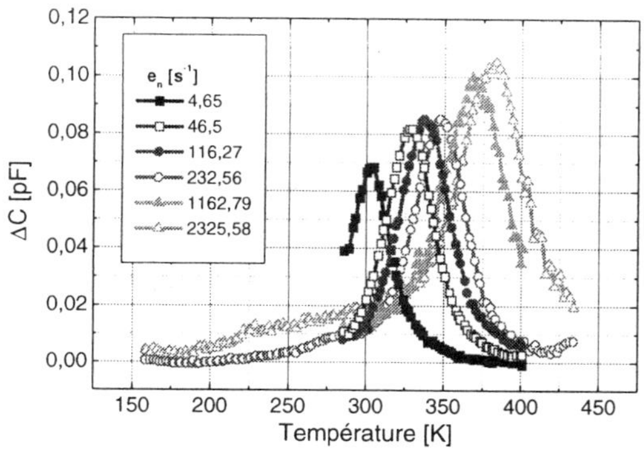

Figure 7: DLTS spectra on SiCOI samples ontained using optimised conditions

4. Applications and adaptation of the structures to these applications

Applications of SiCOI structures can be classified in mainly four categories:

1) Discrete Power devices including RF power devices. SiC layer quality available today without additional epitaxy seems satisfactory for RF devices like MESFETs but specific handle wafers with larger thermal conductivity as for example Poly SiC should be mandatory. Switches and rectifiers for power electronics in most of the case will probably need thick layers with epitaxy after layer transfer.

2) High temperature SiC sensors. SiCOI structures obtained by 3C SiC heteroepitaxy on SOI have been already used for the fabrication of pressure sensors [6]. Use of Smart-Cut and more refractory substrates than Silicon can solve the problems encountered with these previous structures: poor isolation, poor stability of Silicon in harsh environments. Other sensors (flow sensors, catalytic gas sensors) working in harsh environments can take profit from the new SiCOI structures based on Smart-Cut®. For these applications available electrical quality of transferred SiC layer is good enough.

3) Heteroepitaxy of GaN. No specific demonstrations have been obtained yet here but available quality of SiC transferred layers is already good enough for this application. Use of metallic bonding rather than dielectric bonding can be advantageous for this applications to make the diode process more simple.

4) Combination of different devices on the same chip to obtain a complete electronic function for high temperature or specific environments. This include simple logic or analog functions but also power electronic applications. Use of SiCOI with dielectric isolation open the way of obtaining such functions at very large temperature up to 600°C.

5. Conclusions

We have shown that large range of SiCOI structures with different polytypes, different handle substrates and different buried dielectrics can be obtained using the so-called Smart-Cut® process. By using proper process conditions both electrical and physical characteristics are good enough for typical applications. As these applications are diverse SiCOI substrates have to be tailored and adapted to each specific case. Until now no real device has been fabricated on these new structures. This will be the next step in this development.

References

1. M.Bruel, Electron. Lett. 31 (1995) No. 14, pp. 1201-1202

2. L. Di Cioccio, Y. Letiec, F. Letertre, C. Jaussaud and M. Bruel: Electron. Lett. 32 (1996) No. 12, pp.1144-1145

3. B. Aspar, C. Lagahe, H. Moriceau, A. Soubie and M. Bruel, MRS vol 510 (1998) p3812

4. E.Hugonnard-Bruyère, V.Lauer, G. Guillot and C.Jaussaud, Materials Science and Engineering B61-62 (1999) pp 382-388

5. E. Hugonnard-Bruyère, F. Letertre, H. J. von Bardeleben, J.L.Cantin, L. Di Cioccio, T. Ouisse, T. Billon, G. Guillot, ICSCRM Abstracts, Raleigh 1999, p188

6. G.Krötz, W.Legner, Ch.Wagner, H.Möller, H.Sonntag and G.Müller, The 8th International Conference on Solid State sensors and Actuators June 25-29 1995, pp 186-189

ELTRAN® (SOI-Epi Wafer™) Technology

TAKAO YONEHARA and KIFOFUMI SAKAGUCHI
ELTRAN Business Center, Canon Inc.,
6770 Tamura, Hiratsuka, Kanagawa, 254-0013, Japan
http://www.canon.com/eltran/

Abstract

ELTRAN® (Epitaxial Layer TRANsfer), which is the first manufacturable and commercially available product using Porous Si, has been originated, developed and produced in Canon Inc., Japan. The last established technique is highly reproducible splitting in the Porous Si layer by Water Jet and reuse the seed wafer several times. The thicknesses of both SOI and the buried oxide layers are precisely controlled in the very wide range from the extremely thin as 27 nm to as thick as a few μm with the thickness uniformity less than +/- 5 %. The active layer has no COP (Crystal Originated Particle or Pits) by epitaxial growth. The buried oxide is thermally grown on epitaxial Si layers and has no pinholes. We have successfully expanded the wafers to 300-mm (12-inch) diameter, in which SOI-thickness-uniformity of ±1.1% was even better than 8 inch

1. Introduction

SOI (silicon-on-insulator) technology has been at somewhat of a standstill for quite a time, but now we are finally approaching a long-awaited period, with CMOS (complementary metal oxide semiconductor) applications of thin-film SOI in particular about to reach a stage where they can be put to practical use. Thick-film SOI have already been put to practical use in a wide range of SOI-IC (integrated circuit) applications, such as analog devices, photodiode arrays, high-speed bipolar ICs, in-vehicle ICs and PDP (plasma display panel) drivers. Thin-film SOI, on the other hand, has only found a few applications, such as radiation-hard SRAMs (static random access memories), ASICs (application-specific ICs) and ATM (asynchronous transfer mode) ICs. In the future it is anticipated that there will be developments in the area of mainstream CMOS-LSI (large-scale integration) towards devices for low power, high-speed logic and communications. We plan to launch into the stage of mass production trials before the turn of the century so that SOI will be adopted as the new semiconductor IC technology of the 21st century. SOI wafers have a structure whereby a layer made of a large-area single crystal silicon is formed on top of a layer of amorphous silicon dioxide. The history of SOI wafers has involved the challenge of how to construct such an artificial crystalline structure that does not exist in the natural world.

F. Balestra et al. (eds.), Progress in SOI Structures and Devices Operating at Extreme Conditions, 39–86.

SOI-Epi wafer™ and ELTRAN® (Epitaxial Layer TRANsfer) are new wafers that have been originally developed by Canon; they are BESOI (bond and etch-back SOI) wafers that make use of the selective etching of porous silicon [1-4]. The method used involves combining BESOI with epitaxial growth on porous silicon that is capable of being etched with ultra-high selectivity and surface-smoothing by hydrogen annealing. Another special feature of the method is that the bonded wafer is split into two at the porous silicon layer, with one part becoming the ELTRAN® wafer and the other part being reused. Along with product quality, reducing wafer manufacturing costs is one of the most important requirements for further expanding the SOI market. If the starting material, the remains of which were always thrown away in the past after wafer manufacture, can be reused in this way, then production costs can be reduced dramatically.

Research and development into ELTRAN® was started in 1990. Thanks to the good use of ultra-clean technology it subsequently became possible to apply our manufacturing line to 8-inch SOI wafers, with production being commenced in 1997.

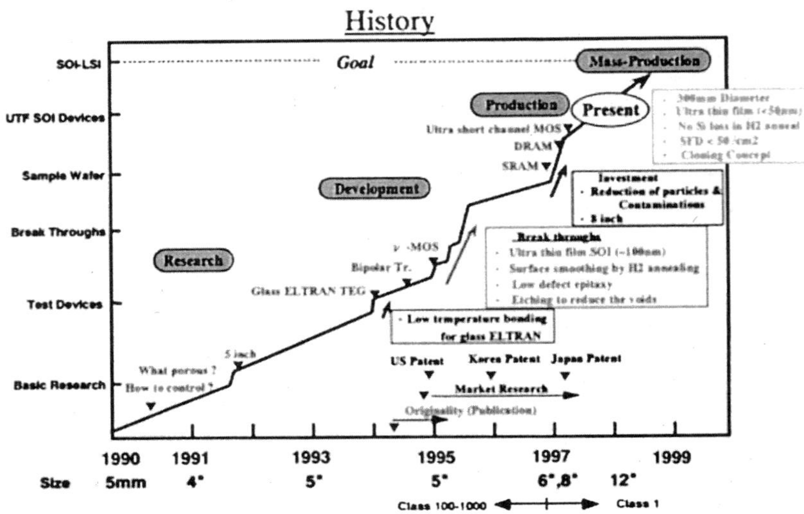

Figure 1. History of ELTRAN R&D to production.

Let us now look back at the history of ELTRAN® R&D with the aid of **Fig. 1**[5]. ELTRAN® was first conceived in 1990 when it came to our attention that porous silicon has an enormous surface area per unit volume (about 200 m^2/cm^3). It was in this year that basic research was started. We were confident that mass production would become possible in the future, and so conducted research into the formation and properties of porous silicon, epitaxial growth, etching properties and optimum etchants, with the intention of protecting our findings as intellectual property. In 1992, we increased the size of our SOI wafers to 5 inches, carried out the trial manufacture and evaluation of individual transistors such as SOI bipolar transistors and MOSFETs (MOS field effect transistors), and developed small displays having integrated drivers. We also succeeded in using low-temperature bonding techniques to form ELTRAN® wafers on top of glass substrates, and tried making corresponding devices. The highlight of this development period was around 1995 with a number of breakthroughs, such as the first formation of a 100nm ultra-thin film and the realization of surface smoothing by hydrogen annealing and low-defect epitaxial growth. Unfortunately, however, the work was only carried out in a class 100 clean room designed for research purposes, and so the expected innate strong advantages of a COP-free (crystal-originated-particle-free) SOI-Epi wafer™ could not be realized. In 1996, so as to get to the stage where we could publish our work, we expanded the diameter of our wafers to 8 inches and at the same time started construction of a new manufacturing line equipped with a class 1 clean room and chemical filter in order to solve the particle problem. In the summer of the following year, we participate in the SEMATECH benchmark and succeeded in confirming the inherent superiority of our SOI-Epi wafers™, i.e. the dramatic reduction in HF defects. We have subsequently looked into a wide range of possible applications for thick-film down to ultra-thin-film SOI wafers and glass substrates, applications such as power devices, micro-mechanisms, memories, ultra-short channel MOSFETs_and displays. At present, with the goal of showing the future potential of our SOI wafers, we are striving towards 300mm-diameter wafers and sub-50nm ultra-thin films, and towards wafer recycling techniques aimed at reducing costs. In addition, since the SOI-LSI market is increasing, we plan to expand our production scale manyfold in the year 2000.

2. Special Features of ELTRAN®

2.1 SOI-EPI (COP-FREE) WAFERS™

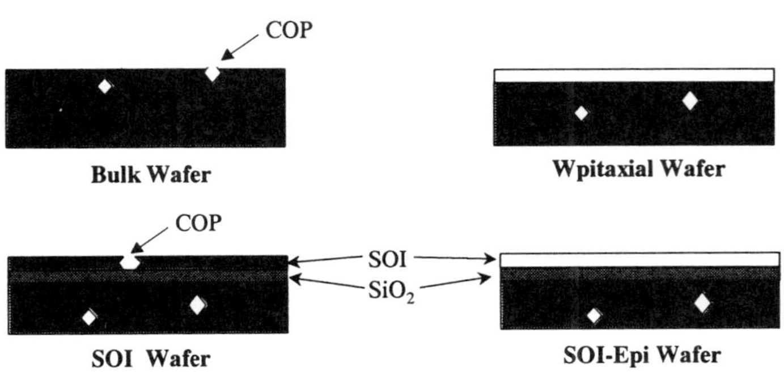

Figure 2. Distribution of COP in bulk wafer, epitaxial wafer, SOI wafer and SOI-Epi wafer.

Let us consider, with the aid of **Fig. 2**, the characteristic features of bulk wafers and epitaxial wafers that are used in current silicon processes, along with SOI wafers and the SOI-Epi wafers™ that we have proposed. With a bulk wafer, when the crystal is pulled from molten silicon using a seed crystal, microscopic voids of size 0.1 $\sim$0.2μm called COPs are formed. It is known that these COPs are regular octahedrons surrounded by $\{111\}$ planes, with the inner wall being covered by an oxide. However, these voids do not propagate by vapor phase epitaxy into the epitaxial layer formed on top of the bulk wafer, and so there are none in the epitaxial layer. This is one of the main reasons why epitaxial wafers are used these days not only in logic devices_but also in memories. The reason that epitaxial wafers have come to be used is the fact that the size of devices has approached that of COPs (0.1~0.2μm), meaning that COPs cause a degradation in the reliability of the gate oxide films of devices. In the case of SOI, the thickness of the thin film SOI_ has become comparable to or even less than the size of COPs, and so if COPs exist there are voids in the corresponding parts of the SOI thin film in place of Si, which is a defect far more fatal to an SOI device than the presence of COPs in the bulk_wafer. If the SOI wafer immerses in HF solution, then the COPs in the SOI thin film become pinholes, and the buried oxide film immediately underneath becomes directly etched, which is easy to detect; such a defect is called an "*HF defect*". Since 0.25μm-rule MPUs (microprocessor units) and DRAM (dynamic random access memory) circuits are manufactured to a 99% yield, the density of HF defects must be no more than about 0.1 /cm^2. It is anticipated that this figure will drop down to around 0.01 /cm^2 with further advances in miniaturization in the future. It is thus considered that technology whereby an epitaxial layer is used as the SOI layer, i.e. COP-free SOI-Epi wafer™ technology, should be useful as far as thin-film SOI devices are concerned. In fact COP-free epitaxial wafers are already used in today's mainstream MPU products. Drawing an analogy from this, it would seem that SOI-Epi wafers™ will also become used in SOI logic devices. In particular, it seems likely that as the miniaturization of the device size progresses, the use of epitaxial wafers as starting materials will become unavoidable regardless of the manufacturing techniques used. In fact, one can already see SOI layers being made epitaxial even in non-ELTRAN® wafers[6].

2.2 SURFACE-SMOOTHING TECHNIQUES

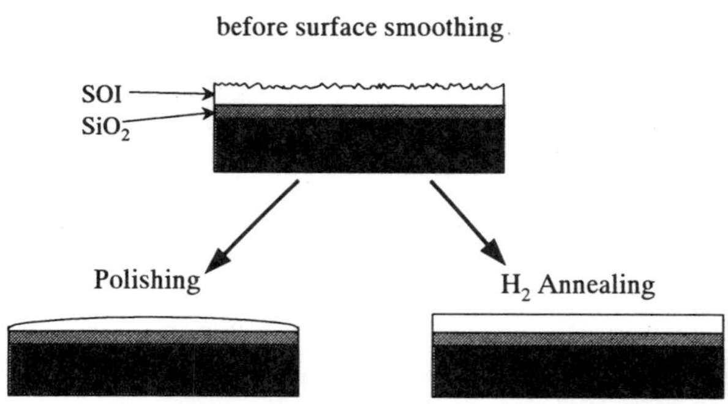

Figure 3. Surface smoothing of SOI wafers by polishing and H_2 annealing.

Let us now discuss the surface smoothness of SOI wafers. We will take an overall look at the surface nature of SOI wafers with the aid of **Fig. 3**. Regardless of the manufacturing technique used, the surface just after the SOI layer has been formed is much rougher than that of a bulk polished wafer. If a device was created with the surface left this rough then the reliability of the gate oxide film would be a big problem. When carrying out ion implantation, not only is the internal crystalline structure damaged, but degradation of the surface nature is also unavoidable. Moreover, when ultra-high-temperature oxidation of the surface is carried out at 1300ºC or above, this degradation becomes even worse. In the case of the propagation technique of micro-cracks by rare gas or hydrogen ion implantation, the crack plane reaches the surface of the SOI layer, and in the case of the porous silicon etching, roughness – which reflects the state of the boundary interface between the porous silicon and the epitaxial layer above it – appears immediately after formation of the SOI layer. This roughness could be removed by CMP (chemical mechanical polishing) in order to achieve a smoothness comparable to that of a bulk polished wafer, but the reduction in the film thickness and the degradation of its in-plane uniformity would become problems, as would control stability during mass production. These problems would be particularly prominent in ultra-thin-film SOI. With ELTRAN®, however, a brand new surface smoothing method totally different to the CMP method has been invented[7)]. This method was devised after discovering the phenomenon whereby the surface pores of the porous silicon close up and disappear due to enhanced surface diffusion of silicon atoms during the hydrogen pre-baking in the CVD (chemical vapor deposition) epitaxial growth on the porous silicon. If about one square micron of the surface of an etched ELTRAN® wafer is studied using an AFM (atomic force microscope) , it is found that the surface roughness (R_{rms}) of 10nm or so is reduced by a factor of about 100 by the hydrogen annealing, meaning that even the atomic steps can be seen. Moreover, with this new surface smoothing method the reduction in film thickness is now less than 1nm[8)]. The new method_has produced truly dramatic effects. We will discuss this suppression of the reduction in film thickness in more details in the section '7. Potential' below.

2.3 HIGH FLEXIBILITY IN THE FILM THICKNESSES

With ELTRAN® wafers, the SOI layer is formed by epitaxial growth and the buried oxide film is formed by thermal oxidation. This means that the thicknesses of the SOI layer and the buried oxide film can be controlled independently from one another and over a wide range. In actual practice both thicknesses can be controlled to be anywhere from 27nm down to a few nm, and both are uniform to within ±5% or better.

3. Processes and Clean Room

3.1 PROCESSES

Porous silicon[9)] is a material that was discovered in 1956[10)]. Although research into its use in FIPOS (Full Isolation by Porous Oxidized Silicon; a technique used in SOI manufacture) and as a luminescent material had been carried out in the past, until now there was no industrial manufacturing technology that made use of it. ELTRAN® provides the only case where a use for porous silicon in industrial manufacturing has been successfully found. The ELTRAN® manufacturing process_is made up of processes that harmoniously unite the 3 special features of porous silicon.

Let us now discuss these special features and processes. The 3 special features of porous silicon are as follows.

1) Sealing of surface pores and surface smoothing by annealing in a hydrogen atmosphere[11)]

   This allows one to achieve a high-quality epitaxial silicon layer on top of the porous silicon.

2) Enormous surface area[12,13)]

   The enormous surface area means that the etching selectivity of porous silicon is as high as 100,000×.

3) Internal stress[14-16)]

   By controlling the internal stress, the splitting plane can be specified.

The ELTRAN® manufacturing process consists of 7 important processes: 3 of them are related to the above-mentioned 3 special features of porous silicon; the other 4 are anodization to form the porous silicon in the first place, wafer bonding, smoothing by hydrogen annealing, and wafer recycling. We will now explain these 7 processes in the order in which they are carried out.

The ELTRAN® process flow is shown in **Fig. 4.**

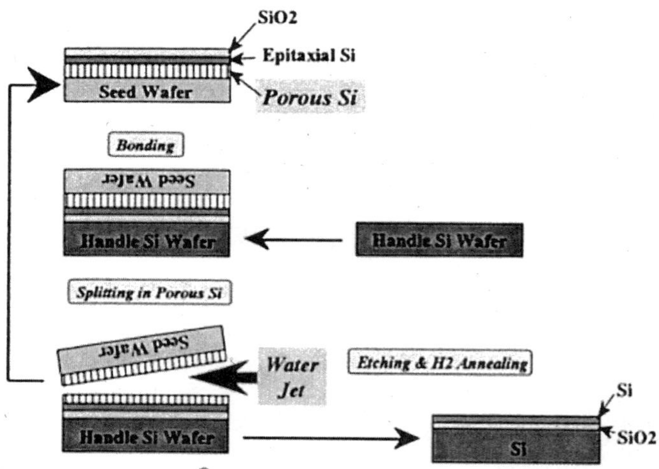

Figure 4. ELTRAN® process flow based on seed wafer reusage.

3.1.1 *Anodization*

Firstly, a single-crystal silicon wafer called a *seed wafer* that can be reused again and again is taken and porous silicon is formed on the surface by means of anodization. The anodization involves passing a current through a solution of HF and ethanol with the single-crystal silicon wafer as the anode in order to form microscopic pores of diameter a few nm on the surface of the wafer at a density of about 10^{11} /cm^2. The reaction occurs at the far end of the pores, meaning that the pores progressively elongate into the inside of the wafer. The structure of the porous silicon can be controlled by the concentration of the solution, the current density and the resistivity of the silicon. Moreover, the thickness of the porous silicon layer can be controlled by the length of time for which the anodization is carried out. The easiest way of controlling the porous structure is to vary the current density. By doing this it is easy to create a porous layer that has a multi-layered structure, an example of which is shown in **Fig. 5**. In this example, the layer of porous silicon closest to the surface was formed using a low current density, and then after this the current density was raised and a second layer of different porosity was formed. It can be seen from the figure that the first layer of porous silicon contains microscopic pores of diameter a few nm, and below this is formed a second layer for which the pore diameter is a few times greater.

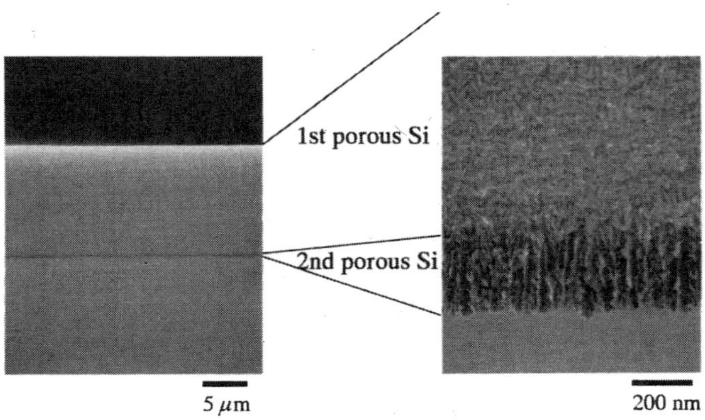

Figure 5. Cross-sectional micrographs of double layered porous Si formed by changing an anodic current.

Dry oxidation of the porous silicon is then carried out at a low temperature of 400ºC[17]. This results in oxidization of about 1~3nm of the inner walls of the pores, preventing the structure of the porous silicon from changing under high-temperature treatment. If this were not done then the porous silicon would be thermally unstable, with agglomeration of the pores occurring particularly at temperatures of 700ºC and above, which would cause a considerable worsening of the subsequent etching properties.

3.1.2 *Epitaxial Growth*[11]

Baking is carried out at about 1000~1100℃ in a hydrogen atmosphere in a chemical vapor deposition (CVD) epitaxial reactor. **Fig. 6** shows SEM micrographs taken from an oblique angle of the surface of the porous silicon before and after the hydrogen pre-baking. It can be seen that the hydrogen pre-baking causes the pores in the porous silicon surface to close up to the extent that the density of these pores goes down from about 10^{11} /cm^2 to less than 10^4 /cm^2, and hence the surface is smoothed. **Fig. 7** shows the mechanism of defect formation and a model of the reduction in defects due to pre-injection. Since during the formation of the porous silicon layer the silicon partially dissolves and pores form, the surface is originally in a state whereby there is a deficiency of silicon atoms. By pre-baking in hydrogen, enhanced surface diffusion takes place, causing the surface pores to close up. The silicon atoms necessary for this must be provided from the side walls of the pores, but since the surfaces of these side walls are partially passivated, there is a limit to the amount of silicon atoms that can be provided, and so pores of relatively large diameter can remain, this being a cause of stacking faults. A pre-injection method[18,19] was thus devised whereby a small additional amount of silicon is provided from the gas phase during the hydrogen pre-baking and surface diffusion is made to occur so that the remaining pores in the surface of the porous silicon close up. In this way a large reduction in defects has been achieved.

Figure 6. Porous Si surface (a) before and (b) after H_2 pre-baking.

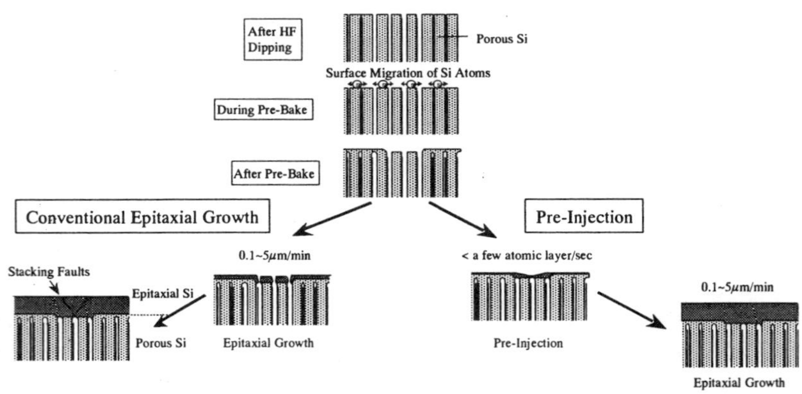

Figure 7. Mechanism of defect reduction by pre-injection.

Fig. 8 shows the history of efforts to reduce defects. With regard to crystal defects in the ELTRAN®, it can be seen that the density of SEDs (secco-etch defects) has been decreased by a factor of about 10 every two years. Big contributing factors to this decrease have been the above-mentioned phenomenon of surface pores being closed up through hydrogen pre-baking, the provision of a small amount of extra silicon atoms (pre-injection), and the use of an epitaxial reactor equipped with a load lock[20]. As of the present, we have achieved a secco-etch defect density of the order of only 10^1 /cm^2.

After the pre-injection, epitaxial growth is carried out at about 900~1000ºC. Next, the surface of the epitaxial silicon layer is thermally oxidized. The resulting oxide film will become the BOX (buried oxide) film of the SOI wafer.

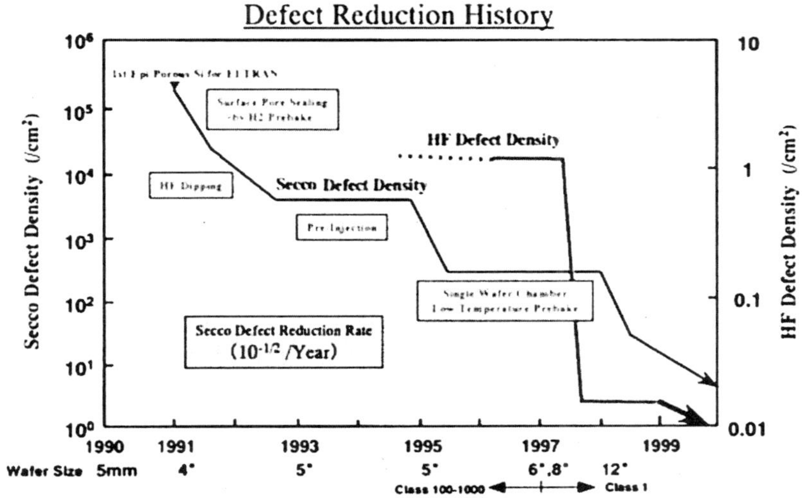

Figure 8. History of defect reduction.

3.1.3 *Bonding*[21)]

The surface of the thermally grown oxide film and the silicon handle wafer that will become the supporting substrate are next cleaned, and then the surfaces of the two wafers are pushed together at room temperature, upon which they bond to one another by van der Waals forces. After that, heat treatment is carried out in order to form covalent bonds and thus strengthen the bonding between the two surfaces. **Fig. 9** shows a cross-sectional TEM micrograph of the bonded wafer. It can be seen from the micrograph that the bonding interface and the SOI/BOX interface are both abrupt. Note that by using a quartz wafer as the handle wafer (the supporting substrate), it is also possible to manufacture light-transmitting ELTRAN® wafers.

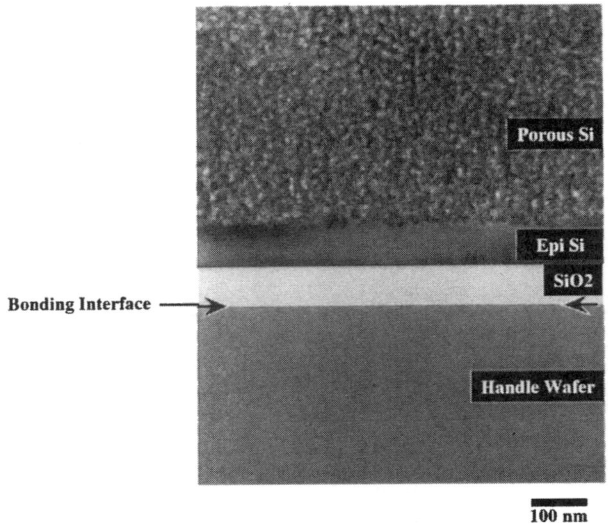

Figure 9. XTEM micrograph to show the bonding configuration.

3.1.4 *Splitting*[14-16, 22)]

The porous silicon layer in the bonded wafer has a double layered structure. The bonded wafer is split parallel to its surface close to the interface between these two layers using a water-jet method. Further details will be given in the section '6. Cost-effectiveness'.

3.1.5 *Etching*[12, 13)]

After the bonded wafer has been split, the first porous silicon layer remains on the handle wafer (supporting substrate) side; this layer is uniform in thickness across the whole wafer. The handle wafer part is then etched using a solution containing a mixture of HF, H_2O_2 and H_2O. The etching characteristics of porous silicon and non-porous silicon are shown in **Fig. 10**. For the sake of comparison, the figure shows the etching characteristics not only for the $HF/H_2O_2/H_2O$ mixture, but also for an $HF/HNO_3/CH_3COOH$ mixture and for an HF/H_2O solution. It can be seen that in the case of the $HF/H_2O_2/H_2O$ mixture, once a certain incubation period has passed, the porous silicon is etched virtually all at once. The selectivity of this etching is as high as 100,000×, meaning that the etching does not cause any degradation of the uniformity of the thickness of the SOI layer. It is inferred from these characteristics that the etching mechanism is as shown in **Fig. 11**. According to this mechanism, the etching solution penetrates into the pores of the porous silicon by capillarity, and then etches into the walls of the pores in a sideways direction. Eventually, the porous structure can no longer support itself and collapses. Since each wall between neighboring pores is etched from both sides, the thickness of epitaxial silicon that must be etched is effectively half of the thickness of the wall, which is at most about 10nm. Moreover, so long as the porous silicon collapses, it does not matter if the etching is not perfect. This means that it is possible to etch porous silicon layer with the thickness up to 10μm or more uniformly across the whole surface of the wafer, and since the selectivity is as high as 100,000×, there is virtually no degradation of the uniformity of the thickness of the SOI layer. When an etching solution that reacts strongly with silicon such as a mixture of HF, HNO_3 and CH_3COOH is used, on the other hand, etching starts to progress from the surface of the porous silicon before the etching solution has had a chance to penetrate into the pores, meaning that the above-mentioned mechanism does not take effect and so selectivity cannot be achieved.

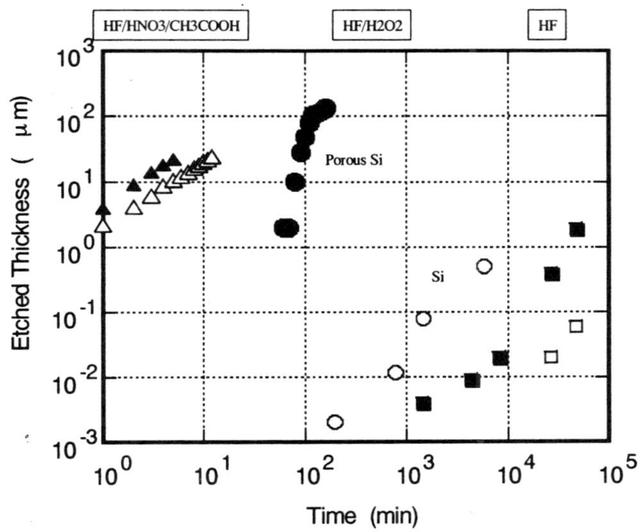

Figure 10. The etching characteristics of porous Si (closed symbols) and nonporous Si (open symbols) by immersing in $HF/HNO_3/CH_3COOH$ (triangles) HF/H_2O_2 (circles) and HF (rectangles) as a function of etching period.

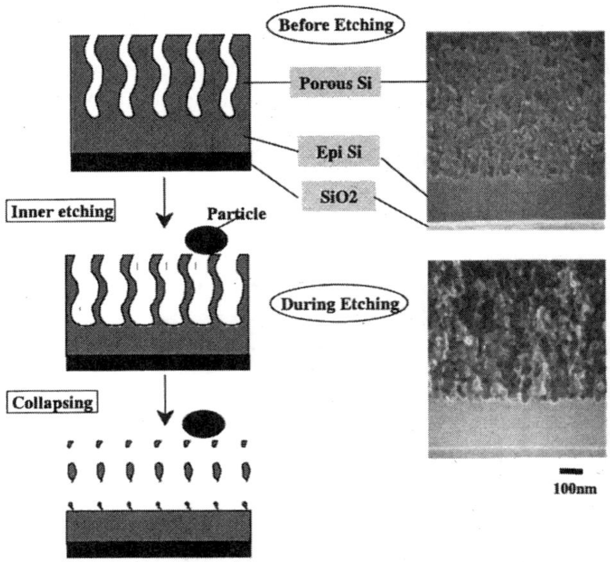

Figure 11. Etching mechanism and cross-sectional micrographs of porous Si before and during etching.

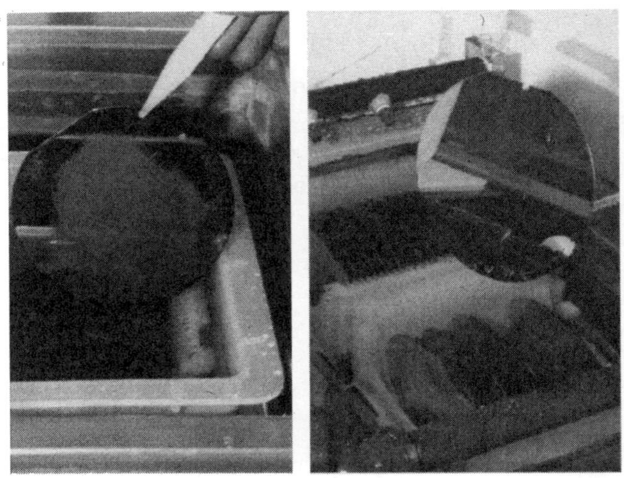

(a) (b)

Figure 12. Photograph of etching process. (a) shows the wafers during the etching process. (b) shows as-etched mirror surface of the wafers.

Furthermore, even if there are foreign particles on the surface of the split wafer, if the porous structure collapses as in **Fig. 11** then these particles can easily be removed by 'lifting off'. **Fig. 12(a)** shows a wafer that is in the process of being etched, while **Fig. 12(b)** shows the wafer after etching has been completed. It can be seen that the porous silicon around the periphery is removed first, and that in **Fig. 12(a)** porous silicon still remains around the center of the wafer. In this way, it is possible to judge when the etching has finished simply by visual inspection, meaning that it is easy to optimize the etching time.

3.1.6 *Hydrogen Annealing*[7)]

After the porous silicon layer has been removed removed by selective etching, the epitaxial silicon / porous silicon interface is left at the SOI surface with a distinctive 'micro-roughness' that results from the etching mechanism described above (see **Fig. 13(a)**). If heat treatment in a hydrogen-containing atmosphere is now carried out, the surface becomes smoothed as in **Fig. 13(b)**. The surface nature was measured using AFM. The observed period of the surface roughness depends on the length of the region scanned, and so the dependence of the surface nature on this length is shown in **Fig. 14**. The measured surface roughness values for the SOI wafer before and after hydrogen annealing are shown, along with those for a commercially available bulk polished wafer for the sake of comparison. It can be seen from the figure that for a $1\mu m \times 1\mu m$ region, the root-mean-square roughness was improved by hydrogen annealing to about 0.1nm, which is at least as good as that for the bulk polished wafer. Moreover, even the atomic steps could be seen, showing that a smoothness down to the atomic scale has been achieved.

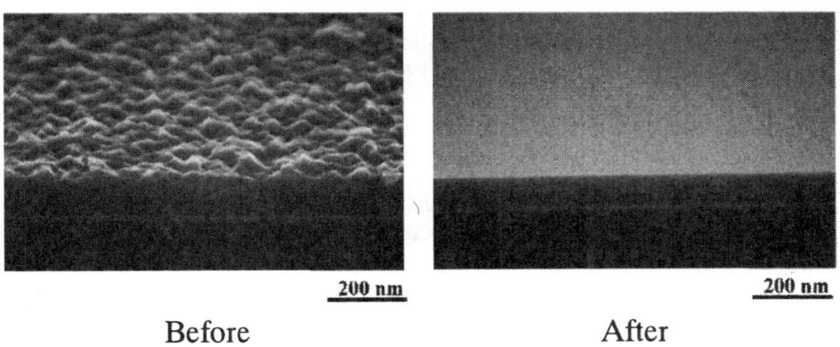

(a) (b)

Figure 13. High resolution micrographs from oblique view of (a) the as-etched ELTRAN® wafers and (b) the H_2 annealed ELTRAN® wafers.

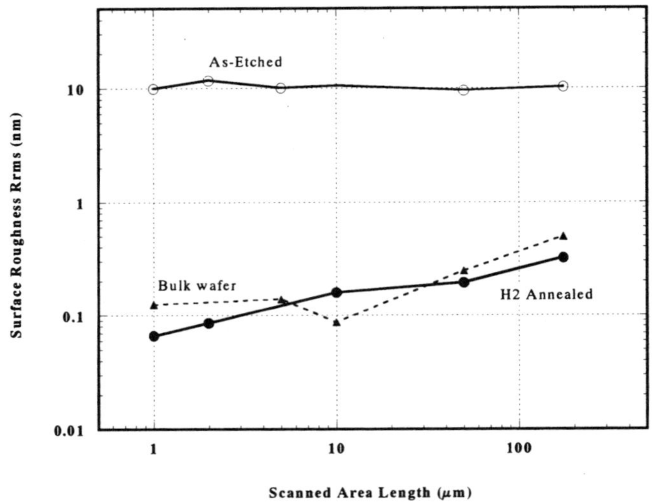

Figure 14. Scanning area size dependence of root-mean-square of the surface roughness (Rrms) measured by AFM. As-etched ELTRAN (open circles), commercially available silicon wafer (triangles) and H_2 annealed ELTRAN (closed circles) are plotted. The annealing conditions are 1150 °C, 80 Torr, 1 h in H_2.

3.1.7 *Wafer Recycling*

After splitting, residual porous silicon is removed from the seed wafer, which can then be recycled and put back into the ELTRAN® manufacturing process again and again.

3.2 CLEAN ROOM

As stated earlier, in 1996 with the aim of announcing our bonded SOI wafers; ELTRAN® to the world, we expanded the diameter of the wafers to 8 inches, and in order to solve the particle problem we constructed a new manufacturing line in a class 1 clean room (see **Fig. 15**). By using a class 1 environment virtually free of particles, we managed to achieve a large reduction in HF defects as shown in **Fig. 8**, thus showing the intrinsic feature of SOI-Epi wafers™ whereby being COP-free results in few HF defects.

Figure 15. Ultra Clean Room for ELTRAN® manufacturing.

Moreover, all of the equipment in the manufacturing line – including the anodization equipment, the bonding equipment, the wafer-splitting equipment and the porous silicon etching equipment, which was developed independently by us – was made to be 'cassette-to-cassette' so that manufacture could be carried out automatically.

In addition, when constructing our class 1 clean room we made use of filter technology in order to try to eliminate boron from the clean room[23,24)].

First of all, SOI wafers were manufactured in a clean room equipped with conventional glass fiber HEPA (high efficient particulate air) filters, and the boron present at the bonding interface was measured using SIMS (secondary ion mass spectroscopy). The results of the measurements are shown in **Fig. 16**. It can be seen that the bonding interface is in the buried oxide film at a distance 0.55μm from the surface, and the amount of boron peaks here at $6x10^{17}$ atoms /cm^3.

Next, 3 countermeasures were taken with the aim of suppressing the amount of boron. The first was to change the material in the ULPA (ultra low penetration air) filters used in the clean room where the bonding was carried out from glass fiber – which was one of the sources of the boron – to PTFE (polytetrafluoroethylene), in order to suppress the generation of boron in the clean room. The second countermeasure was to choose building materials that generate little boron, again in order to curb the amount of boron generated within the clean room. The third countermeasure was to go one step beyond curbing the generation of boron within the clean room and to attach a chemical filter that adsorbs boron to the ULPA filter situated between the cleaning equipment and the bonding equipment, the aim being to decrease the amount of boron adsorbed onto the wafer surface.

In this way, the clean room was environmentally controlled against chemical contamination. The amounts of boron, phosphorous and the like in the clean room were compared to those in other environments. Specifically, the targets of the comparison were the outside air in Hiratsuka City in Kanagawa Prefecture, the air in a clean room equipped with a conventional boron glass fiber HEPA filter, and the air in the newly constructed clean room equipped with the above-mentioned PTFE-ULPA filter and chemical filter. The measurements were carried out using ICP-MS (inductively coupled plasma mass spectroscopy) on samples collected using an impinger. It was found that in the clean room equipped with the boron-containing glass fiber HEPA filter, the boron level in the air was 0.08ng/l, whereas in the clean room equipped with the PTFE-ULPA filter and the chemical filter, it was right down at the limit of determination (0.0003ng/l). Moreover, the amount of boron detected in the outside air was 0.04ng/l; since the amount detected in the conventional clean room was double this (0.08 ng/l), this supports the theory that extra boron was being generated in the conventional clean room.

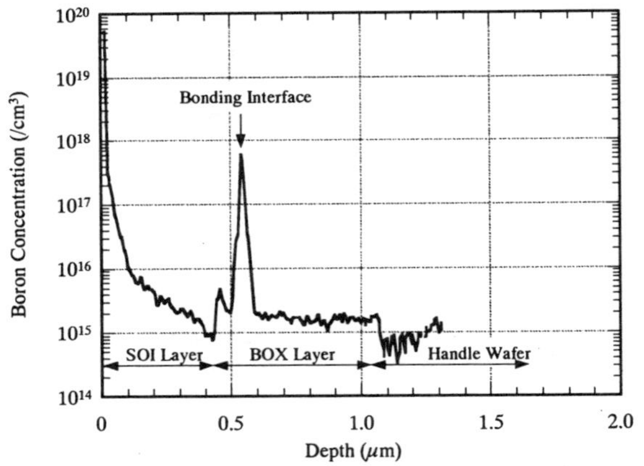

Figure 16. Depth profile of boron concentration around the bonding interface formed in conventional clean room.

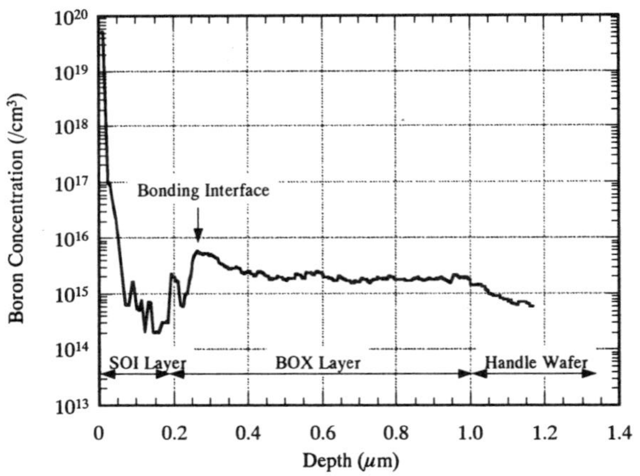

Figure 17. Depth profile of boron concentration around the bonding interface formed in boron-free clean room.

The amount of boron present at the bonding interface of an SOI wafer bonded in the clean room for which the above-mentioned countermeasures had been taken was analyzed in the same way as for **Fig. 16**. Specifically, after adsorbed boron had become trapped in the buried oxide film, the amount of boron was measured in the depth direction using SIMS. The results are shown in **Fig. 17**. It can be seen that the amount of boron at the bonding interface (situated at a depth of 0.35μm) has been successfully reduced to less than one hundredth of that for the case of the conventional clean room; there has been a dramatic decrease in the amount of boron, with a peak of only 5×10^{15} atoms /cm^3 at the bonding interface. It is thought that the reduction in the concentration of boron in the air in the clean room resulted in a reduction in the amount of boron adsorbed onto the wafer surface just before the bonding process, resulting in a reduction in the concentration of boron present at the bonding interface.

Once the amount of boron had been reduced sufficiently, the existence of phosphorous at the bonding interface became apparent. When a high-resistance wafer is used as the handle wafer (supporting substrate), the presence of this phosphorous results in as much as a 100-fold decrease as in the resistivity of the handle wafer at the bonding interface; when a high-frequency circuit is formed on the SOI layer, there is thus a degradation of the circuit characteristics. When bonding is carried out in the current clean room equipped with a chemical filter, this phenomenon cannot be sufficiently prevented. The drop in the resistivity has thus been prevented by technical improvements in the bonding process, and it has been confirmed that the high resistance of 6000Ωcm of the handle wafer is now maintained even after manufacture of the SOI wafer.

As explained above, by using a class 1 environment with virtually no particles and suppressing the generation of boron using a chemical filter and the like, a reduction in both HF defects and interface boron has been achieved.

4. Product Quality

Wafers manufactured using the current method can be made to have any of a wide range of SOI layer and BOX layer thicknesses, and are excellent in terms of surface flatness, uniformity of the layer thicknesses, and other aspects of product quality. The thicknesses of the SOI layer and the BOX layer can be controlled to be anywhere from 27nm down to a few nm, with a uniformity of better than ±5%. In actual practice, 4~8-inch diameter SOI wafers are being produced in a specially designed ultra-clean room.

Table 1 shows a list of parameters that are extremely important to SOI product quality, along with the corresponding values for ELTRAN® wafers and the 1999 ITRS (International Technology Roadmaps for Semiconductors) data of the US SIA (Semiconductor Industry Association) .

TABLE 1. ELTRAN® quality in comparison with 1999 SIA-ITRS.

	Latest SIA-INTRS (Fully Depleted)	ELTRAN
Thickness (SOI/BOX) [nm]	**30-40 /200-400**	**100/100**
LLS >0.09μm[/wafer]	**42**	**21.0**
Δt_{SOI} ± [%]	**< ± 5**	**± 2.37**
Δt_{BOX} ± [%]	**< ± 5**	**± 1.00**
Surface Roughness/ Ra 1μm x 1μm [nm]	**(0.1~0.15)**	**0.107**
HF Defect [cm^{-2}]	**0.064**	**0.032**
Secco Defect [cm^{-2}]	**$2x10^6$**	**434**

The surface micro-roughness Ra is of the order of 0.1nm regardless of whether the splitting method or the grinding method is used to produce ELTRAN® wafers. The surface on ELTRAN® wafer is smoothed by annealing in a hydrogen atmosphere in place of polishing, which is used in the final process of not only conventional silicon wafers but also the other SOI wafers. This is also clear from the SEM photographs of **Fig. 13** taken before and after the hydrogen annealing.

Moreover, it can be seen from **Table 1** that with ELTRAN® SOI wafers an extremely low level of HF defects (which are said to be fatal to such wafers) of less than 0.05 /cm^2 has been achieved. This is because in all types of SOI wafers ELTRAN® is only one SOI wafer which use an epitaxial layer as the active layer, and because there are no COPs (crystal growth defects that occur during the bulk pulling).

The density of SEDs (secco-etch defects) for the ELTRAN® SOI wafers is a few hundred per cm^2, which is significantly better than the SIA-ITRS standard value. This low density of SEDs has been achieved through the improvement in the technique for closing up the pores in the surface of the porous silicon as described earlier in the section '3.1 Processes'.

The minority carrier lifetime was measured for a number of wafers by the PCD method with the finite difference method (see **Fig. 18**)[25)]. For the ELTRAN® Epi-B-SOI wafer, a lifetime of a few hundred microseconds was obtained throughout the whole of the thin-film SOI layer. A drop in the lifetime is observed near the SOI/BOX interface, this being due to the effects of recombination at this interface.

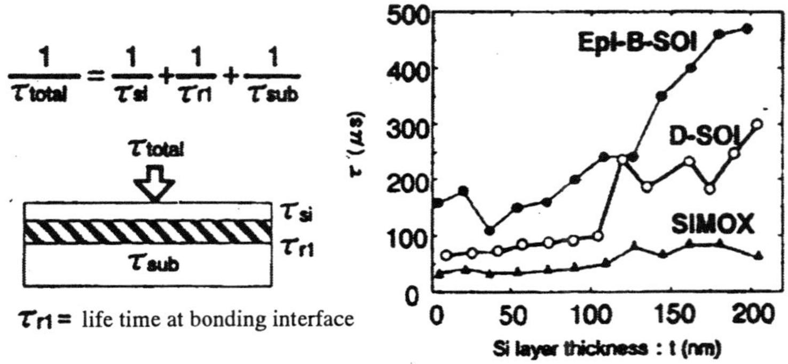

Figure 18. Minority carrier life times various SOI wafers[25].

Most recently, it is reported that a "pure" 1/f noise characteristics is observed in partially depleted floating-body SOI MOSFETs fabricated on ELTRAN® wafers, while the floating-body induced pre-kink excess noise is found in the device simultaneously formed on the other commercially available bonded SOI[26] as shown in **Fig.19**. The difference is stated to reflect the higher quality in the starting material, since the surface and active SOI layers in ELTRAN® are formed by epitaxial growth and hydrogen annealing.

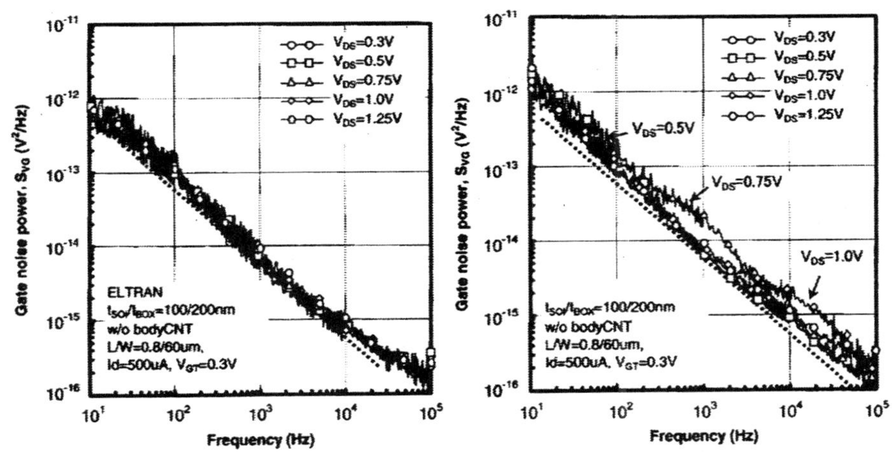

(a) (b)

Figure 19. Drain bias dependence of equivalent input-referred gate noise power spectra in PD floating-body SOI NMOS devices (without body contact) fabricated on (a) ELTRAN® and (b) the other bonded SOI. The dashed line shows an ideal 1/f spectrum.

Let us finish by mentioning the analysis of metal contamination.

With regard to contaminations in silicon wafers and in particular analysis of metals, in the past there has been debate concerning contamination by metals only at the surface. This is surely because with normal silicon wafers, only slicing, lapping, etching, polishing, and surface treatment that does not involve heating are carried out after the crystal pulling, meaning that there is no opportunity for metallic contaminants to find their way into the interior of the wafer. However, in the case of SOI wafers, regardless of the manufacturing method used, there is the possibility of contaminants being drawn into the wafer from the outside; moreover, since heat treatment at a temperature of above 1000ºC is required, there is also the problem of contaminants diffusing into the wafer.

We thus carried out an analysis of metal contaminations inside our SOI wafers using our own special method. The wafer processing procedure used was as follows. Firstly, in order to eliminate the effects of surface contamination, the native oxide film on the SOI wafer surface was removed using an HF solution (usually the state of contamination at the surface is measured using such an HF solution). Next, the following step etching was carried out using an aqueous solution of HF and HNO_3: 1) 80nm of the SOI layer; 2) 20nm of the SOI layer + 100nm of the BOX layer; 3) 80nm of the handle wafer; 4) a further 80nm of the handle wafer. When step etching in this way, control of the thickness etched is very important, since this affects the reproducibility of the measurements and of the minimum limit of detection. The amount of HF necessary for each stage of etching was thus calculated in advance in accordance with the wafer diameter, and then the etching was carried out using an aqueous solution containing this amount of HF along with HNO_3. In this way, once the required thickness had been etched, the HF ran out and the etching automatically came to a halt.

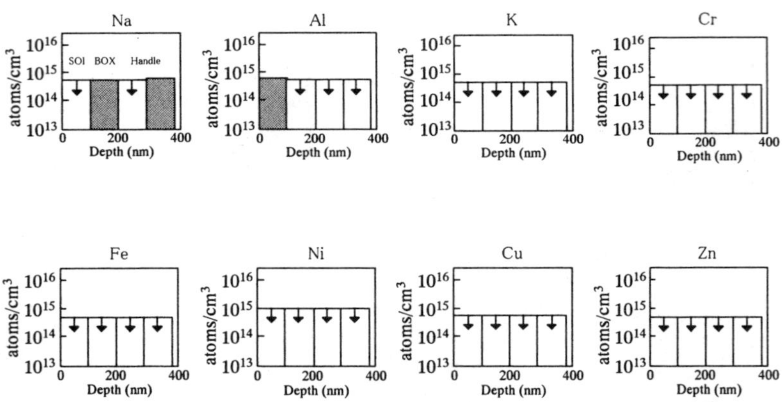

Figure 20. Results of inside metal concentration in ELTRAN®.

Concentration measurements were taken on the etching solution using ICP-MS and frameless AAS (atomic absorption spectroscopy). The results are shown in **Fig. 20**, which shows the distribution of the concentration of each of a number of elements in the depth direction of the wafer. Each bar corresponds to one of the etching stages, and the vertical axis shows the volume concentration. As a rough guideline, a volume concentration of 10^{15} atoms/cm^3 corresponds to an area concentration of about 10^{10} atoms/cm^2. A downward arrow indicates that the concentration for that bar was below the minimum limit of detection. It can be seen from the figure that for both the SOI layer and the BOX layer no element was present in a concentration of more than

10^{15} atoms/cm^3, showing that an extremely high level of purity is achieved with ELTRAN® wafers not just on the surface but also in the interior.

5. Mass Production

As described above, we made good use of ultra-clean technology and adapted our manufacturing line to cope with 8-inch wafers. We then started mass production in 1997, making all of the 8-inch line equipment 'cassette-to-cassette', in order to automate production. With the exception of the four pieces of equipment that we developed ourselves, namely the anodization equipment, the bonding equipment, the wafer-splitting equipment and the porous silicon etching equipment, all of the equipment was that which is used in normal semiconductor processes, meaning that the potential for mass production was good from the outset. The four pieces of equipment that we developed ourselves were all developed at last with the aim of making the production line fully automatic. With the anodization equipment and the etching equipment, the processing capacity was increased by means of batch processing. With both the bonding equipment and the wafer-splitting equipment, since the mechanical accuracy and the location accuracy are important, the single wafer processing was adopted and efforts were made to reduce the tact time accordingly. In this way, the processing capacity of this equipment was made to be as good as or better than that which has already come of age in conventional semiconductor processes, thus making mass production possible.

Another important factor that decides whether mass production is possible or not is the variation between wafers within lots and between lots. However high the processing capacity, if the wafer characteristics vary between wafers in a lot, between lots or even across an individual wafer, then the product yield will drop markedly. Here we will focus in particular on the variation between the wafers between lots of the thickness of the epitaxial silicon layer, the number of LPDs (light point defects) and the density of secco-etch defects.

Fig. 21 shows a graph of the trend in the thickness of the epitaxial silicon layer. The average thickness of each wafer and the variation in thickness across that wafer has been plotted for 450 wafers [produced over a period of time. It can be seen that there was initially a lot of variation in the average thickness in a wafer relative to the target thickness of 180nm, but the most recent 150 wafers show an extremely consistent value. This consistency was achieved by optimizing the surface of the porous silicon layer and improving the epitaxial reactor. The uniformity of the thickness across an individual wafer was extremely consistent right from the outset, with a good value of about 3~8nm for the variation in thickness across a wafer being maintained throughout the set of 450 wafers.

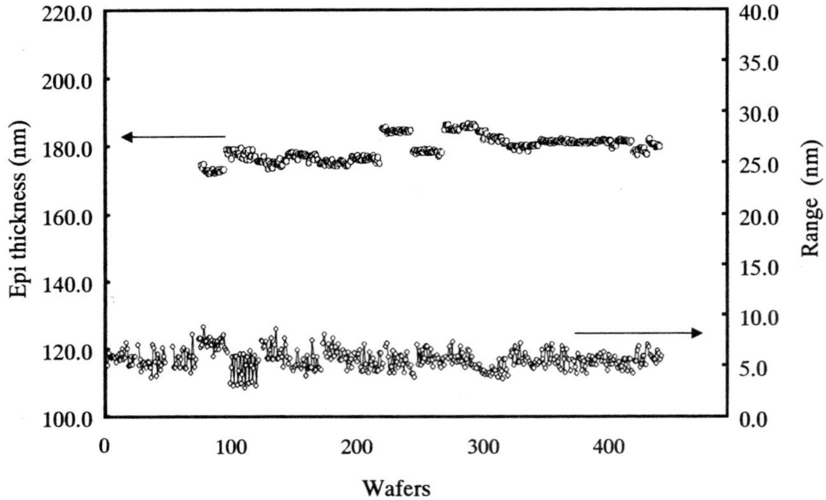

Figure 21. Trend graph of epitaxial Si thickness and their uniformity in *450* wafers.

The number of LPDs is affected greatly by the operational stability of the equipment. When the equipment has just been started up, a lot of particles are produced and these are detected as LPDs on the SOI wafer. As shown in **Fig. 22**, once operation of the equipment has settled down, a good figure of less than 50 LPDs per wafer is maintained.

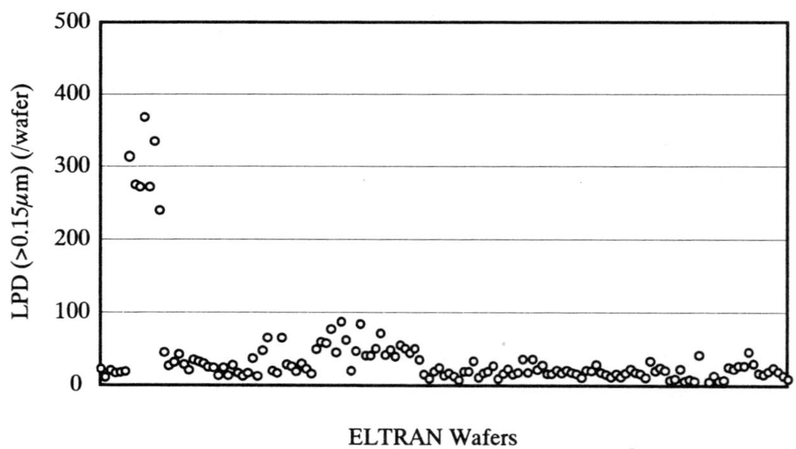

Figure 22. Trend graph of particles on *150* ELTRAN® wafers.

The processing capacities of the various processes must be made such that no bottlenecks occur along the production line. If there is even one process that has a poor processing capacity, then this will determine the overall processing capacity, however much the processing capacities of the other processes are raised. Moreover, making the product quality consistent is extremely important to the users of the SOI wafers, namely device manufacturers. If the wafer thickness or the number of LPDs or other defects varies from year to year or month to month, then device characteristics and yield are liable to be affected. An easy way of judging whether mass production is possible with a certain technology is thus to look at the processing capacities of each of the individual processes involved and the consistency of the product quality.

6. Cost-Effectiveness

In this section we will focus in particular on the wafer-splitting, recycling and reusage techniques.

6.1 COST REDUCTION EFFECTS

Reducing production costs is just as important to expanding the SOI wafer market as product quality. The ELTRAN® manufacturing process used up to the present involves making each SOI wafer by bonding a seed wafer to a handle wafer and then grinding, etching and so on. In order to make each SOI wafer, 2 starting wafers must thus be used. We have established a technique for recycling the seed wafers (which were always thrown away after the grinding and etching processes in the past), thus achieving a large reduction in costs[27)].

The basic ELTRAN® concept of 'split/recycle/reuse' involves reusage of the seed wafer. It can be seen from the ELTRAN® manufacturing process that the seed wafer itself is ultimately not used in any part of the SOI wafer. This is a special feature of ELTRAN® wafers not found with other SOI wafers. It arises because 1) the seed wafer is not used as the handle wafer, and 2) the SOI layer does not come from part of the seed wafer, but rather comes from the epitaxial layer. The special features of SOI-Epi wafers™ come to the fore all the more thanks to this technique for reducing the manufacturing costs. Turning to the handle wafer, a new handle wafer is put into the manufacturing process each time an SOI wafer is made, the epitaxial layer is also grown anew each time, and of course the thermally grown oxide film that becomes the BOX is formed anew each time by oxidizing the epitaxial layer.

In this way, because each completed SOI wafer is always made entirely of new materials, even if the seed wafer is reused again and again there is no degradation in the product quality of the SOI wafers.

Let us now examine the extent of the reduction in wafer material costs due to such repeated reuse of the seed wafers. The following equation represents the wafer material costs when the 'seed wafer reusage' concept is used[7)].

$$Y = H + \frac{S + R\cdot(n-1)}{n} \qquad \textbf{(Eqn. 1)}$$

Here, Y is the SOI wafer material cost, H is the price of the handle wafer, S is the price of the seed wafer, R is the cost of reclaiming the seed wafer, and n is the number of times that the seed wafer is used.

As n is increased, according to the above equation:

$$Y \rightarrow H + R \qquad \textbf{(Eqn. 2)}$$

In other words, the SOI wafer material cost becomes the price of the handle wafer plus the cost of reclaiming the seed wafer, and the original price of the seed wafer is essentially irrelevant.

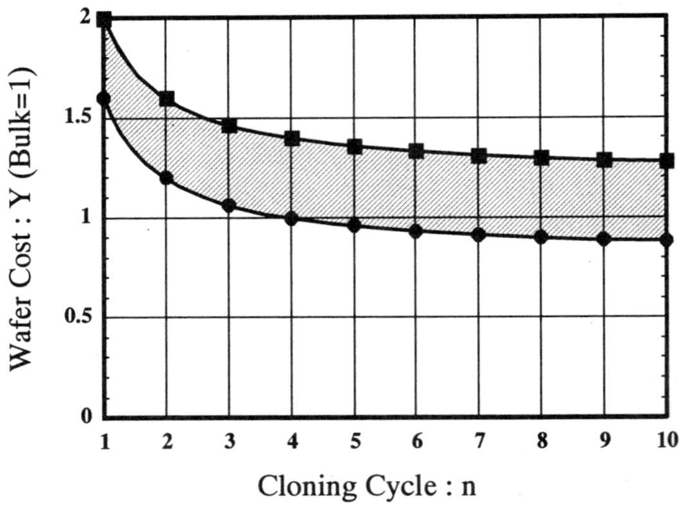

Figure 23. Wafer material cost dependent on usage number of seed wafers. The cost greatly depend on handle wafer material grade. Assuming that SOI wafer production yield is *100 %*.

Fig. 23 shows the graph of **Eqn. 1**. Note, however, that the following assumptions have been made.

a) The SOI wafer production yield is 100%.
b) Process costs are not included.
c) Things like depreciation and labor costs are also not included.

In other words, only the cost of the starting wafers has been considered. The SOI wafer material cost depends greatly on the handle wafer specifications; this causes a change in the gray part of the graph. Regardless of the type of handle wafer used, the cost reduction effect is very large for each time the seed wafer is used up to 4 times. After that, the cost reduction effect upon reusing the seed wafer becomes relatively small, with the cost asymptotically approaching the total of the price of the handle wafer and the cost of reclaiming the seed wafer as in **Eqn. 2**.

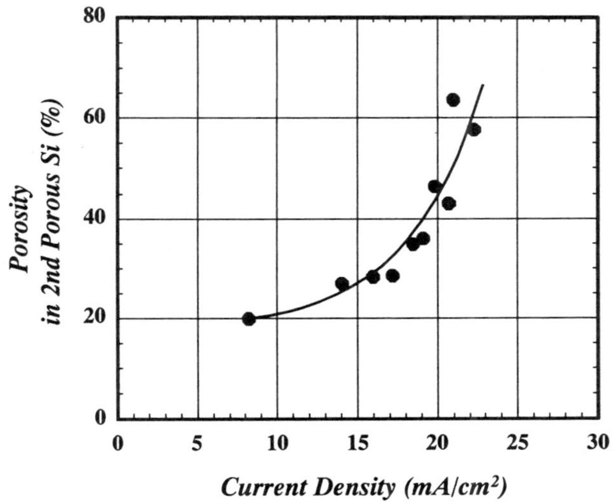

Figure 24. The porosity in the 2nd porous Si layer versus the current density during anodization.

6.2 DOUBLE POROUS SILICON LAYERS[14-16]

We have already explained in section 3 'Processes and use of a clean room' that a multi-layered porous silicon layer that consists of two layers of different porosity can easily be created by increasing the current in the middle of the anodizing process. The first porous silicon layer must have low porosity in order that the epitaxial silicon layer will be of high quality; the porosity of this porous silicon layer cannot be varied in order to improve the splitting. In other words, it is fair to say that optimizing the porous silicon layer for splitting is purely down to optimizing the structure (i.e. porosity) of the second porous silicon layer. **Fig. 24** shows the dependence of the porosity of the second porous silicon layer on the current density during anodization. It can be seen that it is possible to control the porosity of the second porous silicon layer to anywhere within the range 20~60% by changing the current density.

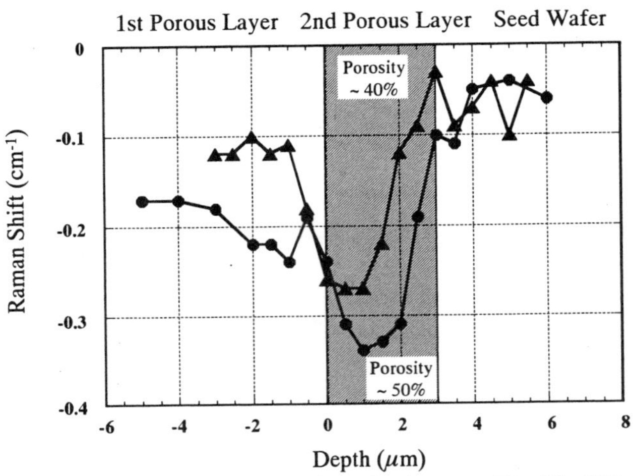

Figure 25. Raman shift depth profile around the 2nd porous Si with different porosity of *40 %* and *50 %* just before splitting.

The strain around the second porous silicon layer in the bonded wafer just before splitting was measured using micro Raman spectroscopy for two wafers with second porous silicon layers of different porosity (40% and 50%). The porosity of the first porous silicon layer was the same in both cases. The Ramon shift depth profiles observed in both cases was similar. The results are shown in **Fig. 25**. It can be seen that the Raman shift showed the maximum negative value in the second porous silicon layer close to the interface between the two layers. Moreover, this shift was larger in magnitude when the porosity of the second layer was the higher of the two values.

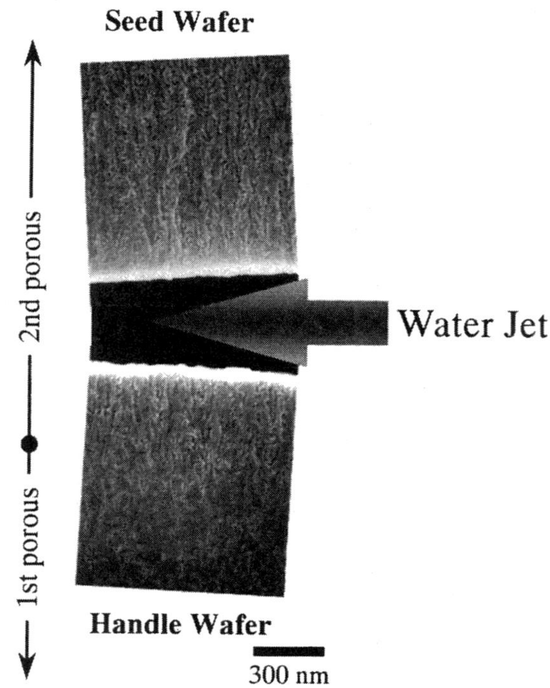

Figure 26. Cross sectional micrographs of as-split wafers. The splitting plane is located around the interface between 1st and 2nd porous Si layers.

Cross-sectional SEMs of the region around the splitting plane for the two wafers (the seed wafer and the handle wafer) after splitting the bonded wafer are shown in **Fig. 26**. Roughly speaking, the first porous silicon layer ends up on the handle wafer side, while the second porous silicon layer remains on the seed wafer side. However, splitting occurs not exactly at the interface between the two layers but rather slightly into the second layer. The splitting plane corresponds to the position of maximum negative Raman shift in the depth direction as shown in **Fig. 25**. In other words, splitting occurs at the plane where the stress is at a maximum. The magnitude of the Raman shift increases when the porosity of the second porous silicon layer is increased, which suggests that the higher this porosity the easier it should be to consistently split the bonded wafer right at the place where the magnitude of the Raman shift is a maximum.

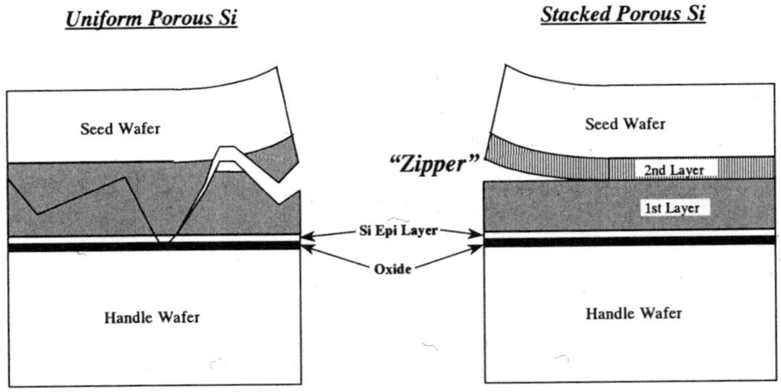

Figure 27. Schematic drawings of crack propagation in double layered porous Si and single layered porous Si.

In this way, the porosity of the second porous silicon layer is an extremely important factor in successfully restricting the splitting plane to be within a certain limited region. **Fig. 27** shows schematic drawings of the progression of the splitting plane for a single-layered porous silicon of uniform porosity and a double-layered porous silicon made up of two layers of different porosity. If the porosity is uniform throughout the porous silicon layer then splitting can occur at any depth, meaning that the progression of the splitting plane cannot be controlled and cracks may reach as far as the seed wafer or the epitaxial layer. However, in the case of a double-layered porous silicon, splitting occurs close to the interface between the two layers, and so these layers act to protect the seed wafer and the epitaxial silicon layer respectively, thus providing a definitive means for raising yield.

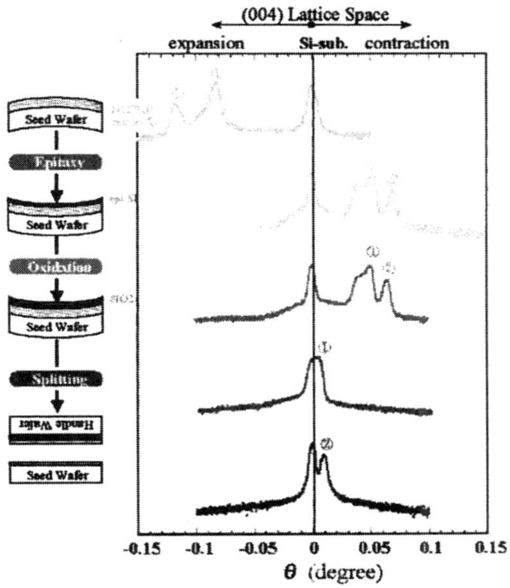

Figure 28. X-ray diffraction rocking curves of each wafers before and after ELTRAN® major processes.

Next, with the aim of understanding the mechanism by which stress is generated in the double-layered porous silicon consisting of two layers and the splitting mechanism, the stress values in the two layers were evaluated. An extremely effective way of evaluating the stress value in a single-crystal layer is to measure the lattice spacing by X-ray diffraction. A double-crystal X-ray rocking curve has been already used in analyzing the stress in porous silicon[9)]. In the case of a double-layered porous silicon with different porosity, it is also necessary to detect the lattice strain between the layers, and so we used a 5-crystal X-ray diffract meter having improved resolution. Above all else, in the case of ELTRAN® by splitting, gaining an understanding of the dynamic stress changes during the ELTRAN® process is the key to improving the splitting yield.

First of all, an investigation was carried out into the way in which the profile of the X-ray rocking curve (i.e. the lattice stress/strain) changes during the ELTRAN® manufacturing process. The results are shown in **Fig. 28**. In the porous silicon consisting of two layers of different porosity, before epitaxial growth the lattice spacing is extended in the direction perpendicular to the wafer surface, with the lattice spacing expansion factor in this direction for the second porous silicon layer (2.97×10^{-3}) being about 1.5 times that for the first porous silicon layer (2.09×10^{-3}). After the epitaxial growth, the lattice of the porous silicon shrinks and the corresponding peaks in the rocking curve shift to the high angle side relative to the substrate peak; the contraction factor is now 1.32×10^{-3} for the first layer and 1.76×10^{-3} for the second layer. The fact that the expansion of the lattice of porous silicon changes to a contraction on epitaxial growth was already known[28-32)], but this is the first time that observations have been carried out on a double porous silicon layers with different porosity. Before the high-temperature heat treatment, the second porous silicon layer with the higher porosity is more expanded than that with the lower porosity and the lattice spacing is greater, but when the porous silicon is made to undergo the high-temperature heat treatment (in order to cause epitaxial silicon growth), the second porous silicon layer with the higher porosity contracts more than that with the lower porosity, with the lattice spacing of the former shrinking by about 1.3 times as much as that of the latter. This lattice expansion and contraction has been explained as follows[33]. Since the porous silicon is formed by anodizing in an HF solution, during the anodization hydrogen diffuses into the crystalline region of the porous silicon. This hydrogen causes stretching of the Si-Si bonds, explaining the lattice expansion of the porous silicon. When the porous silicon is subsequently heat-treated, the first thing that happens is that the hydrogen is desorbed and so the lattice spacing returns to its original value. Up to this point, the expansion and contraction of the lattice can be explained purely by the adsorption and desorption of hydrogen. In order for there to be further contraction, however, another force must be at work. It is thought that this is probably the surface tension acting on the surfaces of the pores that accompanies minimization of the surface energy. In such a case, a lattice contraction of around $2\sim3\times10^{-3}$ has been measured, which agrees pretty well with the contraction factor observed in our case.

The peak just to the low angle side of the peak corresponding to the first porous silicon layer in the rocking curve after the epitaxial growth is a signal from part of this layer; it is not the diffraction peak of the epitaxial silicon layer. It has been found that as the epitaxial layer is made thicker, the diffraction peak at the position corresponding to the substrate gets larger. In other words, the epitaxial layer must have the same lattice spacing as the substrate and so must be subject to no strain.

There is virtually no change in the rocking curve profile upon surface oxidation of the epitaxial layer. Bonding and splitting are then carried out, and the first and second porous silicon layers go to the handle wafer side and the seed wafer side respectively. Evaluation must now be carried out for the two wafers separately. For both porous silicon layers, the diffraction peak due to the layers approaches that of the respective substrate, showing that there is very little residual stress. It is thought that this is because the internal stress in each of the porous silicon layers is released when the porous silicon layers are split from one another, and as a result the lattice strain is eased. It was found that after this release of stress, the 'in-plane stress' (here and in the following 'in-plane stress' refers to the stress in a plane parallel to the wafer surface) has dropped from 1.35×10^9 dyn/cm^2 to 1.31×10^8 dyn/cm^2 for the first layer and from 1.90×10^9 dyn/cm^2 to 2.88×10^8 dyn/cm^2 for the second layer.

Let us now examine the in-plane stress of the porous silicon layer. The porous silicon is formed on the surface of a single-crystal silicon bulk wafer, and so there is virtually no freedom for the lattice to expand or contract in-plane. If one looks at the X-ray diffraction reciprocal lattice map, it can be seen that the in-plane lattice spacing varies little across the interface between the bulk wafer and the porous silicon layer. Expansion/contraction of the lattice is thus limited to the direction perpendicular to the wafer surface. It was concluded from such X-ray diffraction measurement results that the following kind of stress acts in the two porous silicon layers.

If the porous silicon was in a free state not connected to the single-crystal silicon bulk wafer, then it would expand/contract equally in all directions. However, in actual practice the in-plane lattice spacing of the porous silicon is regulated by the single-crystal silicon bulk wafer, meaning that the porous silicon receives in-plane stress from the single-crystal silicon bulk, and the in-plane lattice spacing becomes the same as that of the single-crystal bulk silicon. According to material mechanics, the relationship between the in-plane stress and the lattice spacing in the perpendicular direction can be expressed in terms of the Young's modulus and the Poisson ratio of the material in question. In other words, if the lattice spacing in the perpendicular direction is determined from the angle of the X-ray diffraction peak, and in addition the Young's modulus and Poisson ratio for the crystalline part of the porous silicon are known, then the in-plane stress can be calculated. Up to the present time, particularly in the case of porous silicon, the Young's modulus has usually been discussed in terms of the relationship between the curvature of the warp and the thickness[32,34]. In such a discussion, it is necessary to treat the porous silicon as being a homogenous medium. In other words, it is the average Young's modulus for the pores and the crystalline silicon part that has been considered in the past. However, the stress acts only on the crystalline silicon part, and so in the crystallographic discussion here we have decided to ignore the porosity and use the Young's modulus of single-crystal silicon as the Young's modulus of the porous silicon. The Poisson ratio, on the other hand, has been derived from X-ray diffraction results even in past discussions, and so we have decided to consider use the relationship between the Poisson ratio and the porosity as follows[32].

$$\nu = 0.278 - 0.348\rho \qquad \textbf{(Eqn. 3)}$$

Here, ν is the Poisson ratio and ρ is the porosity.

Moreover, the lattice spacing can be determined from the X-ray diffraction angle as follows.

$$d_z = d_0 \cdot \left(1 - \frac{\theta_1 - \theta_0}{\tan\theta_0}\right) \qquad \textbf{(Eqn. 4)}$$

Here, d_z is the spacing in the porous silicon layer in the direction perpendicular to the surface, d_0 is the spacing in the single-crystal silicon bulk wafer in this direction, θ_1 is the diffraction angle for the porous silicon, and θ_0 is the diffraction angle for the single-crystal silicon bulk wafer.

The stress σ is represented using the basic equation of materials mechanics as follows.

$$\sigma = E \times \frac{d_{xy} - d_z}{2\nu d_{xy} + (1-\nu)d_z} \qquad \textbf{(Eqn. 5)}$$

Here σ is the biaxial in-plane stress of the porous silicon layer, E is the Young's modulus of the porous silicon layer, ν is the Poisson ratio of the porous silicon layer, d_{xy} is the in-plane lattice spacing of the porous silicon layer, which is equal to d_0, and d_z is the lattice spacing for the porous silicon layer in the direction perpendicular to the surface.

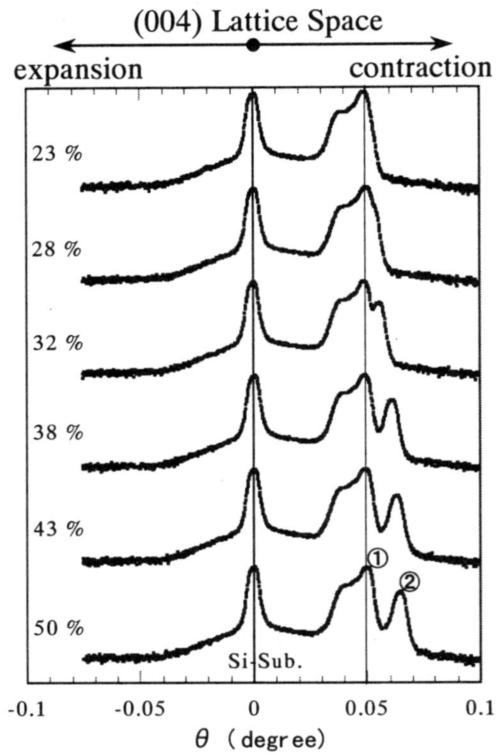

Figure 29. X-ray diffraction rocking curves of each wafers with the different porosity of *23 % - 50 %* after the epitaxial surface oxidation process.

For the Young's modulus, the value for single-crystal silicon is used. The Poisson ratio is determined from the porosity using **Eqn. 3** above. For d_{xy}, the value d_0 for the single-crystal silicon bulk wafer is used as explained above. d_z is calculated from the X-ray rocking curve. In this way, it is thus possible to calculate the stress acting in-plane from X-ray diffraction angle measurements.

The next thing investigated was the change in the stress for each of the porous silicon layers with the porosity of the second layer. **Fig. 29** shows the rocking curves after the epitaxial silicon surface oxidation process for each of a number of second layer porosities. It can be seen that when the porosity of the second layer is low at 23%, the peaks for the first and second layers more-or-less overlap, with very little separation between the two. However, as the porosity of the second porous silicon layer is raised, the peak for this second layer shifts to a higher angle; in this case, there is virtually no change in the peak for the first porous silicon layer. For each of the second layer porosities, the in-plane stress for each of the porous silicon layers was calculated from the rocking curve using the above-mentioned method. The results are shown in **Fig. 30**, where the in-plane stress for each of the two layers is plotted against the porosity of the second porous silicon layer. It can be seen that the in-plane stress for the second porous silicon layer increases linearly with the porosity of this layer. However, the in-plane stress for the first porous silicon layer is unaffected by the porosity of the second layer. This is because the porosity of the first layer is unaffected by the second porous silicon layer and is constant. This in turn is due to the special nature of the porous silicon reaction process. In the porous silicon, pores of diameter around 10nm form at a spacing of about ten to a few tens of nm. During the pore formation, the regions of single-crystal silicon between pores become electrically depleted, meaning that the formation current is only provided to the far ends of the pores and the anodization reaction occurs here only. Once a region of porous silicon has been formed, it is no longer affected by the continuing anodization process. The structure, porosity and stress of each of the two porous silicon layers can thus be controlled *independently of one another* by means of the anodization current.

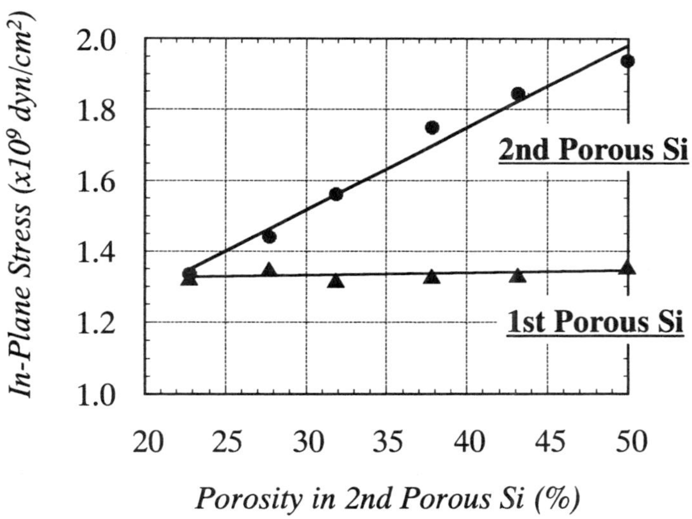

Figure 30. In-plane stress in 1st and 2nd porous Si layers as a function of the porosity in 2nd porous Si layers.

The difference in stress between the first and second porous silicon layers causes distortion in a confined region around the interface between the two, causing strain energy to accumulate and thus a high-energy state to be generated. Once splitting starts to occur in this region, the leading edge of the split progresses in such a way as to reduce the accumulated strain energy, and so the splitting plane remains close to the interface between the two layers. In other words, it is the 'desire' to reduce this strain energy that is the 'driving force' that makes splitting progress uniformly with the splitting plane remaining close to said interface.

6.3 SPLITTING WITH A WATER JET[15,16,22)]

The bonded wafer is split near to the interface between the first and second porous silicon layers using a water jet. We have tried a whole range of different splitting methods: thermal stress, oxidation from the edge of the porous silicon, application of ultrasonic waves, insertion of a solid wedge, and insertion of a fluid wedge using a water jet. Thermal stress had virtually no effect. Oxidation caused serious warping of the wafer. With ultrasonic waves, splitting across the whole wafer was possible, but reproducibility and controllability were poor. With both the solid wedge and the fluid wedge, splitting across the whole wafer was possible, but the solid wedge caused damage to the wafer as will be described below. Considering all of the above, it was decided to adopt the water jet method. It is thought that when the bonded wafer is split using a water jet, the primary mechanism is not cutting but rather peeling by the fluid wedge produced by the water jet. The setup that we are currently using in actual practice involves a straight water jet used only with high purity water designed to be used in the manufacture of semiconductors. If the nozzle bore is made to be 0.1mm then splitting with a jet pressure of 20~60MPa is possible.

Let us now explain the theoretical background to water jets[35)].

The basis of understanding a water jet is Bernoulli's theorem. This is the energy conservation law of hydrodynamics, and can be expressed as in the following equation.

$$\frac{1}{2}av^2 + agh + p = const. \qquad \textbf{(Eqn. 6)}$$

Here, a is the density of water, v is the flow velocity of the water, p is the water pressure, h is the height, and g is gravitational acceleration. According to this equation, the water hummer pressure can be expressed as follows.

$$\begin{aligned} p_w &= acv \\ &= ac\sqrt{\frac{2p}{a}} \\ &= c\sqrt{2ap} \end{aligned} \qquad \textbf{(Eqn. 7)}$$

Here p_w is the water hummer pressure and c is the speed of sound in water. Moreover, the quantity of flow is given as follows.

$$\begin{aligned} Q &= sv \\ &= s\sqrt{\frac{2p}{a}} \end{aligned} \qquad \textbf{(Eqn. 8)}$$

Here Q is the quantity of flow and s is the cross-sectional area of the nozzle.

Using the 'fluid wedge effect' of a water jet in the splitting of a bonded wafer can be expected to result in the superiority described below. The above-mentioned equations for the water hummer pressure and the quantity of flow are very important in that they lend theoretical support to this superiority.

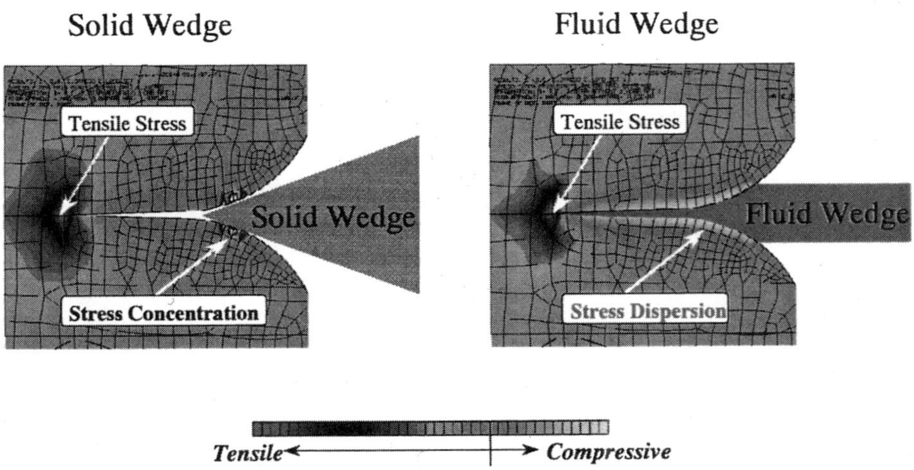

Figure 31. Stress simulation of bonded pairs just before splitting by (a) solid wedge and (b) fluid wedge.

According to **Eqn. 7**, the water hummer pressure when the diameter of the nozzle is 0.1mm is of the order of 0.24~0.43N. This is insufficient to cut through a silicon wafer, which is why it is thought that the fluid wedge peeling effect is dominant in working to split the bonded wafer. Due to the ability of a fluid to change its shape, the water jet penetrates as far as the periphery of the bonding interface and causes a buildup of pressure, meaning that the water hummer pressure is felt across a broad region of the wafer periphery. Unlike with a solid wedge, there is not a single point of contact with the wafer edge, and so splitting can take place without stress being concentrated in one particular place, meaning that cracking of the wafer periphery can be prevented.

Fig. 31 shows stress measurements for the water jet and a solid wedge carried out using a finite element method simulator for the envisaged situation at the start of wafer splitting. In this model, the splitting force and the water hummer pressure exerted on the wafer by the solid wedge and the water jet respectively at the start of splitting were both made to be as close to the true situation as possible by using values obtained from experiments. Looking at **Fig. 31**, it can be seen that both the solid wedge and the water jet give a tensile stress of about 70MPa around the periphery of the bonding interface, meaning that both generate more-or-less the same splitting force. Looking now at the extent to which the stress is concentrated on the wafer edge, however, it can be seen that because the solid wedge comes into contact with the wafer at essentially a single point but the water jet comes into contact with the wafer over a wider area, there is a point where the stress generated by the solid wedge is 10 times that generated by the water jet, showing that there is a much greater chance of cracking with the solid wedge.

This pressure-dispersing effect of the water jet continues to apply throughout the splitting. When the fluid wedge is inserted, the force at which the water jet strikes the wedge is kept extremely low so that splitting just about manages to proceed along the plane where the internal stress is locally concentrated (i.e. close to the interface between the two porous silicon layers). This means that a pressure that is dispersed more-or-less equally across the whole of the splitting plane can be applied, and so splitting can be achieved without the exertion of excessive force.

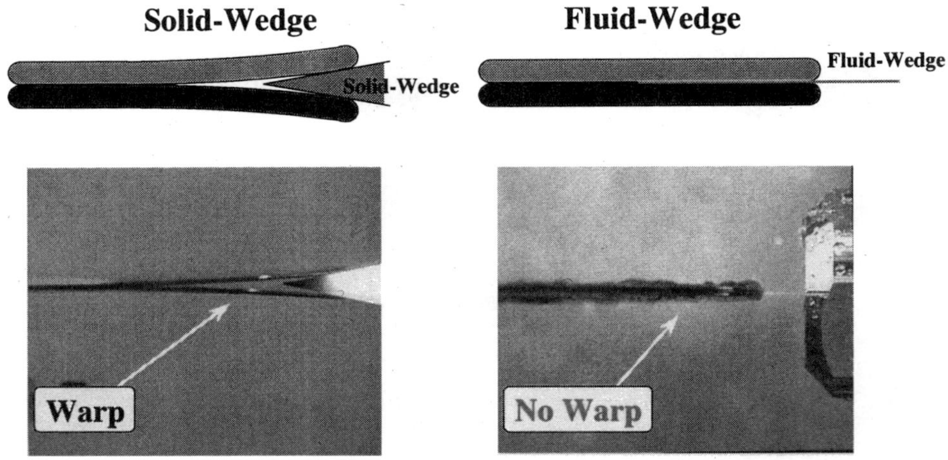

Figure 32. Schematic drawings and side view photographs during splitting the bonded pairs by (a) solid wedge and (b) fluid wedge.

Another effect of the fluid being able to change its shape is that splitting can be achieved without causing warping of the wafer. As shown in **Fig. 32**, with the solid wedge the mechanism is such that splitting is made to proceed by applying a force to the wafer edge. This means that as the splitting proceeds, the wafer is forced to change shape excessively, and so there is a risk of damage, in particular to the thin-film SOI layer and the buried oxide film layer. In the worst cases, the wafer may even crack during splitting. Moreover, when the solid wedge is inserted, it is bound to scrape the wafer at the place of contact, leading to problems such as scratches and the introduction of particles. With the water jet, on the other hand, the mechanism is such that the wedge is constantly changing in shape and so the pressure is exerted across the whole of the region where the splitting is proceeding. This means that the splitting can be made to proceed with virtually no deformation of the wafer. This will be extremely advantageous when the diameter of wafers is increased to 300mm as expected in the future. Furthermore, because pure water can be used in the water jet, not only is there no risk of the wafer being scratched as in the case of the solid wedge, but the problem of particles becoming stuck to the wafer is also solved.

The amount of water used is extremely small, meaning that running costs can also be dramatically reduced. According to **Eqn. 8**, only about 366cc of pure water is required to split an 8-inch bonded wafer.

Due to the above, we concluded that a water jet offered the optimum method for the bonded wafer splitting process, focusing on the 'fluid wedge' effect of the water jet in reaching this decision.

Let us now discuss the bonded wafer splitting process.

Since our wafers are disc-shaped, from the point of view of the efficiency of the splitting process we decided to adopt a setup whereby the bonded wafer is clamped in a rotating holder and the water jet is projected while rotating the wafer. **Fig. 33** shows a schematic diagram of this setup, while **Fig. 34** shows a photograph taken from the side during splitting. Due to the rotation, the splitting plane progresses in a spiral fashion. In order to control the progression of the splitting plane, the nozzle is gradually moved from the wafer periphery towards the center. This means that the splitting region can be prevented from progressing unnecessarily, and so by incorporating a sequence whereby the nozzle is gradually moved towards the center of the wafer while controlling the splitting region, spiral splitting of high reproducibility can be achieved. **Fig. 35** shows a graph of measurements of the size (diameter) of the region yet to be split while moving the nozzle from the periphery of the bonded wafer towards the center. At any given time the region yet to be split is circular. It can be seen from **Fig. 35** that due to the way in which the nozzle is made to progress the diameter of the region yet to be split decreases linearly; in this way the progression of the splitting plane is controlled.

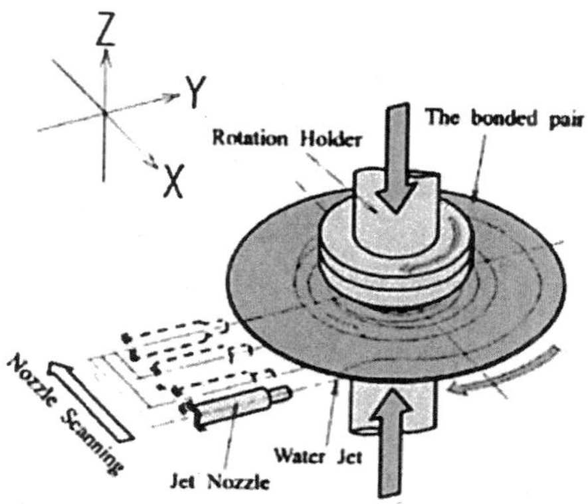

Figure 33. Schematic drawings of splitting sequence.

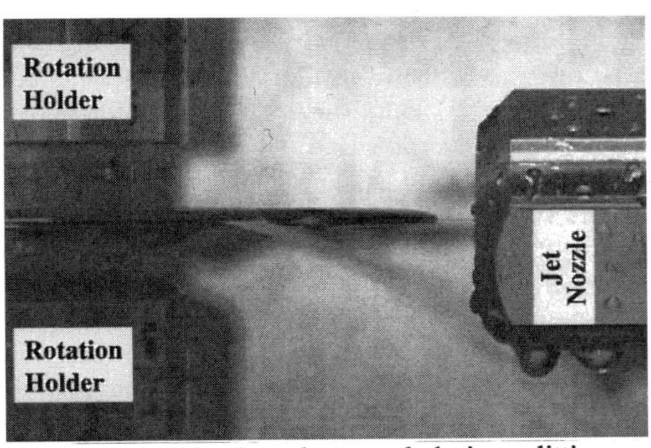

Figure 34. Side view photograph during splitting.

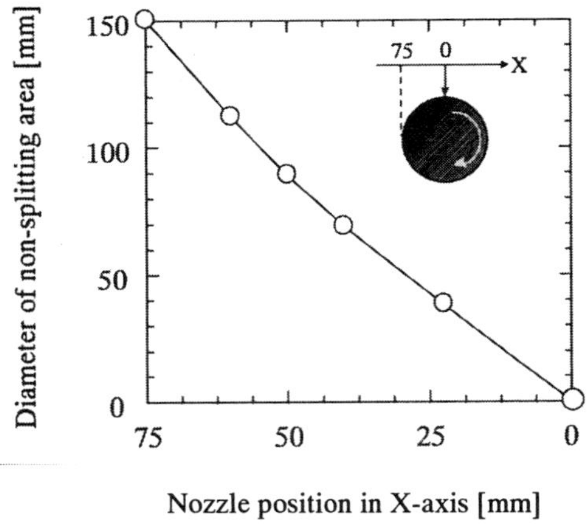

Figure 35. Relationship between nozzle position and non-splitting area.

Since such a splitting process was to be introduced into the ELTRAN® manufacturing process, the equipment involved was developed while focusing on the following 2 points.

a. The accuracy of the equipment must be such that the wafer splitting is reproducible.
b. The equipment must be suitable for use in the ELTRAN® manufacturing process in terms of cleanliness.

When carrying out automatic processing of a number of wafers, if the way in which the water jet strikes the bonded wafer varies greatly, then the splitting may not be reproducible. In terms of the equipment accuracy, controlling the position of the nozzle and the bonded wafer in the direction perpendicular to the wafer surface is thus very important.

Moreover, of processes in the field of semiconductors, the ELTRAN® manufacturing process is particularly sensitive to particles. Most of the individual processes involved in the ELTRAN® manufacturing process are thus carried out in a class 1 clean room. When putting the equipment into the ELTRAN® clean room, various measures against particles and contamination were taken.

The mass production version of the water jet equipment (which uses the 'cassette-to-cassette' method) is shown in **Fig. 36**. The equipment consists of a wafer aligner, a wafer-flipping part, a wafer-cleaning part, a splitting chamber, a loader for bonded wafers and two unloaders for each of the split wafers, with a double-arm robot. The chamber contains a splitting part equipped with a rotating holder as described above. By combining such measures as high accuracy and cleanliness, it was possible to make the water jet splitting equipment such that it could be withstand being used in the ELTRAN® manufacturing process.

Thanks to the precise control of the stress in the porous silicon and the fluid wedge effect of the water jet described above, a bonded wafer splitting yield of a consistently high value has been realized even for the case that each seed wafer is used 3 times, and in fact it has even become possible to use each seed wafer a fourth time.

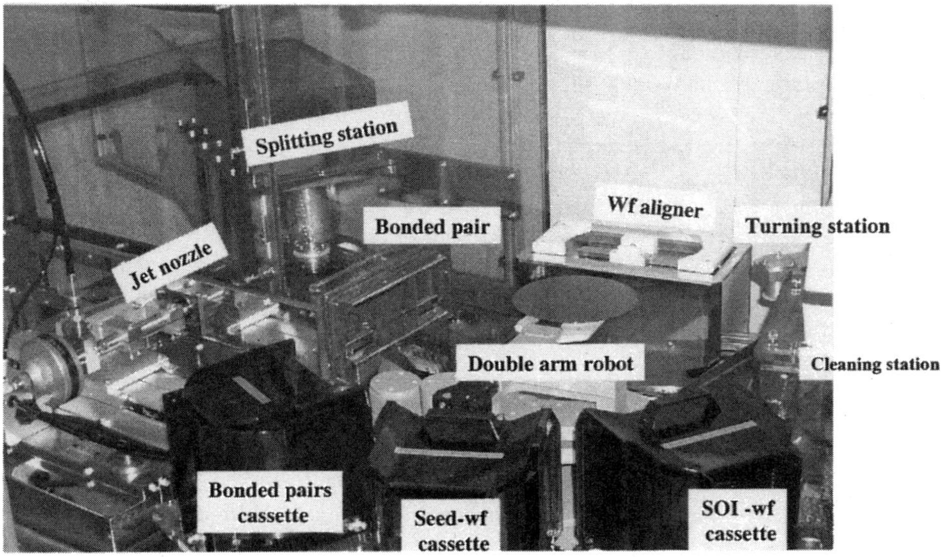

Figure 36. Custom-made automated water jet splitting machine.

TABLE 2. ELTRAN® quality by water jet splitting up to 3rd generation.

	ELTRAN®	Water Jet		
		1st	2nd	3rd
Thickness (SOI/BOX) [nm]	100/100	100/100	100/100	100/100
LLS [/wafer]	21.0	19.4	32.8	21.4
Δt_{SOI} ± [%]	± 2.37	± 2.21	± 2.68	± 2.67
Δt_{BOX} ± [%]	± 1.00	± 1.08	± 1.05	± 0.80
Surface Roughness/ Ra 1μm x 1μm [nm]	0.107	0.106	0.101	0.121
HF Defect [cm^{-2}]	0.032	0.037	0.043	0.023
Secco Defect [cm^{-2}]	434	440	764	692

Table 2 shows a list of parameters that are extremely important to the product quality of SOI wafers and the corresponding values for ELTRAN® wafers. The ELTRAN® product quality achieved in the past is shown along with that for 3 generations of ELTRAN® wafers made using the current splitting/recycling process. It

can be seen that the numbers of HF defects for the 3 generations of wafers made using the current process (0.037, 0.043 and 0.023 /cm^2 respectively) are all under 0.05 /cm^2. Moreover, the numbers of secco-etch defects are 440, 764 and 692 /cm^2 respectively, figures that are comparable to that obtained in the past. It can thus be seen from the table that the product quality of the ELTRAN® wafers produced using the current splitting/recycling process is by no means inferior to that achieved in the past. It can also be seen that there is no degradation in product quality even if splitting and recycling of the seed wafer is carried out repeatedly. This is due to the basic 'seed wafer reusage' concept unique to ELTRAN®. Because ELTRAN® wafers, regardless of the generation, always consist of a handle wafer that is introduced into the manufacturing process anew every time and an epitaxial silicon layer that is grown anew every time (along with an SiO_2 layer formed by thermal oxidation of the surface of this epitaxial silicon layer), it has been possible to realize manufacture of ELTRAN® wafers with a consistently high product quality.

7. Potential[36)]

We first reported that we had successfully smoothed our SOI wafers by hydrogen annealing to achieve a surface roughness comparable to that of commercially available silicon wafers in 1994[7)]. However, silicon etching during the hydrogen annealing causes degradation of the uniformity of the thickness of the SOI layer, and so it is necessary to restrict the amount of such etching. The rate of such etching is greatly affected by things like the equipment setup and the purity of the gas. **Fig. 37** shows a comparison between the case where the hydrogen annealing was carried out in an epitaxial reactor (single wafer type or barrel shaped batch type) and the case where it was carried out in a vertical type hydrogen annealing furnace. It can be seen that the rate of etching during the hydrogen annealing was dramatically suppressed in the case of the vertical type hydrogen annealing furnace compared with the case of the epitaxial reactors, specifically to 0.0013 nm/min at 1050ºC and to 0.0022 nm/min at 1100ºC. This difference in the rate of etching can be explained by the difference in gas flow across the surface of the SOI layer for the two types of device. In the epitaxial growth reactors, the hydrogen constantly flows parallel to the wafer surface, meaning that the products from the reaction between the silicon of SOI surface and the hydrogen or the small amount of oxygen and water present in the hydrogen are rapidly removed from the wafer surface by this gas flow. This means that the silicon etching reaction is always proceeding at the wafer surface. With the vertical type annealing furnace, on the other hand, the gas only flows past the sides of the wafer, with there being no intentional gas flow over the wafer surface. This means that the concentration of the reaction products rises up to the saturated concentration around the wafer surface, after which the reaction products diffuse away from the wafer only on account of the concentration gradient. This is the reason why the rate of silicon etching can be kept down so low with the vertical type annealing furnace. As a result, the reduction in the thickness of the SOI layer during the hydrogen annealing is kept down to below 1nm, and as shown in **Fig. 38** it is possible to manufacture a 6-inch wafer with an extremely thin SOI layer of average thickness about 27nm_such that this thickness is uniform to the extent that the standard deviationσ is only 1nm. The uniformity of the SOI layer thickness is thus determined almost exclusively by the uniformity of the thickness of the epitaxial silicon layer, i.e. it is more-or-less unchanged during the hydrogen annealing. The thickness uniformity of the epitaxial silicon layers produced by epitaxial reactors is being continually improved, due to the demands not only of SOI but of all silicon processes, and there has been a report that a film thickness uniformity better than ±1% has been achieved with an epitaxial reactor designed to be used with 300-mm silicon wafers[37]. It is expected that as wafers of larger and larger diameter are produced, there will be further improvements in the thickness uniformity.

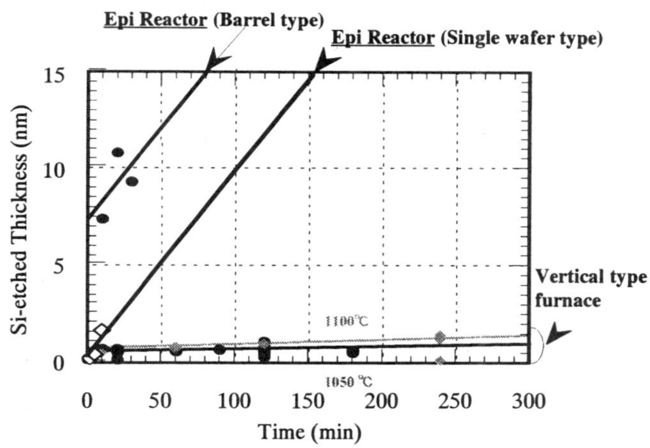

Figure 37. Si consumption by H_2 annealing in the epitaxial reactor vs the vertical type furnace.

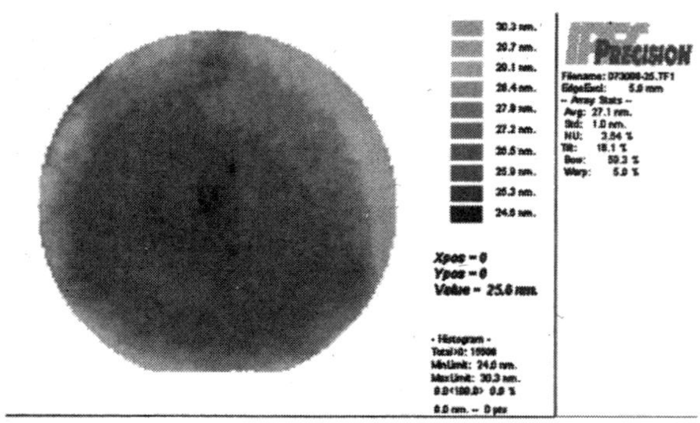

t_{SOI} = 27.1nm, σ = 1 nm over 15000 points in a 6inch wafer

Figure 38. The thickness uniformity in *27* nm-thick SOI wafer (*150* mm).

As described above, ELTRAN® wafers are SOI-Epi wafers™ having the special feature of being COP-free. The low density of HF defects attests to this special feature. Nevertheless, The number of HF defects increases when the HF solution permeates into any weak points in the SOI layer, and so it is expected that the number of such defects should vary according to the thickness of the SOI layer. **Fig. 39** shows the change in the number of HF defects with the thickness of the SOI layer of the ELTRAN® wafer. It can be seen that an extremely high product quality of only about 0.03 HF defects /cm^2 is maintained for an SOI layer thickness right down to 100nm. It can also be seen that even in the case of an ultra-thin SOI layer of thickness only 50nm, a low HF defect density of 0.06 /cm^2 is maintained. This shows that we have already achieved a product quality – in terms of both thickness uniformity and an HF defect density of 0.04~0.06 /cm^2 or less – that will answer to the needs of the new era of ultra-thin films of thickness 25~40nm, this thickness being the 1999 target for fully depleted type indicated in the SIA Roadmap and the 2009 target for partially depleted type.

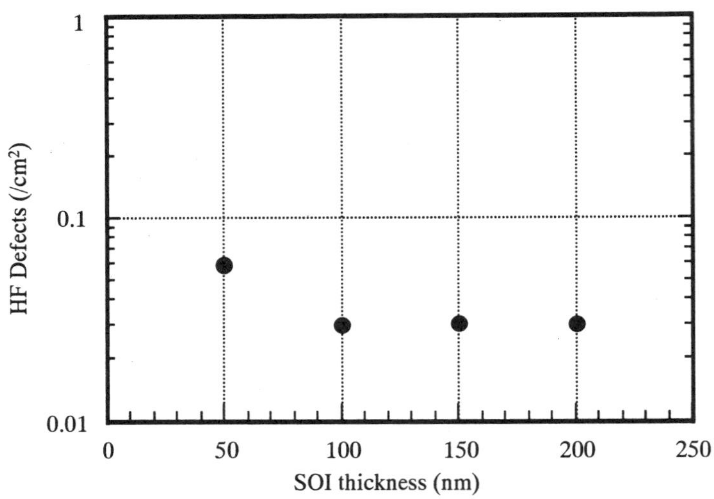

Figure 39. SOI thickness dependence of HF defects.

Another aspect of the SOI crystal product quality is the density of secco-etch defects. We have achieved a satisfactory level of such defects even when the SOI layer thickness is only 50nm.

Device characteristics that attest to the extremely good SOI layer thickness uniformity and high crystal product quality of ELTRAN® wafers are shown in **Fig. 40**[37]. When the gate length drops below 0.1μm, the threshold value starts to roll off and the S-value starts to worsen due to short channel effects. One way of suppressing these things is ultra-thin film MOSFETs. **Fig. 40** shows the change in the threshold voltage and the S-value plotted against the gate length for a 4nm ultra-thin film MOSFET. It can be seen that, because of the ultra-thin film MOS, the rolling off of the threshold voltage and the deterioration in the S-value as the gate length is reduced are suppressed. In order to achieve these goals, it becomes very important that the ultra-thin film is highly uniform in thickness.

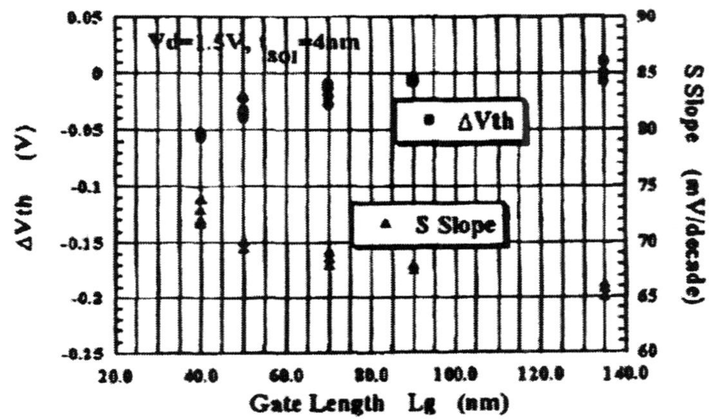

Figure 40. ΔVth and S-slope as a function of the gate length Lg (*40-135* nm) for the fabricated *4* nm-thick SOI n-MOSFETs, where Vd=*1.5* V[38].

The usage of conventional equipment – an epitaxial reactor and a heat treatment furnace – means that it easy for the ELTRAN® manufacturing method to be applied to larger-diameter wafers. As explained earlier, the anodization equipment, the bonding equipment, the splitting equipment and the etching equipment were developed by us independently. We have already made it such that this equipment can handle 300mm-diameter wafers, and in fact have already demonstrated that the equipment can be used with such large wafers by actually manufacturing 300mm ELTRAN® wafers (see **Fig. 41**). Moreover, we achieved a high thickness uniformity 147nm±1.6nm as shown in Figure 42[39], that is better than that of our 200mmwafers.

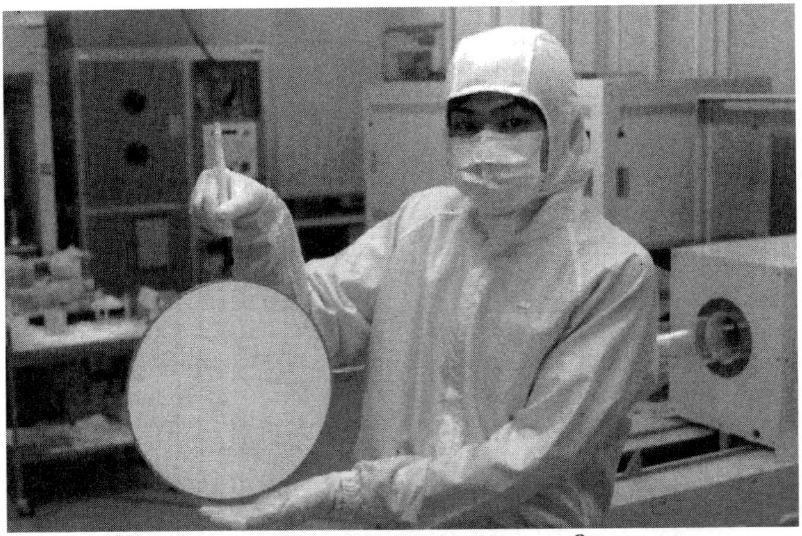

Figure 41. 300mm-diametered ELTRAN® Wafer.

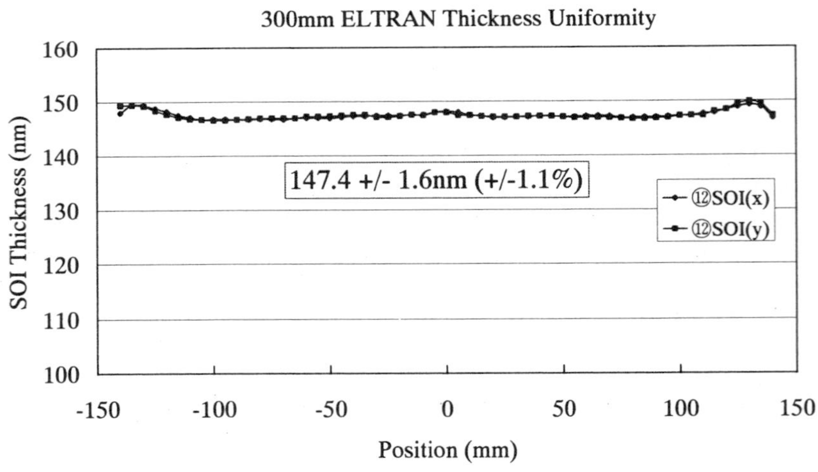

Figure 42. SOI thickness (tSOI) of 300mm ELTRAN®

8. Conclusion

Since ELTRAN® wafers are COP-free SOI-Epi wafers™, they have an extremely low density of HF defects. Moreover, due to a combination of the extremely good epitaxial silicon layer thickness uniformity, selective etching of the porous silicon with remarkably high selectivity, and a dramatic reduction in the amount of silicon etching that occurs during the hydrogen annealing of the SOI surface, ELTRAN® wafers also have SOI layers of highly uniform thickness. As a result, even wafers with an SOI thickness of only 27nm show a good thickness uniformity. We have put the equipment used in the ELTRAN® manufacturing process in a boron-free class 1 clean room, and have managed to automate all of the equipment and completed other preparation necessary for mass production. In this way, we have managed to prevent variation in product quality. Moreover, there are no new technical problems in scaling up the manufacturing process to 300mm-diameter wafers.

R&D into ELTRAN®/SOI-Epi wafers™ was started in 1990, and now we are on the verge of reaching the mass production stage. At the mass production stage, product quality and costs will be important factors. We have thus developed wafer splitting/recycling technology based on a 'seed wafer reusage' concept unique to ELTRAN® wafers. In addition, we have taken a fresh look at porous silicon – which has always been a key material of ELTRAN® wafers – from the standpoint of internal stress, and succeeded in creating stress/strain energy in a limited region close to the interface between two porous silicon layers of different porosity. We have also established new wafer-splitting technology, first devising and developing a special splitting method that makes use of a water jet and then carrying out our own equipment development.

In the future, in order for SOI to enter into the domain of mainstream CMOS and become widely used in logic devices, analog devices, memories and systems on chips, a product quality comparable to that of bulk and epitaxial wafers will be required, and it will be necessary to set the price at a reasonable level whereby sufficient device development can be carried out. At the same time, it will be necessary to allow the risk associated with the R&D activities and prior investment of the SOI wafer producers. A setup for the supply of wafers is indispensable to the mass production of SOI devices, and with regard to such a setup we are now approaching a time for careful consideration of timing, scale and form from a business strategy viewpoint. In other words, it is SOI wafer manufacturers and device manufacturers joining forces and working together to build a mass production setup that will lead to an expansion of the SOI market.

9. References

1. http://www.canon.co.jp/eltran
2. http://www.digital-contents.com
3. Yonehara, T., Sakaguchi, K. and Sato, N. (1994) Epitaxial Layer Transfer by Bond and Etch Back of Porous Si, *Appl. Phys. Lett.* ***64***, 2108-2110.
4. Yonehara, T., Sakaguchi, K. and Sato, N. (1999) ELTRAN®; Epitaxial Layer Transfer, in P.H.L.Hemment (ed.), *Proc. 9th Int. Symp. on Silicon-on-Insulator Tech. and Devices* ***99-3***, The Electrochemical Society, Seattle, pp.111-120.
5. Yonehara, T. (1999) SOI-Epi Wafers, *Break Through*, April 1999 edition, Realize Inc., Tokyo, pp.15-18 (in Japanese).
6. Mitani, K. (1998) Technical Challenge in High Volume Unibond SOI Wafer Manufacturing, *SEMI® Silicon-on-Insulator (SOI) Manufacturing Technology*, SEMICON West 98, San Francisco, pp.H-1-H-12.

7. Sato, N. and Yonehara, T(1994) Hydrogen Annealed Silicon-on-Insulator, *Appl. Phys. Lett.* ***65***, 1924-1926.
8. Sato, N., Ito, M., Nakayama, J. and Yonehara, T. (1998) Extremely Low Si Etching (<1nm) during Hydrogen Annealing of Silicon-on-Insulator, *Ext. Abst. 1998 Int. Conf. SSDM*, The Japan Society of Applied Physics, Hiroshima, 298-299.
9. Edited by Canham, L. (1997) Properties of Porous Si, INSPEC, United Kingdom.
10. Uhlir, A. (1956) Electrolytic Shaping of Germanium and Silicon, *Bell System Tech. J.* ***35***, 333-347.
11. Sato, N., Sakaguchi, K., Yamagata, K., Fujiyama, Y., and Yonehara, T. (1995) Epitaxial Growth on Porous Si for a New Bond and Etchback Silicon-on-Insulator, *J. Electrochem. Soc.* ***142***, 3116-3122.
12. Sakaguchi, K., Sato, N., Yamagata, K., Fujiyama, Y., and Yonehara, T. (1994) Extremely High selective Etching of Porous Si for Single Etch-Stop Bond-and-Etch-Back Silicon-on-Insulator, *Ext. Abst. 1995 Int. Conf. SSDM*, The Japan Society of Applied Physics, Yokohama, 259-261.
13. Sakaguchi, K., Sato, N., Yamagata, K., Fujiyama, Y. and Yonehara, T. (1995) Extremely High selective Etching of Porous Si for Single Etch-Stop Bond-and-Etch-Back Silicon-on-Insulator, *Jpn. J. Appl. Phys.,* ***34***, 842-847.
14. Sakaguchi, K., Yanagita, K., Kurisu, H., Suzuki, H., Ohmi, K. and Yonehara, T. (1999) ELTRAN® by Splitting Porous Si Layers, in P.H.L.Hemment (ed.), *Proc. 9th Int. Symp. on Silicon-on-Insulator Tech. and Devices,* ***99-3***, The Electrochemical Society, Seattle, pp.117-121.
15. Ohmi, K., Sakaguchi, K., Yanagita, K., Kurisu, H., Suzuki, H. and Yonehara, T. (1999) Water Jet Splitting of Thin Porous Si for ELTRAN®, *Ext. Abst. 1999 Int. Conf. SSDM*, The Japan Society of Applied Physics, Tokyo, 354-355.
16. Sakaguchi, K., Yanagita, K., Kurisu, H., Suzuki, H., Ohmi, K. and Yonehara, T. (1999) ELTRAN® by Water Jet Splitting in Stress-Controlled Porous Si, *Proc. 1999 IEEE Int. SOI Conf.*, IEEE, Rohnert Park, 110-111.
17. Herino, R., Perio, A., Barla, K. and Bomchil, G. (1984) Microstructure of Porous Silicon and its Evolution with Temperature", Mater. *Lett.* ***2***, 519-523.
18. Sato, N., Sakaguchi, K., Yamagata, K., Fujiyama, Y., Nakayama, J. and Yonehara, T. (1995) Advanced Quality in Epitaxial Layer Transfer by Bond and Etch-Back of Porous Si, *Ext. Abst. 1995 Int. Conf. SSDM*, The Japan Society of Applied Physics, Osaka, pp.845-847.
19. Sato, N., Sakaguchi, K., Yamagata, K., Fujiyama, Y., Nakayama, J., and Yonehara, T. (1996) Advanced Quality in Epitaxial Layer Transfer by Bond and Etch-Back of Porous Si, *Jpn. J. Appl. Phys.* ***35***, 842-847.
20. Sato, N., Ishii, S., Matsumura, S., Ito, M., Nakayama, J. and Yonehara, T. (1998) Defect Engineering in Epitaxial Layers over Porous Silicon for ELTRAN® SOI Wafers, *Ext. Abst. 1998 Int. Conf. SSDM,* The Japan Society of Applied Physics, Hiroshima, pp.302-303.
21. Sakaguchi, K., Sato, N., Yamagata, K., Atoji, T., Fujiyama, Y., Nakayama, J. and Yonehara, T. (1997) Current Progress in Epitaxial Layer Transfer (ELTRAN®), *IEICE Trans. Electron.* ***E80-C***, 378-387.
22. Yanagita, K., Sakaguchi, K., Kurisu, H., Suzuki, H., Ohmi, K. and Yonehara, T.

(1999) An Application of the Water Jet for Splitting Bonded Wafers, *Proc. Int. Symp. on New Applications of Water Jet Technology*, WJTSJ, Ishinomaki, pp.137-144.
23. Yonehara, T. (1998) Latest Technical Trends in 'ELTRAN' SOI-Epi Wafers and Future Prospects, *Denshi-Zairyo (Electronic Parts and Materials),* June edition, pp.4547 (in Japanese).
24. Nakayama, J. and Yonehara, T. (1999) Boron and Phosphorus Contamination of Bonded SOI Wafers, *The Status Quo of Contamination in ULSI Manufacturing*, Realize Inc., Tokyo, pp.389-394 (in Japanese).
25. Ohshima, S., Sutoh, M., Kuroda, T. and Hara, T. (1998) Measurement of Minority Carrier Lifetime for SOI Device Si Layer (II), *Ext. Abst. (The 45th Spring Meeting, 1998), No. 2,* The Japan Society of Applied Physics, Tokyo, pp.805 (in Japanese).
26. Ushiki, T. and Ohmi, T. (2000) Effect of Starting SOI Material Quality on Low-Frequency Noise Characteristics in Partially Depleted Floating-Body SOI MOSFETs, *IEEE Electron Device Lett.* ***21***, 610-612.
27. (1999) SOI: Low-Cost Mass Production, *Nikkei Sangyo Shimbun (Nikkei Business News)*, 24 September 1999, page 1 (in Japanese).
28. Labunov, V., Bondarenko, V., Glinenko, L., Dorofeev, A. and Tabulina, L. (1986) Heat Treatment Effect on Porous Silicon, *Thin Solid Films* ***137***, 123-134.
29. Sugiyama, H. and Nittono, O. (1989) Annealing Effect on Lattice Distortion in Anodized Porous Silicon Layers, *Jpn. J. Appl. Phys.* ***28***, L2013-L2016.
30. Young, I.M., Beale, M.I.J. and Benjamin, J.D. (1985) X-Ray Double Crystal Diffraction Study of Porous Silicon, *Appl. Phys. Lett.* ***46***, pp.1133-1135.
31. Ito, T., Yasumatsu, T., Watabe, H. and Hiraki, A. (1990) Structure Change of Crystalline Porous Silicon with Chemisorption, *Jpn. J. Appl. Phys.* ***29***, L201-L204.
32. Barla, K., Herino, R., Bomchil, G., Pfister, J.C. and Freund, A. (1984) Determination of Lattice Parameter and Elastic Properties of Porous Silicon by X-Ray Diffraction, *J. Crystal Growth* ***68***, 727-732.
33. Sugiyama, H. and Nittono, O. (1991) Microstructure and Lattice Distortion of Anodized Porous Silicon Layers, *Bulletin of the Japan Institute of Metals* ***30***, 268-275 (in Japanese).
34. Bellet, D., Lamagnere, P., Vincent, A. and Brechet, Y. (1996) Nanoindentation Investigation of the Young's Modulus of Porous Silicon, *J. Appl. Phys.* ***80***, 3772-3776.
35. Edited by The Water Jet Technology Society of Japan(1993) *Water Jet Technology Handbook*, Maruzen, Tokyo (in Japanese).
36. Sato, N. (1999) Recent ELTRAN® Advances, *Investigative Research Report on Trends in Multilayer Integration Technology VI*, Japan Electronic Industry Development Association, pp.14-28 (in Japanese).
37. Hashimoto, M. (1998) MOSLSI Epitaxial Reactors, *MOS Device Epitaxial Wafer*, Realize Inc., Tokyo, pp.46-55 (in Japanese).
38. Suzuki, E., Ishii, K., Kanematsu, S., Maeda, T., Tsutsumi, T., Nagai, K., Sekigawa, T. and Hiroshima, H. (1999) Highly Suppressed Threshold Voltage Roll-off Characteristics of the 4 nm-Thick SOI N-MOSFETs in the 40-135 nm Gate Length Regime, in P.H.L.Hemment (ed.), *Proc. 9th Int. Symp. on Silicon-on-Insulator Tech.*

*and Devices **99-3***, The Electrochemical Society, Seattle, pp.260-265.
39. Ito, M., Yamagata, K., Miyabayashi, H. and Yonehara, T. (2000) Scalability Potential in ELTRAN® SOI-Epi Wafer™, *Proc. 2000 IEEE Int. SOI Conf.*, IEEE, Massachusetts,

LOW DIMENSION PROPERTIES OF NANOSTRUCTURES ON ULTRATHIN LAYERS OF SILICON FORMED BY OXIDATION OF ION CUT SOI WAFERS AND ELECTRON LITHOGRAPHY

V.P. Popov, A.L. Aseev, I.V. Antonova, Yu.V. Nastaushev,
T.A. Gavrilova, O.V. Naumova, A.A. Franzusov, G.N. Feafanov,
V.A. Kolosanov
Institute of Semiconductor Physics, Lavrentieva 13, Novosibirsk 630090,
Russia, E-mail: popov@isp.nsc.ru

1. Introduction

High quality silicon-on-insulator (SOI) wafers with appreciable properties for quantum devices can be produced by very long time high temperature annealing of commercial SIMOX wafers [1] or by Internal Thermal Oxidation (ITOX) of low-dose SIMOX wafers [2]. A promising and elegant method to fabricate SOI is by ion cut technology [3]. High dose hydrogen implantation with subsequent annealing allows one to delaminate a thin layer of silicon from a thick substrate and to connect it to an oxidized silicon wafer. The main advantages of this technology are high quality buried oxide, top silicon layer, and high flatness of interface between layers [4].

The aim of this effort was to produce ultrathin SOI structures with different dopant concentration by oxidation, to fabricate nanoscale transistor using electron lithography and investigate the properties of these structures.

2. Experimental details

A thickness of initial silicon top layer is managed by energy of implanted hydrogen ions and can be in the range of 200-2000 nm for ordinary implanter. A scatter of the silicon layer thickness in wafers with diameter of 100 mm is lower than 5 nm. Further reduction of the SOI film thickness has been attempted by dry oxidation of wafer followed by stripping in diluted HF. The thickness and uniformity of the remaining SOI films and interface layers were determined by spectroscopic ellipsometry measurements. This conventional technological process provides for example the thickness of SOI of about 3nm with a dispersion of 1 nm on 100 mm wafer. Transmission Electron Microscopy (TEM) and High Resolution Electron Microscopy (HREM) investigations proved the high perfection of final silicon layer and silicon/oxide interface [4].

Top Si layers in SOI were then doped with B^+ (130 keV, $D = 2 \times 10^{15}$cm^{-2}) either P^+ (300 keV, $D = 6 \times 10^{15}$ cm^{-2}) ions with subsequent activation of dopant at 1100°C for 2 hours. As a result, nanostructures with three different doping levels were fabricated: low doped n-type ($n = 3 \times 10^{15}$ cm^{-3}), highly boron doped p-type ($p^+ = 2 \times 10^{18}$cm^{-3}) and highly phosphorous doped n-type ($n^+ = 10^{20}$cm^{-3}).

In-plane gate field-effect transistors (IPGFET) are formed using electron lithography, chemical etching of photoresist, which was exposed to electron beam with diameter of 50 nm, and then a plasma-chemical etching of Si overlayer. The electron lithography resolution was about 20 nm at the thickness of p^+type Si film 40 nm for electron microscope BS-350 (TESLA) with image generator PROXY WRITER (Raith GmbH) used as an e-beam source (22 keV, 1 nA). Plasma etching with CCl_2F_2/Ar gas mixture was used for nanostructures fabrication. Polymetilmetacrilat (PMMA) was utilized as a resist for electron lithography in IPGFET process. The Ti/Au contacts were deposited by vacuum evaporation.

F. Balestra et al. (eds.), Progress in SOI Structures and Devices Operating at Extreme Conditions, 87–91.

TEM, HREM, spectroscopic ellipsometry, point contact pseudo-MOSFET measurements, capacity –voltage, current –voltage and Hall effect measurements were used for investigation. of the nanostructures.

3. Results and Discussion

The results of simulation of the current –voltage characteristics for the point contact pseudo-MOSFET are presented in Fig.1. As it is seen in the figure, quantum confinement effects become essential for the thickness of top silicon layer lower than 10 nm. The point contact measurements show also enhanced decrease in conductivity below 10 nm.

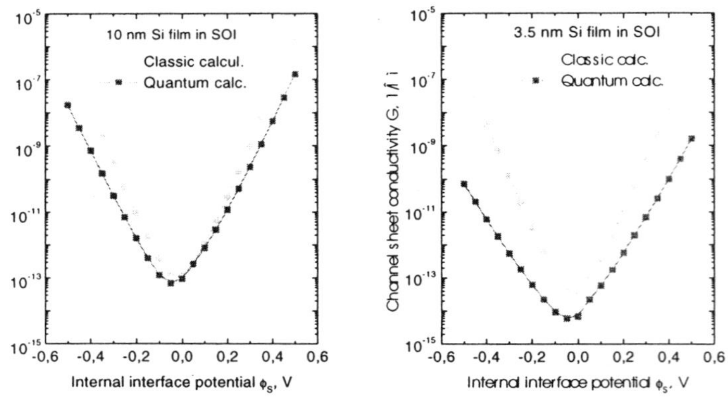

Fig.1. Classic and quantum confinement calculations of SOI FET channel conductivity

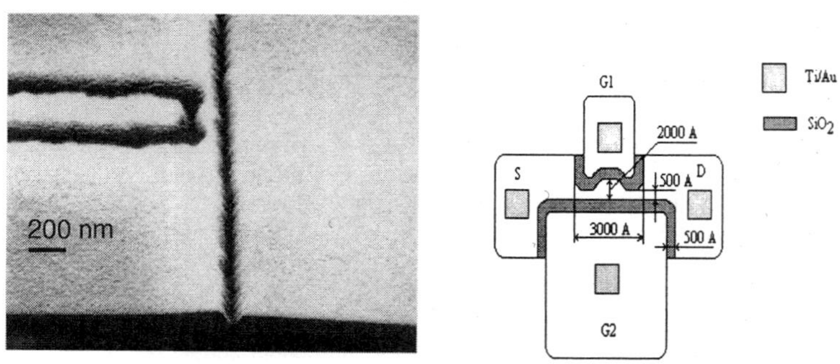

Fig.2. SEM and schematic pictures of in-plane-gate SOI FET with two splitting gates and a dot between the source and drain.

The electron and hole mobilities were equal to 300-550 cm^2/Vs and 100-300 cm^2/Vs, respectively depending on film thickness and doping. The interface state density was lower than 1×10^{12} $cm^{-2}eV^{-1}$. The resistance between nanostructures obtained in such a manner and substrate was higher than 10^{10} Ohm. The drain-source resistance at room temperature for this SOI nanostructure was about 10^7 Ohm and practically independent of drain and gate voltages due to electron gas degeneration. The resistance is nonlinear at cryogenic temperatures. The results of measurements of the I-V characteristics at liquid helium temperature for IPGFET fabricated on Si films with thickness of 45 nm are presented in Fig. 3. A Coulomb blockade (zero conductance) near voltage 0 V is seen in the top figure.

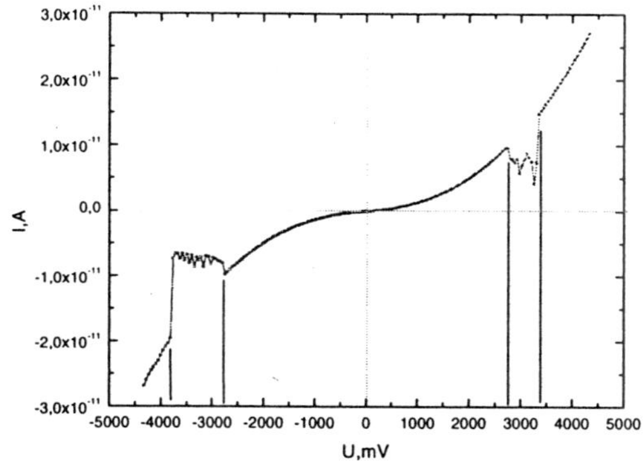

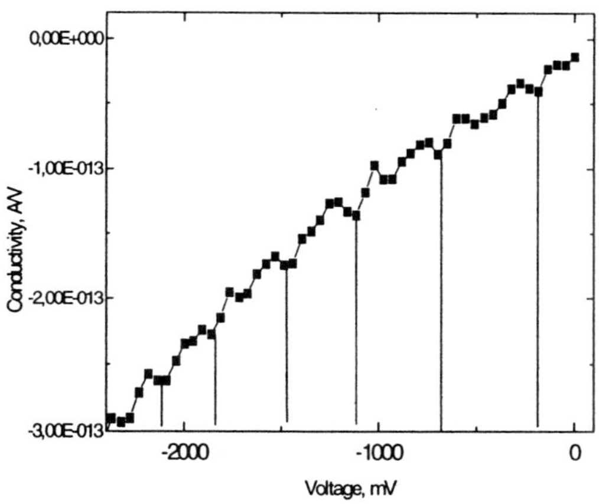

Fig.3. Drain current dependence on the drain voltage at 4.2 K (V_{gate} = 0 V)

The features on the I-V curve at $\pm$2.8 V are connected with electron transport through electron traps in the buried oxide or at the interface [5]. Oscillations with a period of about 0.1-0.3 V are seen in Fig. 3 (down). If a size of quantum dot determines these oscillations, the distance between levels should be about 20 meV. Thus in our case the oscillations of conductance with different periods are caused by fluctuation of potential in nanowire, for example, due to interface or oxide traps. The 45 nm size of channel is too large for single electron effects.

The single electron oscillations might be observed in IPGFET at low temperatures provided that top silicon layer is thinned down to 10 nm by chemical enhanced oxidation.

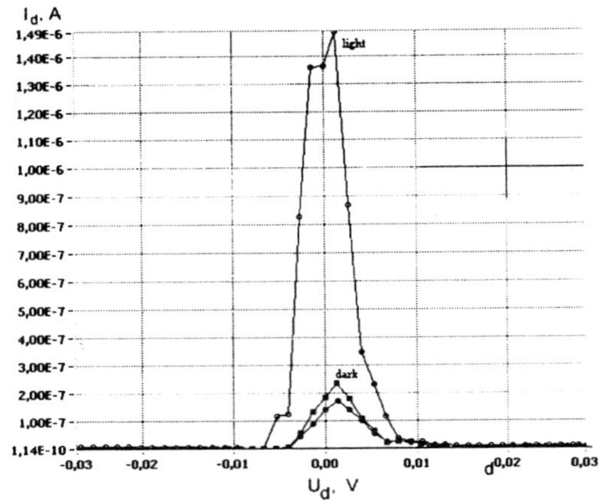

Fig.4. Drain current versus drain voltage for IPGFET with 36nm Si film under different level of illumination

$I_d(V_d)$ characteristics for IPGFET with 36 nm low doped Si film are presented in Fig. 4. These characteristics are measured at different illumination (dark or light) and room temperature. A strong dependence of drain current on illumination during measurements is most likely caused by presence of potential barriers in the low doped channel. At least two opposite connected potential barriers with heights about few mV are present in the channel most likely due to inhomogeneities of impurity or interface (oxide) traps distributions.

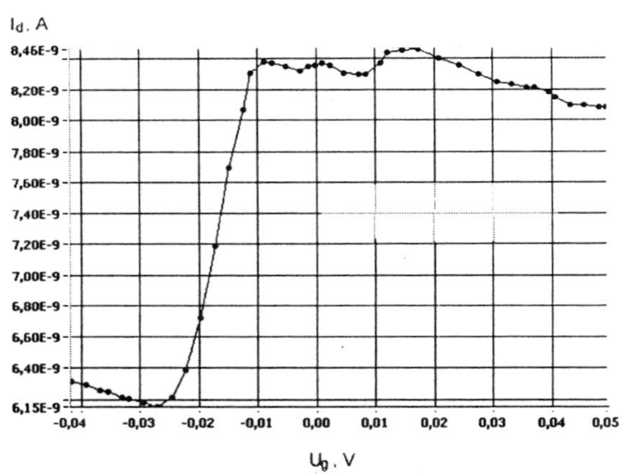

Fig.5. Drain - Source current vs. second gate G2 voltage in lightly doped IPGFET at room temperature

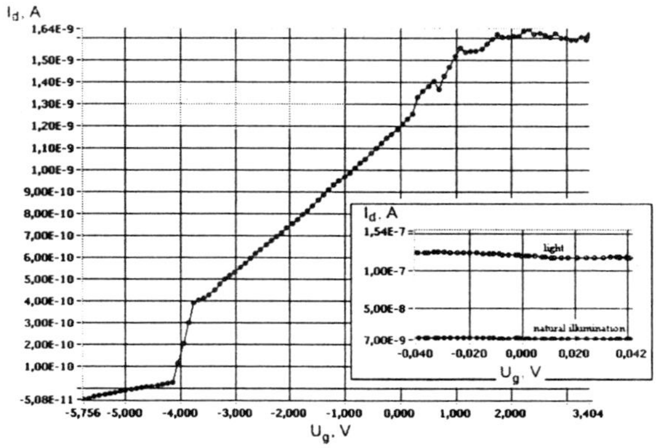

Fig.6. Drain - Source current vs. first gate G1 voltage in lightly doped IPGFET at room temperature

Drain current as a function of the first and second gate voltage in lightly doped IPGFET measured at room temperature is given in Fig. 5 and 6. Drain voltage was equal to 30 mV in both cases. IPGFET channel is on grounded substrate, in a partially depleted regime (accumulation regime on the G2 side and depletion regime on the G1 side). The maximal drain current is observed for U_{G2} > - 10 mV and U_{G1} > 1.8 V. Breakdown voltage for G1 is equal to – 4 V.

4. Conclusion

Step by step thinning of high quality Si films in SOI structures is a convenient way for fabrication of ultrathin silicon films with measurable quantum effects. Degradation of electrical properties was not observed during preparation procedure.

Quantum effects manifest for heavily doped channels only at liquid He temperature. Oscillation of conductivity is connected with potential fluctuation inside the doped Si film, a still unresolved problem for these structures.

Our results indicate a perspective for using thinned ion cut SOI structures for quantum device production.

Acknowledgment

The present work was partially supported by ISTC grant N 563, project no. 204-9(00)-П of Federal Program "Physics and technique of nanostructures" and Russian-Ukrainian joint program "Nanophysics and nanoelectronics".

References

[1] T. Sakamoto, H. Kawaura, T. Baba. Appl. Phys. Lett. 72, 795 (1998).
[2] L. Zhuang, L. Guo, S.Y. Chou. Appl. Phys. Lett. 72, 1089 (1998).
[3] M. Bruel. Electron. Lett. 31, 1201 (1995).
[4] V.P. Popov, I.V. Antonova, V.F. Stas, L.V. Mironova, A.K. Gutakovskii, E.V. Spesivtsev, A.A. Franzusov, A.S. Mardegov, G.N. Feofanov. J. Mater. Sci. Eng. B, 73, 82 (2000).
[5] A. Stesmans, V.V. Afanas'ev, J. Phys: Condens. Mat. 10, L19 (1998).

SOI FOR HARSH ENVIRONMENT APPLICATIONS IN THE USA

Cynthia A. COLINGE
Dept. of Electrical and Electronic Engineering
California State University, Sacramento, USA

1. Introduction

SOI Technology has received much attention since the announcement of an SOI line of microprocessor products by IBM in 1998. Semiconductor manufacturers such as Motorola have followed IBM's lead and have announced SOI products [1], while other major US semiconductor vendors such as AMD are currently developing SOI products. Most of these companies develop SOI products for low-power, low-voltage applications, however, a small number of vendors have specialized in the field of harsh-environment electronics, which will be the topic of this paper.

2. SOI Material Availability

Until recently the major impediment to the use of SOI material for producing commercial integrated circuits was the lack of wafer supply in large number. Furthermore, the fabrication of fully depleted devices was slowed down due to unpredictable threshold voltage caused by variations in thickness of the silicon overlayer. These problems seem to be a thing of the past now. All major SOI wafer suppliers report a silicon thickness uniformity better than 5 nm although, in practice, the uniformity is usually better than 2 and even 1 nm (Table 1).

TABLE 1. SOI wafer characteristics

Material	Company	Si uniformity	Wafer size	Reference
SIMOX	Ibis	5 nm	100-200 mm	[2]
Unibond®	Soitec	5 nm	100-200 mm	[3]
Eltran	Canon	5 nm	100-200 mm	[4]

The SOITEC and Shin-Etsu Hendotai partnership projects a combined annual production of 1,000,000 Unibond® wafers. Canon is expanding its facilities to manufacture 120,000 Eltran wafers per year [5] and Ibis has recently developed a new implanter, the IBIS-2000 for mass producing SIMOX substrates.

Peregrine Semiconductor is using thin-film SOS material for the fabrication of fully depleted RF and rad-hard products. Good quality of the silicon layer is obtained by

F. Balestra et al. (eds.), Progress in SOI Structures and Devices Operating at Extreme Conditions, 93–104.

amorphization and low-temperature regrowth of the bottom of the film, followed by sacrificial oxidation of the defective top of the layer (Figures 1 and 2).

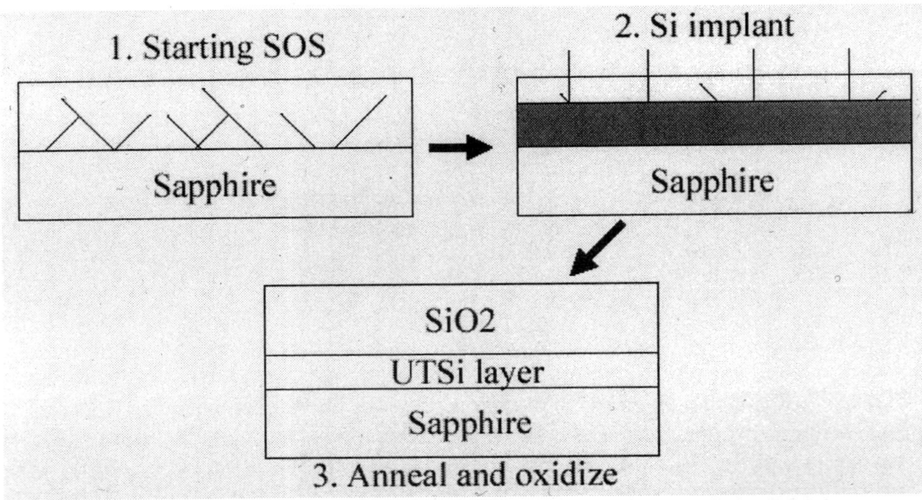

Figure 1. Amorphization, regrowth and sacrificial oxidation of the SOS layer.

Figure 2. TEM cross section of as-grown and improved SOS layer. [6]

3. Commercial SOI Circuits

In 1998 IBM announced what it believes to be the first commercially-viable implementation of silicon-on-insulator (SOI). Their announcement is significant

because, while others, including IBM, have been successful in developing SOI technology, IBM is the first to be able to apply it in building fully-functional mainstream microprocessors. They report that the use SOI technology improves performance over bulk CMOS technology by 25 to 35%, equivalent to two years of bulk CMOS advances. SOI technology also brings power use advantages of 1.7 to 3 times. IBM is currently working with many circuit designers and product groups that are designing with SOI technology. The company expects SOI to eventually replace bulk CMOS as the most commonly used substrate for advanced CMOS in mainstream microprocessors and other emerging wireless electronic devices requiring low power. [7] IBM's first 0.22 μm devices with these features will be used in Apple Computer's Macintosh systems and in IBM servers. At the same time, IBM is developing a 1-GHz PowerPC that will be built on a 0.18 μm process which compares well with Intel's second-generation IA-64 chip made on bulk CMOS. [8]
In the wake of IBM's announcement, Motorola announced that the company would now gear its core semiconductor process technology toward low-power, integrated devices for mobile system applications. And instead of rolling out multiple process technologies, Motorola for the first time will craft one base process with plug-in modules for silicon-on-insulator, mixed-signal-RF, non-volatile, embedded DRAM (eDRAM), and other capabilities. [9] The new Motorola G4 and G5 microprocessors used in Macintosh computers are now made in SOI as well. Reported microprocessor speeds show and increase of up to 35 percent over conventional bulk-CMOS processes, and reduced power consumption of up to 65 percent at the same speed is obtained.[10]

In April 2000 Hewlett-Packard Co. revealed details of its 64-bit PA-8700 processor, HP's first chip fabricated with copper interconnects. Built on a 0.18-micron silicon-on-insulator (SOI) process, the new RISC processor is designed to operate at and above 800 MHz. The PA-8700 was released to manufacturing in late March and is expected to ship in servers and workstations in the first half of 2001. HP made the announcement at InterWorks 2000, an HP enterprise computing users.[11]

Other companies, such as AMD [12], Texas Instruments [13], Cypress Semiconductor and Peregrine, are actively developing SOI processes for low-voltage logic, analog and RF applications. A list of US companies using SOI for commercial products is presented in Table 2.

TABLE 2. US Companies using SOI for commercial products.

Company	Part./Full. Depleted	Main Application
IBM	PD	Microprocessor
Motorola	PD/FD	Microprocessor, RF, DRAM
AMD	PD/FD	Microprocessor
Cypress	PD/FD	SRAM/RF
TI	PD	Various
HP	PD	Microprocessor
Peregrine	FD (SOS)	RF, logic, EEPROM, analog

4. Radiation-hardened circuits

Rad-hard SOI circuits are traditionally made using partially depleted SOI devices. The reason for this is that the creation of charges in the thick buried oxide influences the front threshold voltage in fully depleted devices, which is highly undesirable for rad-hard applications. With advancement in today's SOI materials it is now possible to fabricate rad-hard, fully depleted devices. Some of the companies manufacturing rad-hard SOI circuits are presented in Table 3.

TABLE 3. Rad-hard and high-temperature circuits

Company	Part./Full. Depleted	Application
Synova	PD	Rad-Hard
Honeywell	PD	Hi-T°; Rad-hard
Peregrine	FD	Rad-hard
Lincoln Lab	FD	Low-power, Rad-hard

NASA is supporting research in the field of SOI for its ability to withstand radiation, as well as low and high temperatures. Silicon on Insulator (SOI) technology is emerging as a major contender for digital and mixed signal device and circuit applications. The technology features small device geometry's (0.18 - 0.25 μm) and provides for low power dissipation, the potential for high radiation tolerance, and is of great interest to system designers specifically for potential space applications. Of particular interest are the characteristics of these devices at low temperatures (-100 C) for applications on the Martian surface and at cryogenic temperatures for applications in instruments and sensors. Missions to Mercury require high-temperature electronics, and a mission to Europa (one of Jupiter's moons) requires the spacecraft to withstand a total dose approaching 100 MRad. The purpose of the effort developing SOI for space applications is be to perform a comprehensive technology characterization of SOI processes in direct partnership with MIT/Lincoln Labs, Honeywell, Allied Signal, and the National Security Agency (NSA). This characterization will use simplified process and reliability test structures and existing circuit designs suitable for process and reliability evaluation. The results of this evaluation will provide critical information for design engineers and circuit designers relating to the characteristics and limitations of this technology for high reliability applications. Additional benefit will also be realized by establishing the necessary process and reliability characterization information desired by our industry partners and tremendously beneficial for future NASA applications.

Honeywell has developed a line of SOI products circuits ranging including a 256k ROM, a 4 Mb SRAM and a 5 Mb SRAM module and a gate array.[14] The CMOS fabrication process is rather conventional and makes use of partially depleted SOI MOSFETs. Honeywell and Motorola have announced the joint development of a rad-hard PowerPC microprocessor for aerospace applications.[15] The characteristics of some Honeywell SOI circuits are presented in Table 4.

TABLE 4. Some Honeywell rad-hard products

Circuit	L (μm)	V_{DD}	Max. dose	SEU rate	Transient
Gate array	0.28	2.5 V	1 MRad(SiO_2)	10^{-10} error/bit/day	
256k SRAM	0.75	5 V	1 MRad(SiO_2)	10^{-10} error/bit/day	10^9 Rad(Si)/s
4M SRAM	0.35	3.3 V	1 MRad(SiO_2)	10^{-10} error/bit/day	10^{12} Rad(Si)/s

Synova has developed a rad-hard (1 Mrad) 32-bit MIPS processor called the Mongoose-V.[16] Mongoose-V is a radiation hardened MIPS R3000 32-bit microprocessor that is fabricated in CMOS Silicon-on-Insulator (SOI). The single chip device offers a highly integrated solution to many spacecraft processor applications, such as embedded instrument controllers. Mongoose-V incorporates on-chip cache memory, on-chip peripheral functions and full hardware support for IEEE-754 floating point. The Mongoose-V development was sponsored by NASA Goddard Space Flight Center. The Mongoose-V has a linear energy transfer (LET) larger than 80 MeV.cm^2 mg^{-1}, which makes it virtually SEU-free for space applications, and has a dose hardness of 1 Mrad.

The Lincoln Laboratory is currently working on a fully depleted SOI process for low-power applications, aiming at threshold voltages of 400 mV and a supply voltage of 900 mV. The minimum gate length is 0.18 μm and the silicon film thickness is 50 nm. The source and drains are silicided and the process comprises three metal levels. The process is characterized by the use of mesa isolation. The sidewalls of the mesas are oxidized, and anti edge leakage implants are performed after sidewall oxide growth (N-type for P-channel devices, and P-type for N-channel transistors). These implants require 2 mask steps. A key parameter for the improvement of the radiation hardness has been the reduction of the sidewall oxide thickness from 25 to 8 nm (Figure 3).[17] Recent improvements to the process have made it possible to reduce the threshold voltage for a dose of 1 MRad(Si) to less than 140 mV. The Lincoln Laboratory fully depleted process is developed in collaboration with organizations interested in radiation hardness such as the NASA Jet Propulsion Lab, Honeywell and Rockwell as well as with companies interested in low-power, high-speed circuits such as Lucent, Boeing and DEC.

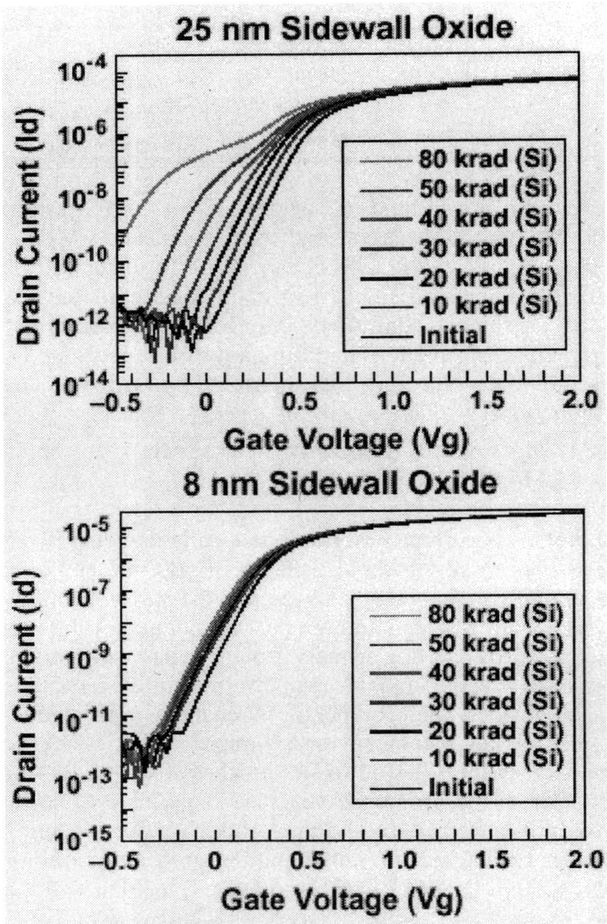

Figure 3. Improvement of dose hardness
brought about by the reduction of sidewall oxide thickness.

The low-power, high-speed capability of the process is illustrated by the compressive receiver test circuit of Figure 4 where an operational frequency of 0.55 and 1 GHz is obtained for a supply voltage of 1.1 and 2 volts, respectively. More recently the Lincoln Laboratory reported the fabrication of sub-100 nm FD SOI transistors. [[18]]

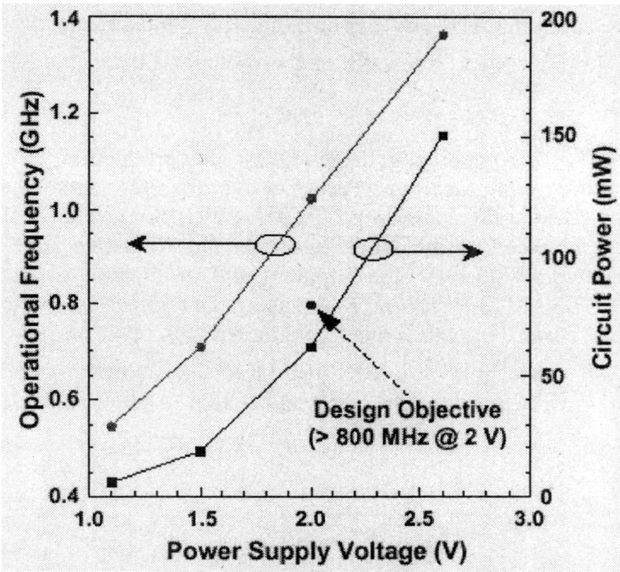

Figure 4. Performance of a compressive receiver circuit (0.25 µm SOI CMOS)

Peregrine Semiconductor designs and manufactures high-performance integrated circuits based on its patented, commercial Ultra Thin Silicon (UTSi®) CMOS/SOS process technology. Offering excellent benefits in integration, speed, power consumption, linearity and cost, the UTSi® based on SOS material is a very suitable process technology for wireless and satellite communications. Peregrine Semiconductor's UTSi® process uses LOCOS isolation between devices followed by sidewall boron implantation of the nMOS devices.

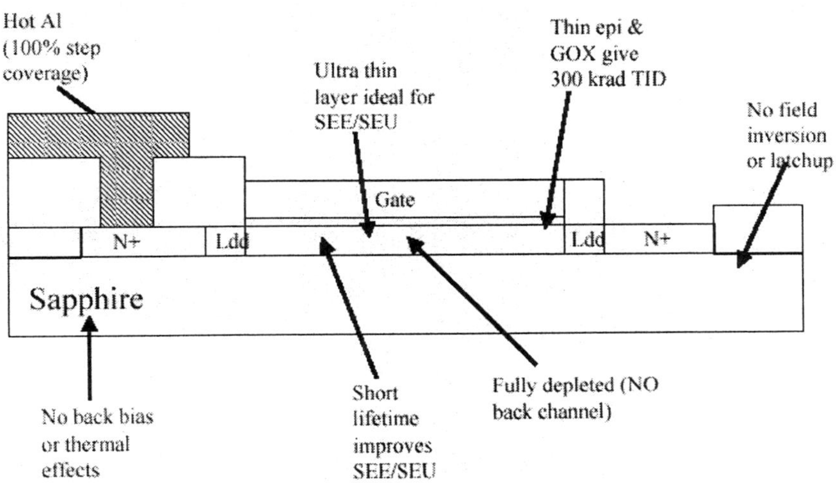

Figure 5. Cross-section of the UTSi® process highlighting some rad-hard features.

The low minority carrier lifetime in the thin SOS produces excellent single-event-upset (SEU) hardness, and the use of sapphire material underneath the active silicon combined with the P^+-doped sidewalls of the n-channel devices ensures good total-dose hardness (Figure 5).

The UTSi® process was originally designed for the production of RF circuits for mobile and satellite communications. Peregrine circuits are currently used in mobile products sold by Qualcomm, Samsung, Alcatel, NEC, Sony, etc. In 1999 Peregrine Semiconductor announced at the Nuclear Space and Radiation Effects Conference (NSREC) that it would start the development and production of rad-hard circuits, including FPGA, SRAM, EEPROM, DSP and A/D converters. Figures 6a and 6b present the drain current vs. gate voltage characteristics of n- and p-channel 0.5 μm UTSi® devices. The threshold shift after 100 kRad(Si) irradiation is less than 50 mV and the increase of leakage current (< 100 pA/μm) after irradiation is barely noticeable.

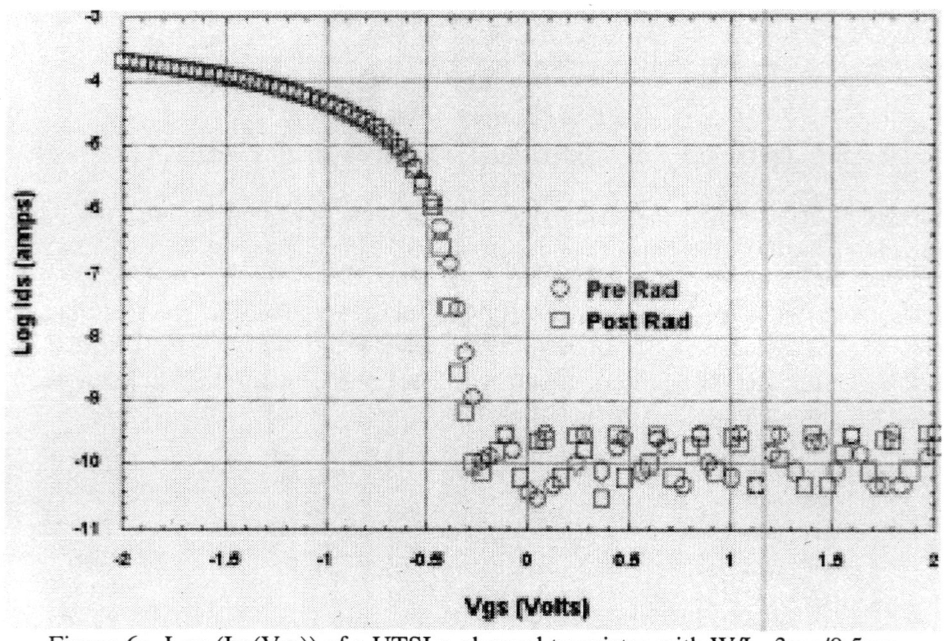

Figure 6a. Log ($I_D(V_G)$) of a UTSI n-channel transistor with W/L=3μm/0.5μm before and after irradiation at 100 kRad(Si). V_{GS} = 3V during irradiation.

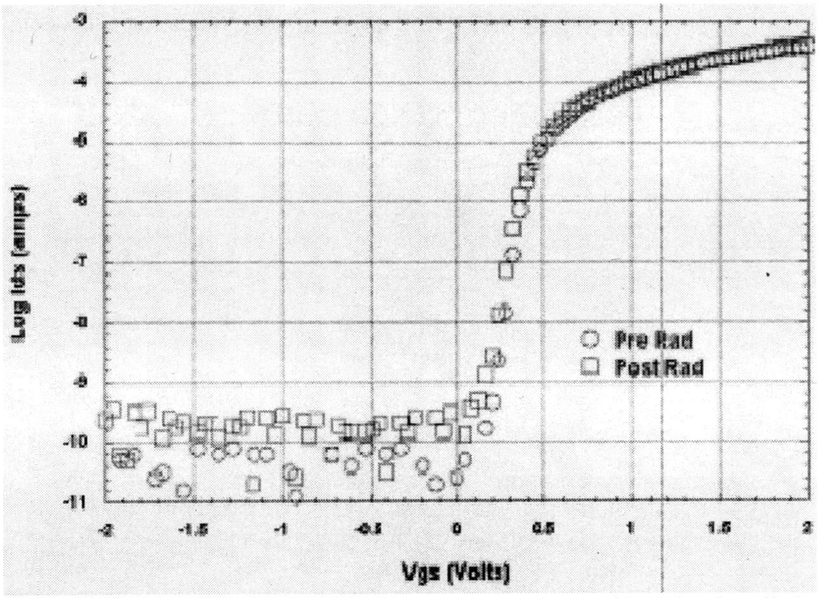

Figure 6b. Log ($I_D(V_G)$) of a UTSI p-channel transistor with W/L=3μm/0.5μm before and after irradiation at 100 kRad(Si). V_{GS} = -3V during irradiation.

5. High-temperature Circuits

The benefits of SOI for high-temperature applications are well known: a 100X or better reduction in leakage current, latchup immunity, lowered parasitic capacitance which provides 20% faster switching speed, reduced cross talk on adjacent devices, low device leakage, operation to >300°C and reduced shift threshold voltage shift with temperature in FD devices.[19,20] As an example, Figure 7 presents the leakage current and gate delay in SOI and bulk devices fabricated by Honeywell.

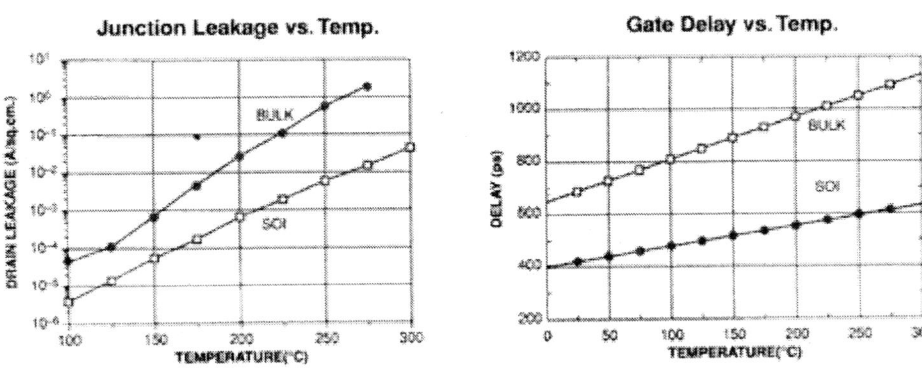

Figure 7. Some temperature features of the Honeywell 0.8 μm SOI process.[21]

Honeywell has a product line (called the HT (High-temperature) series) rated at 225°C for continuous operation and at 300°C for up to a year. Some typical circuits are presented in Table 5.

TABLE 5: Some Honeywell high-temperature circuits. [22]

Circuit	Temp.	Max. Temp.	Properties
Op. amp.	225°C	300°C	gain=100 dB; input offset drift: 100 μV/year
12-bit A/D converter	225°C	300°C	11-bit linear
32k × 8 SRAM	225°C	300°C	50 ns access time
Magnetic sensor	225°C	300°C	sensitivity: 3.2 mV/V/Gauss
Voltage reference	225°C	300°C	drift: 3 mV/1000 hour

These circuits find application in oil drilling and oil well monitoring, geothermal energy, nuclear power plants, and the automotive industry.

Figures 8 and 9 show the gain of Honeywell's HT1104 Quad operational amplifier vs. temperature. The reduction of transconductance caused by the reduction of mobility is probably compensated by the improvement of MOSFET output conductance.[9] It has been shown in the literature that the use of ZTC (zero temperature coefficient) bias conditions allow analog devices to operate at temperatures as high as 400°C [23,24] and that the decrease of dc gain with temperature is solely due to the decrease of carrier mobility with temperature.

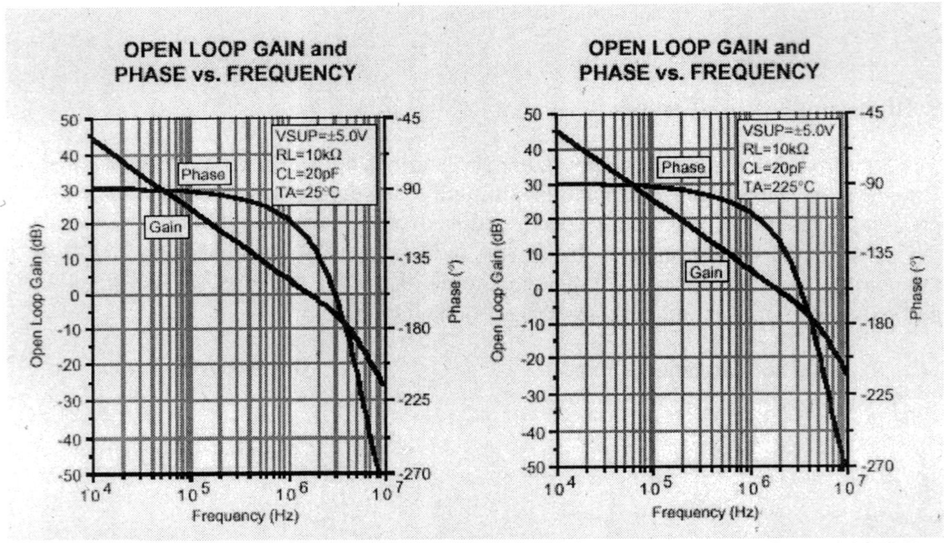

Figure 8. Open loop gain and phase vs. frequency for the HT1104 at 25 and 225°C.

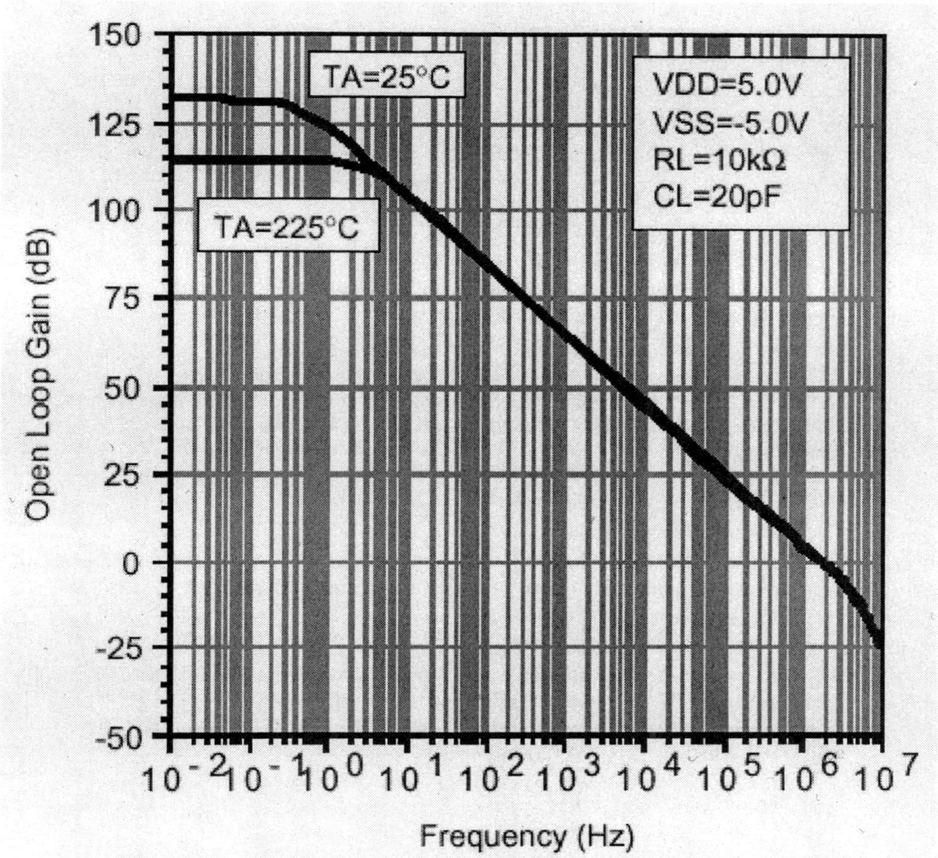

Figure 9. Open loop gain vs. frequency for the HT1104 at 25 and 225°C, over a wide range of frequencies.

6. Conclusion

SOI Technology is now employed by several US companies to fabricate low-power, low-voltage integrated circuits. Some vendors have specialized in the field of harsh-environment electronics (radiation and high temperature) using SOI or SOS substrates. A list of products as well as their performance, is presented.

References

1 http://www.soisolutions.com

2 http://www.ibis.com/wafer.htm

3 http://www.soitec.com/

4 http://www.canon.co.jp/Eltran/elt_index-e.html

5 K. Sakaguchi and T. Yonehara, "SOI wafers based on epitaxial technology", Solid-State Technology, pp. 88-92, June 2000

6 http://www.peregrine-semi.com/Tech/UTSiProc.htm

7 http://www.chips.ibm.com/bluelogic/showcase/soi/

8 http://www.planetit.com/techcenters/docs/desktop_computing/news/ PIT19990618S0050

9 http://www.edtn.com/story/prod/OEG20000815S0049-R

10 http://www.zdnet.com/zdnn/stories/news/0,4586,2336967,00.html

11 http://www.eocenter.com/edtn/out.asp?n=33586386&ty=all&dr1year=1999&dr2=on&dr2day=26&i=SOIbop=AND&ty2=all&relrange=0&dr2year=2000&dType=range&range=between&dr1month=1&dr1=on&dr1day=01&dr2month=5&url=http%3A%2F%2Fwww%2Esemibiznews%2Ecom%2Fstory%2FOEG20000411S0053&title=HP+unveils+64%2Dbit+RISC+processor+made+with+copper%2C+SOI+technologies

12 http://www.zdnet.com/zdnn/stories/news/0,4586,2594500,00.html

13 http://www.ti.com/research/docs/deviflow.htm

14 http://www.ssec.honeywell.com

15 http://content.honeywell.com/space/news/release/mot-hi.htm

16 http://www.synova.com/proc/processors.html

17 http://www.ll.mit.edu/AST/soipage.html

18 C.K. Chen *et al.*, "SLOTFET fabrication of self-aligned sub-100-nm fully depleted SOI CMOS", Proceedings of the IEEE International SOI Conference, p. 82, 2000

19 A.J. Auberton-Hervé *et al.*, "High-temperature applications of SIMOX technology", Japanese Solid State Technology, pp. 12-17, Dec. 1993 (In Japanese)

20 J.P. Colinge, *Silicon-on-Insulator Technology: Materials to VLSI*, 2nd Edition", Kluwer Academic Publishers, pp. 23-236, 1977

21 http://www.ssec.honeywell.com/hightemp/datasheets/htmo/htmo.html

22 http://www.ssec.honeywell.com/hightemp/datasheets.html

23 B. Gentinne, J.P. Eggermont and J.P. Colinge, "Performances of SOI CMOS OTA combining ZTC and gain-boosting techniques", Electronics Letters **32-24**, 2092-2093, 1995

24 B. Gentinne, J.P. Eggermont, D. Flandre, J.P. Colinge, "Fully depleted SOI-CMOS technology for high temperature IC applications", Materials Science and Engineering **B46**, 1-7, 1997

PERFORMANCE AND RELIABILITY OF DEEP SUBMICRON SOI MOSFETS IN A WIDE TEMPERATURE RANGE

F. Balestra
Institut de Microélectronique, Electromagnétisme et Photonique
CNRS / INP Grenoble
ENSERG, 23 Avenue des Martyrs
BP 257, 38016 Grenoble, France.

1. Introduction

The SOI technology is well known for its advantages as compared to bulk devices, in particular in the field of low voltage/low power and high frequency applications [1-3]. It has also been shown that SOI devices are interesting for space and military applications in radiation environments. Important issues are the optimum SOI MOSFET architectures which have to be used in the deep submicron range and the performance and reliability of these transistors under extreme temperature conditions.

In this paper, the impact of device architecture on the performance of ultra-short channel SOI transistors will be reviewed. The interest in using ultra-thin Si films, body contact, low doping and/or double gate MOSFETs will be discussed. The electrical properties (subthreshold and strong inversion operation, floating body phenomena, short channel and hot carrier effects) of these various SOI structures as a function of temperature, down to cryogenic operation, will also be thoroughly addressed.

2. Impact of device architecture on floating body effects

2.1. Kink and transient effects

The buried insulator of the SOI structure leads to floating body effects. Indeed, contrary to the case of bulk silicon devices, there is usually no substrate (or body) contact to determine the potential of the active Si layer and to collect free carriers induced by various mechanisms (for instance carriers created by impact ionization). When a body contact is available, the efficiency of this contact depends on the thickness of the Si film and on device architecture, owing to possible series resistance effect and/or potential barrier between the active silicon layer and the body contact.

F. Balestra et al. (eds.), Progress in SOI Structures and Devices Operating at Extreme Conditions, 105–127.

For partially depleted SOI transistors, a kink effect is obtained at high Vd. For a fully depleted SOI MOSFET realized on ultra-thin films, the depletion capacitance Cd vanishes which suppresses the shift of the threshold voltage, the excess drain current and, thereby, the kink effect [4].

Fig. 1. Output Id(Vd) characteristics of n-channel thin film SIMOX MOSFET showing influence of body connection.

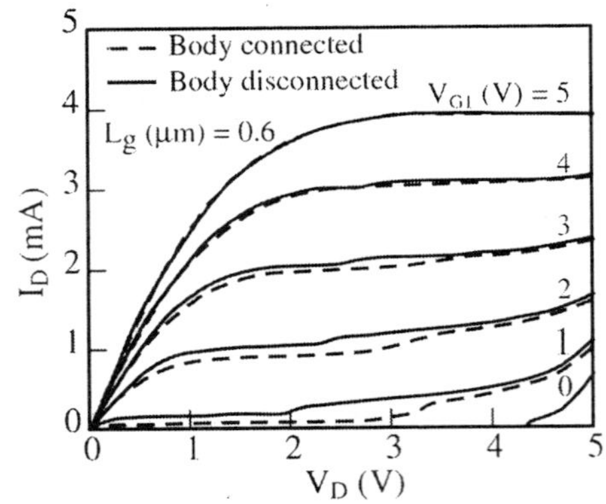

However, it has been shown that a moderate kink effect can still be obtained for fully depleted SOI MOSFETs with intermediate Si film thickness (Fig. 1) [5]. This effect can be modulated by connecting the body terminal (Fig. 1), if available. This kink is due to the substantial barrier height of the source/thin film diode in moderately fully depleted transistors. The influence of temperature on the kink effect will be shown below.

On the other hand, parasitic current transients can also be suppressed in fully depleted SOI MOSFETs where the source potential barrier is small and the threshold voltage is not sensitive to a small amount of excess holes.

2.2. Latch and Breakdown Phenomena

Another floating body effect is the parasitic bipolar transistor (PBT) action that can be observed for SOI MOSFETs in a certain voltage range. This PBT induces latch and premature breakdown as compared to bulk Si devices [6]. As the kink effect, this parasitic phenomenon is triggered by the impact ionization current which leads to an internal forward bias of the (floating) base/emitter (thin Si film/source) junction of the PBT. A large collector (drain) current is thus created giving a substantial enhancement of the drain current. This PBT action portends power-consumption problems in SOI CMOS, hysteresis in the subthreshold characteristics and eventually the loss of gate control.

Breakdown and latch phenomena have been reported for both N- and P-channel enhancement- and depletion-mode devices [7-10]. For P-channel transistors, these effects are attenuated by the lower impact ionization rate of holes. On the other hand, a body contact can reduce or suppress this phenomenon. The influence of temperature on the PBT action will be presented below.

2.3. Noise

Flicker noise in microelectronics devices, observed at low frequency with a variation inversely proportional to the frequency (1/f), is very harmful for analog or RF applications. It is attributed to carrier mobility or carrier number fluctuations. In MOS transistors, the main fluctuations are due to dynamic exchange of channel carriers with interface or oxide traps (carrier number fluctuations).

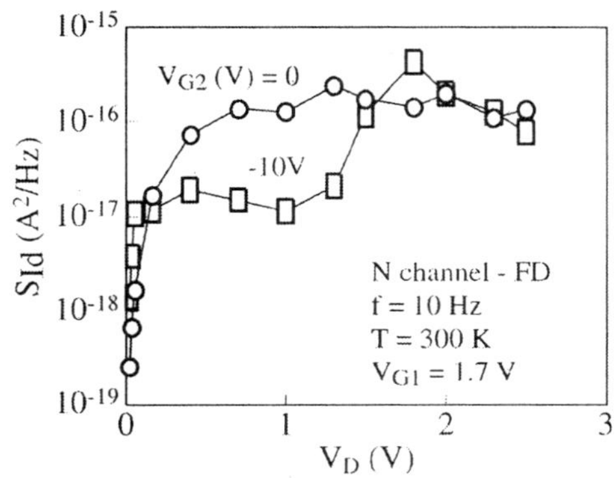

Fig. 2. Experimental drain current noise as a function of drain voltage, for various back gate biases (N-channel thin film (80nm) enhancement-mode SOI MOSFET); a kink effect and an excess noise is obtained when the device is partially depleted (Vg2=-10V).

In SOI MOSFETs, supplementary fluctuations are obtained owing to floating body effects. For a thin film fully depleted transistor, a conventional drain current noise (bulk-like) is obtained as a function of Vd (Fig. 2 with Vg=0). However, when a negative bias is applied to the substrate (back gate), the device becomes partially depleted (accumulation of the back interface) and a kink effect is obtained leading to an excess noise around the kink (Fig. 2 with Vg2=-10V) [11]. This excess noise can be attributed to the dynamic trapping of carriers created by impact ionization in the Si layer or at the interfaces of the SOI structure as well as the impact of shot noise at the source/body junction.

The parasitic bipolar transistor action, which can be observed in both partially and fully depleted SOI MOS transistors, also leads to a peculiar low frequency noise behavior. The noise measurements are shown in Fig. 3. For small drain current, below the threshold of the parasitic bipolar transistor, similar noise magnitudes are obtained for the various V_d values. However, for large I_d, the PBT regime is reached for a sufficiently high drain bias ($V_d \geq 3.3$ V) leading to a significant increase of the noise. Similar results for the drain current noise are obtained in the case of forward- (Vd=3.35V) and reverse (Vd = 3.3 V) -V_g-scan latch [12].

Fig. 3. Normalized drain current noise vs. Drain current for N-channel partially-depleted SOI MOSFET (tsi=100nm, tox1=4.5nm, tox2=80nm); an excess noise is observed in the presence of both forward- and reverse-Vg-scan latch.

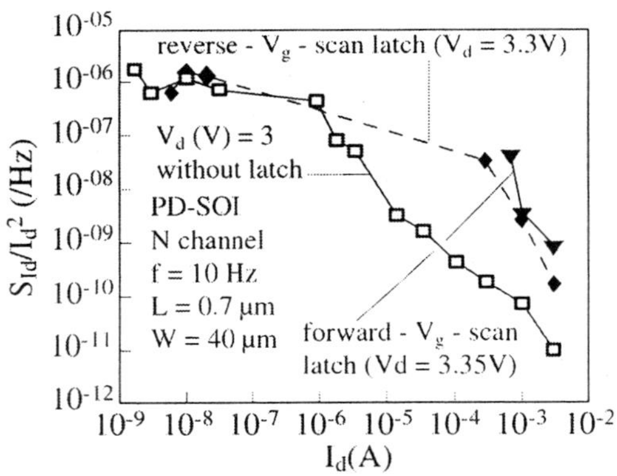

This excess noise is attributed to the same phenomena as the noise obtained in the kink regime. Note also that the impact-ionization current leads to a biasing of the base of the parasitic bipolar transistor, and further fluctuations can be caused by the PBT carrier transport between emitter (source) and collector (drain).

On the other hand, an interesting behavior in the low frequency noise has been observed in volume inversion with a double gate operation (see below). In this case, an important part of the carrier transport is carried out in the Si volume far from the interfaces, inducing a reduction of the dynamic exchange of carriers with the oxide traps and a screening of the oxide charge fluctuations, thus enabling a significant reduction of the noise of the MOS transistors [11].

2.4. Self-Heating Effects

SOI MOSFETs suffer from self-heating effects conveyed by the low thermal conductivity of the buried oxide. At high power levels, one observes the onset of negative output conductance in the saturation region [13]. This behavior is mainly attributed to the reduction of the mobility with increasing channel temperature by self-heating. However, other device parameters (threshold voltage, saturation velocity, etc.), have to be taken into account for accurate modeling. Self-heating also leads to an increase of the interconnect temperature which is critical for electromigration considerations.

The temperature rise is proportional to the power, and is much larger in a SOI device than in a bulk Si transistor. As the silicon layer is thinner, the channel temperature substantially increases. The channel temperature is also raised with increasing the buried oxide thickness and the channel-metal contact separation. The reduction of this important parasitic effect in SOI technology may require the device structure as well as the film and buried oxide thicknesses be optimized. Fortunately, lower self-heating effects are obtained under dynamic operation. For thin buried oxides, the self-heating phenomena are also substantially reduced (Fig. 4).

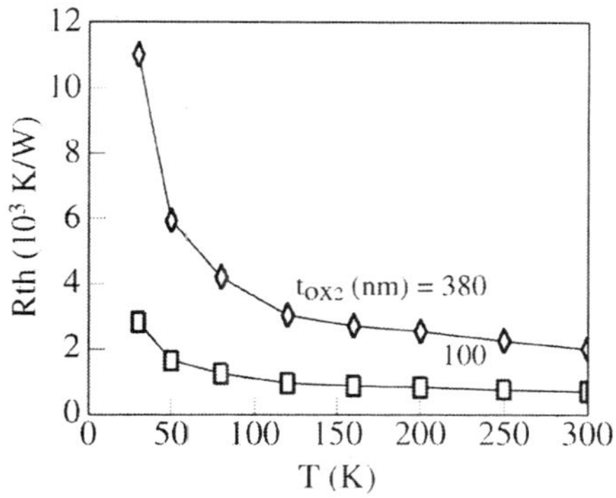

Fig. 4. Simulated variations of the thermal resistance versus temperature with various buried oxide thicknesses for SIMOX MOSFETs.

3. Influence of device architecture on short channel effects

Short channel effects can become harmful in the deep submicron range. In this respect, two phenomena have to be optimized in order to obtain a reliable device and circuit operation. The charge sharing effect is due to the increased influence of the depletion region at the source and drain junctions with scaling down the MOSFETs. This leads to a reduction of the depletion charge controlled by the gate and thereby a decrease of the threshold voltage of the transistor that can induce substantial leakage currents. The drain-induced barrier lowering (DIBL) is due to the electrostatic influence of the drain potential on the source/Si film barrier height at high Vd. This phenomenon has been shown to jeopardize deep submicron device operation with, in particular, a significant drain leakage current.

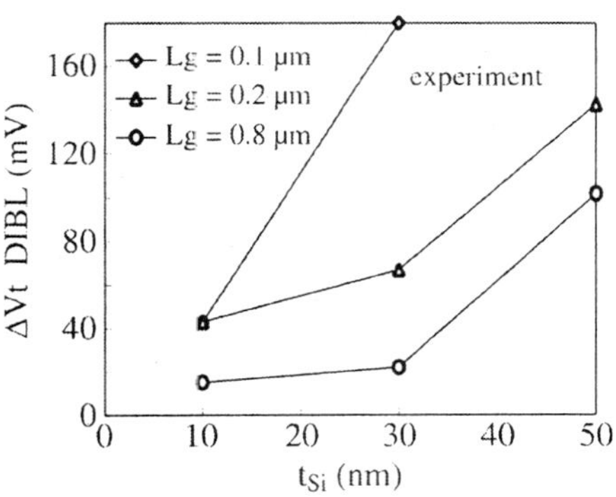

Fig. 5. Experimental variations of DIBL as a function of Si layer thickness for various gate lengths.

SOI structures offer unique options for the reduction of short channel effects. However, a careful adjustment of the SOI parameters is necessary to improve the performance of deep submicron devices. For instance, Fig. 5 shows typical DIBL experimental results observed for various gate lengths. A substantial decrease of this short channel effect is obtained with reducing the Si layer thickness, especially for 0.1µm devices [14]. The low temperature performance will be investigated in section 6.

4. Impact of device architecture on hot carrier effects

Hot carrier effects (HCE) are created by the large lateral electric field observed at high Vd close to the drain/Si film junction inducing high-energy carriers in the conduction channel. A substantial reduction of the normalized gate current is found for decreasing the Si film thickness. 2-D simulations (including the energy balance equation) show that the inversion layer thickness increases with reducing the film thickness in fully depleted SOI transistors, which results in a decrease in the maximum electron temperature (Fig. 6) [15]. However, another possible effect in order to explain the decrease of hot carrier effects is the lowering of the secondary impact ionization at the drain/substrate junction. Indeed, in thin film SOI, the area of this junction decreases as compared to thick Si layers or bulk MOSFETs, and therefore a reduction of the number of high-energy carriers is obtained. Moreover, the reduction of the transverse electric field also leads to a smaller injection probability into the gate.

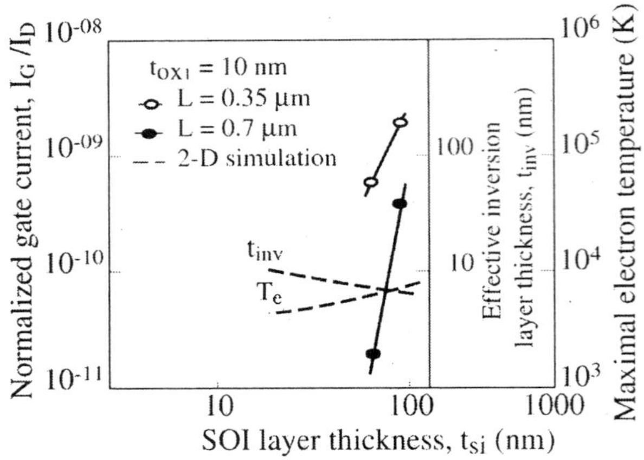

Fig. 6. Variation of the experimental normalized gate current as a function of Si film thickness compared with simulated inversion layer thickness and maximum electron temperature (after [15]).

On the other hand, photon emission measurements have shown that a number of hot carriers are also created in the parasitic bipolar regime for small gate biases ($0 \leq Vg \leq Vt$) and high Vd [16]. Substantial degradation have been pointed out in this gate voltage range. Indeed, the worst case degradation of fully depleted SOI nMOSFET is often obtained in the PBT regime in off state operation.

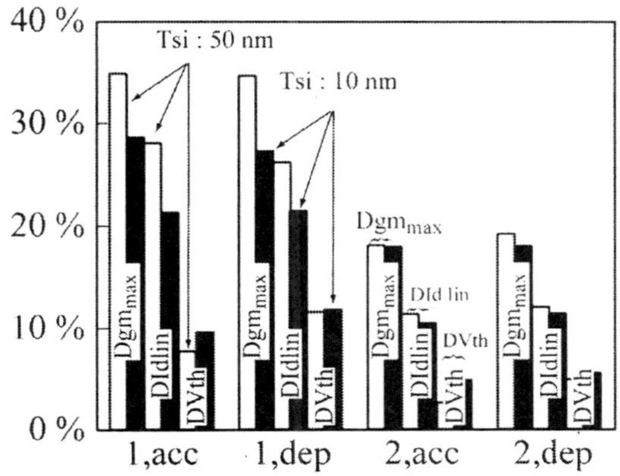

Fig. 7. Degradation of 0.15 μm SIMOX nMOS electrical properties after 50 000s of electrical aging @ Vd = 3.0 V and Vg=0 V, dark bar: Tsi = 10 nm and white bar: Tsi = 50 nm.

For thick films (50 nm), the degradation is enhanced as compared to ultra-thin films (Fig. 7). The electrical properties at the front interface with a back interface accumulation (1,Acc) or with a back interface depletion (1,Dep), and the electrical properties at the back interface with a front interface accumulation (2,Acc) or a front interface depletion (2,Dep) are investigated in order to determine the impact of interface coupling effects. The degradations of the transconductance and the drain current controlled by the front gate (1,Acc or 1,Dep) are substantially lower for a device realized on a 10 nm silicon layer. A small increase of the threshold voltage degradation is observed for a 10nm Si film. The back interface degradation (2,Acc or 2,Dep) of 10nm and 50nm Si films are similar (with a low reduction of drain current degradation for a 10nm Si layer). This behavior can be explained by the reduction of carrier energy in thinner films which leads to a decrease of the front interface degradation and a similar aging of the back interface whatever the Si-film thickness is, due to the enhancement of the coupling effect. These very interesting results show that ultra-thin Si films lead to an improvement of long term device reliability [17]. The low temperature reliability will be studied in section 6.

5. Ultimate device architectures

The inversion channel induced by the gate of a MOSFET is located at the Si/SiO_2 interface with a typical length of a few nanometers. The double-gate control of a SOI MOSFET allows forcing the whole silicon film (interface layers and volume) in strong inversion and gives rise to the "volume inversion (VI)" concept [18]. The fact that the current drive of the VI-MOSFET is governed by two gates and carriers are no longer confined at one interface presents remarkable advantages: enhancement of the number of minority carriers, increase in carrier mobility and velocity due to reduced influence of scattering associated with oxide charges and surface roughness, increase in drain current and transconductance, ideal subthreshold slope [18].

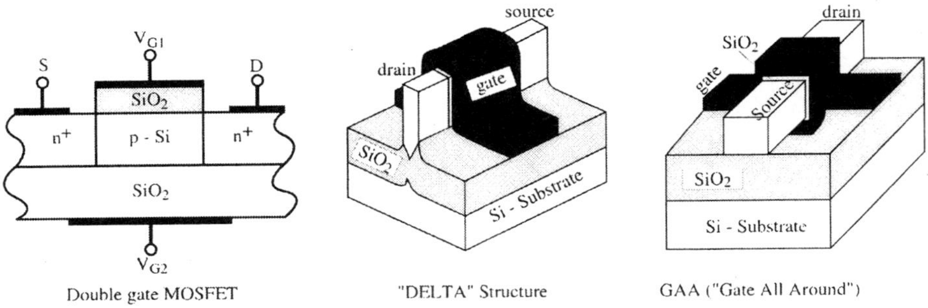

Fig. 8. Various SOI structures using the concept of volume inversion.

Various SOI structures (double-gate [19], DELTA [20], GAA (Gate-All-Around) [21] have been proposed in order to take advantage of this original feature (Fig. 8). Fig. 9 shows an example of the performance improvement during volume inversion operation as compared to a conventional single-gate SOI operation [21] : the GAA device exhibits a transconductance up to 3 times larger.

Fig. 9. Transconductance in a conventional SOI MOSFET, a double-gate transistor without volume inversion, and a GAA device with volume inversion (W/L = 3μm / 3μm, Vd=100mV) (after [21]).

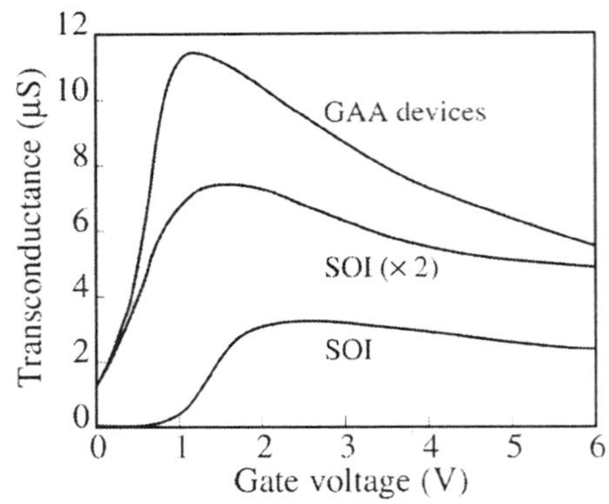

The VI-MOSFET seems also to be an ideal device for alleviating short-channel effects in ultimate ultra-short channel MOS transistors. A substantial reduction of the DIBL (Fig. 10) and charge sharing phenomena in volume inversion operation has been observed [22]. These advantages are obtained together with very large driving current and very small leakage currents for drain bias up to 1.5V in a wide Si film thickness range. For 0.05μm SOI-MOSFETs, the short channel effects (DIBL and charge sharing) are also reduced with decreasing the silicon film thickness (down to 5nm) and doping whatever the architecture is (single gate and double gate (Fig. 10)). Besides, a reduced sensitivity on the silicon film thickness and doping is observed for the double gate devices which is very interesting for the optimization of their electrical properties.

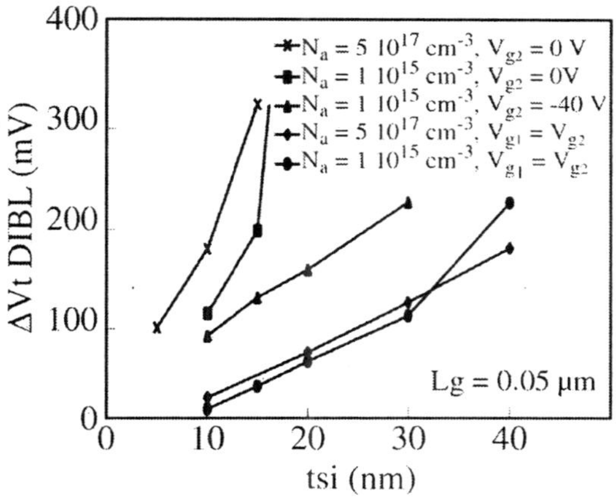

Fig. 10. DIBL effect vs. silicon film thickness for 0.05μm single gate SOI MOSFETs (tox2=380nm, tox1=3nm) with high doping ($Na=5\times10^{17}cm^{-3}$, Vg2=0V), low doping ($Na=10^{15}cm^{-3}$, Vg2=0V), back channel accumulation ($Na=10^{15}cm^{-3}$, Vg2=-40V), and for double gate SOI MOSFETs (tox1=tox2=3nm, $Na=10^{15}cm^{-3}$ and $Na=5\times10^{17}cm^{-3}$ with Vg1=Vg2).

It is also worth noticing that the electron temperature is reduced with decreasing the Si film doping leading to a better long-term device reliability. The lowest electron temperature has been obtained in the case of a double gate SOI MOSFET.

This transistor seems to be the best candidate for the ultimate integration of silicon and could be proposed for devices down to at least 20nm gate length without the need of an alternative gate dielectrics. The electrical properties of these double gate devices at low temperature will be discussed in section 6.

6. Impact of temperature on MOSFET performance, floating body phenomena, hot carrier and short channel effects

6.1. Influence of temperature on MOSFET performance

a) Subthreshold swing

The dependence of the subthreshold swing S on temperature is shown in Fig 11a for fully depleted SOI MOSFETs. S for the front channel, which is around 70mV/dec at 300K, can reach very small values, about 20mV/dec, at very low temperature [23].

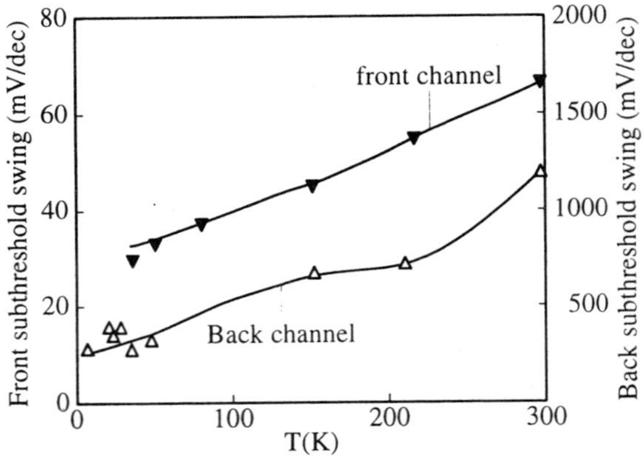

Fig. 11a. Subthreshold swing versus temperature for front and back channels in fully depleted SIMOX MOSFETs; L=2μm, Vd=30mV.

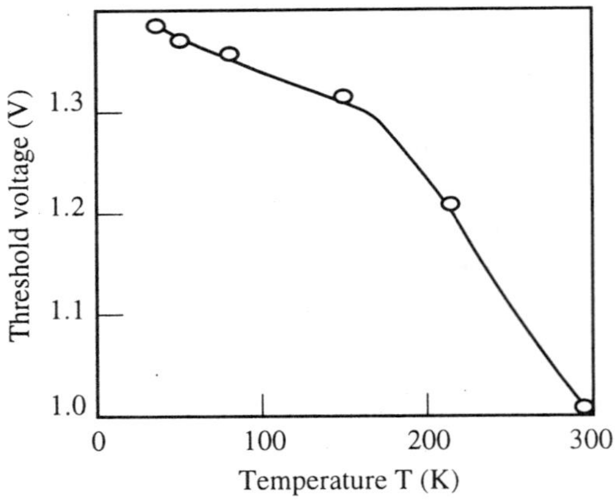

Fig. 11b. Threshold voltage against temperature for a partially-depleted (at 300K) SIMOX MOSFET becoming fully-depleted at low temperature (L=2μm, Vd=30mV).

b) Threshold voltage

The low or high temperature properties of SOI transistors are rather different from those of bulk devices as a consequence of the specific mechanisms existing in SOI structures. The variation of the threshold voltage as a function of temperature is illustrated in Fig. 11b [23].

Two slopes can be clearly seen in this characteristic. The device is partially depleted in the high temperature range and becomes fully depleted in the low temperature range. The reduction of the Vt(T) slope at low temperature occurs when the full depletion of the silicon film prevents the further extension of the depletion region under the gate with decreasing the temperature [23].

The variation of the threshold voltage with temperature can be expressed as

$$\frac{dV_t}{dT} = \frac{d\phi_F}{dT}\left[\alpha\left(\frac{q\varepsilon_{si}N_a}{\phi_F C_{ox}^{\ 2}}\right)^{1/2} + 2 + \frac{qN_{ss}}{C_{ox}}\right]$$

where α=1 for a partially depleted film and α=0 for a fully depleted film, ϕ_F is the Fermi potential, Na the Si film doping, Cox the gate oxide capacitance and Nss the interface state density.

Therefore, a substantial reduction of the variation of Vt versus temperature can be achieved in fully depleted SOI MOSFET because of the suppression of the variation of the depletion charge with temperature. This behavior is very attractive for device operation in a large temperature range.

c) Transport parameters

Fig. 12 exemplifies the variation of the maximum field effect mobility as a function of temperature [23].

FD SOI MOSFETs offer the highest mobility both at room and low temperature for front and back channel operations. The excellent mobility observed below 77K for fully depleted devices is attributed to the stronger coupling between the two interfaces at very low temperature causing the minority carriers to extend into the Si film volume. Carriers flowing in the volume and contributing to the whole conduction have much higher mobility than the surface carriers.

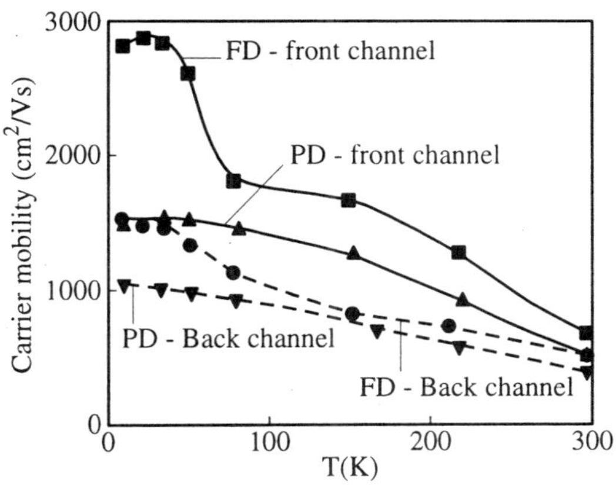

Fig.12. Maximum field-effect mobility for front (——) and back (----) channels in PD and FD SIMOX MOSFETs; L=2μm, Vd=30mV.

6.2. Influence of temperature on floating body effects

a) kink effect

In Fig. 13 the kink effect in N-channel enhancement- and accumulation-mode SOI MOSFETs are compared at very low temperature. A clear kink effect is obtained for partially depleted enhancement-mode devices for low gate biases (at high Vg, the kink is strongly reduced due to self-heating effects). However, for accumulation-mode transistors, no kink effect is obtained whatever the gate voltage is [23]. This behavior is very interesting for cryogenic applications.

The kink effect can also be suppressed in fully depleted enhancement-mode SOI MOSFET down to liquid helium temperature.

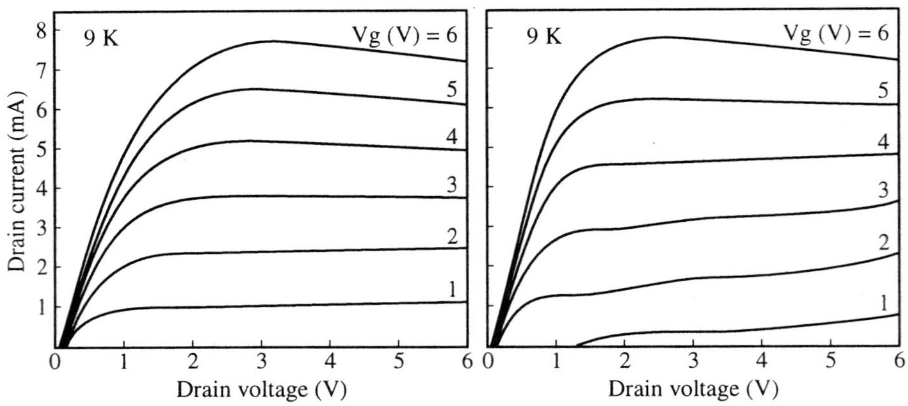

Fig. 13. Id(Vd) output characteristics in (a) accumulation-mode and (b) partially depleted enhancement-mode n-channel SIMOX MOSFETs.

b) Latch and Breakdown Phenomena

The latch and breakdown phenomena associated with the parasitic bipolar transistor action are substantially temperature dependent. SOI MOSFETs present some interests for high temperature applications (up to 300°C) due to the reduction of junction leakage currents. However, the enhancement of the gain of the parasitic bipolar transistor (PBT) has to be taken into account for the optimization of the transistors (Fig. 14).

Fig. 14. Gain of the PBT versus temperature up to 300°C.

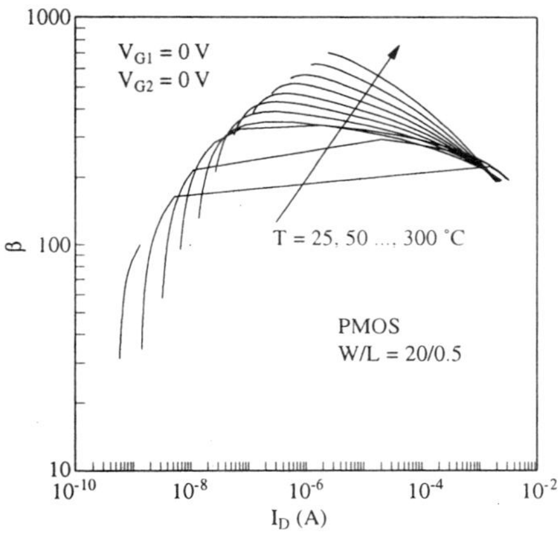

Fig. 15 presents a typical off-state (Vg=0) breakdown for a thin-film (80 nm) LDD SOI n-MOSFET at low temperature [10]. At 300K, a conventional breakdown is observed at low drain bias (<4V), unlike at liquid nitrogen temperature where breakdown is avoided for Vd below 6V.

Fig. 15. Off-state breakdown (Vg=0) at 300 and 77K for a n-channel thin-film accumulation-mode SIMOX MOSFET (tsi=80nm, W=40μm, Lg=0.8μm).

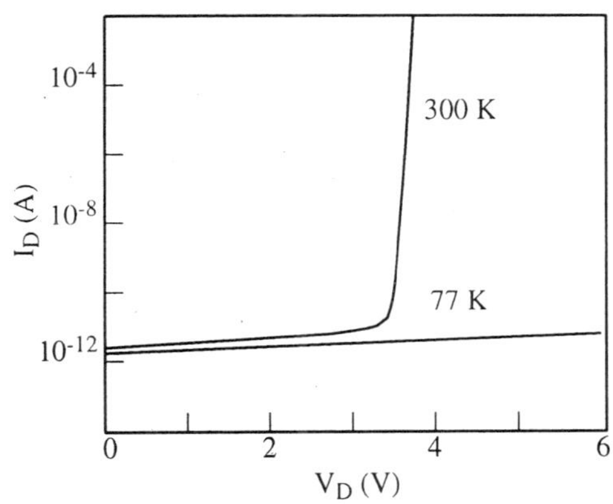

Fig. 16a,b,c presents the transfer Id(Vg) characteristics obtained for a thin film (80 nm) accumulation-mode SIMOX MOSFET (Leff=0.3μm). A clear latch phenomenon is observed at 300K. The forward Vg-scan latch is problematical in CMOS circuits due to prohibitive power consumption. At 77K, this forward-Vg-scan latch is suppressed, and at 4.2K the reverse-Vg-scan latch also disappears. For enhancement-mode devices (N+PN+ or P+NP+), the reduction of the PBT is stronger than that observed for accumulation-mode ones at low temperature [10].

Fig. 16d,e shows the transfer characteristics (forward and reverse Vg-scans) for advanced 0.2μm gate length NMOS SIMOX-MOSFET (4.5nm gate oxide) with a 40nm Si film thickness (enhancement-mode transistor). At 300K (Fig. 16d), for a drain bias larger than 3V, a clear latch phenomenon is obtained for small gate biases with a very large leakage current in accumulation. Fig. 16e presents the behavior of the same device at liquid nitrogen temperature. A substantial reduction of this parasitic effect is observed up to a drain voltage of 3.5V. Nevertheless, the latch phenomenon is not completely suppressed and can be seen for Vd=3.5V [24].

These considerable advantages at low temperature are attributed to various mechanisms. Firstly, the PBT gain β decreases quasi-exponentially with reducing the temperature ($\beta \propto \exp(-\Delta E_g/kT)$), due the difference in bandgap ΔE_g between emitter (source) and base (thin Si film). In addition, the lightly doped regions at the source and drain freeze-out at low temperature. Since the lightly doped drain is useful to reduce the impact ionization by lowering the lateral field, the freeze-out of the LDD region induces a further decrease of the ionization. The benefit of the lightly doped source is to reduce the PBT gain by maximizing the carrier recombination in the source (emitter) region; the freeze-out of this region produces a further decrease in gain.

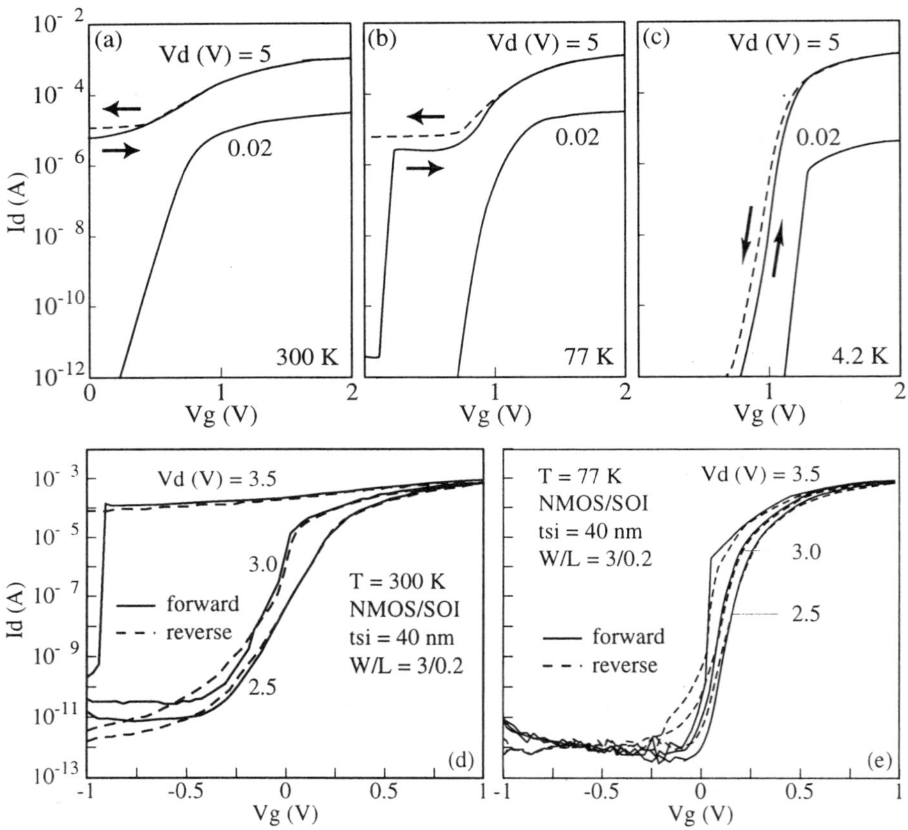

Fig. 16. Id(Vg) characteristics of N-channel thin-film accumulation-mode SIMOX MOSFET (W = 20 μm, Leff = 0.3 μm, tsi = 80 nm) at:
(a) 300 K, (b) 77 K and (c) 4.2 K, for high and low Vd in the case of forward- and reverse-Vg-scan.
Transfer characteristics for a N-channel enhancement-mode SIMOX MOSFET at:
(d) 300K and (e) 77K with forward- and reverse-Vg scans (tsi=40nm, W/L=3/0.2).

In Fig. 17 are plotted the variations of Id(Vg) obtained with a grounded substrate. A significant decrease of the latch effect is found compared with Fig. 16d at 300K (Fig. 17a), showing that the body terminal is useful for hole collection. However, the action of the parasitic bipolar transistor is not completely suppressed, even in the 77K range (Fig. 17b), which demonstrates experimentally the limits of this contact for ultra-thin SOI layers. For a 20nm Si film enhancement-mode N-channel devices, similar improvements of the PBT effect have been found at low temperature [24].

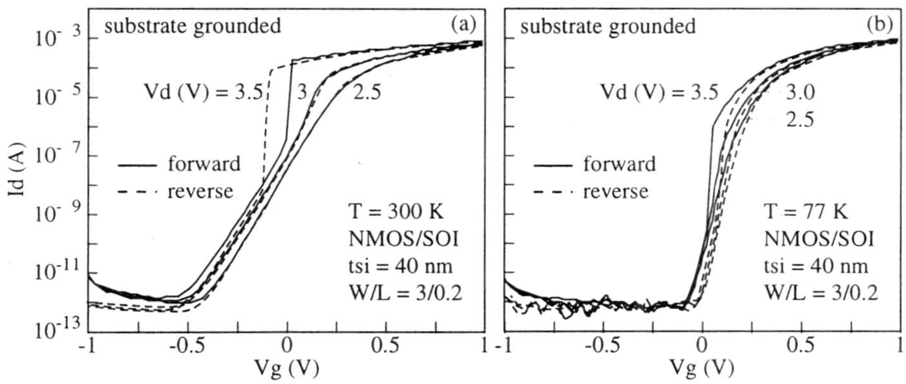

Fig. 17. Transfer characteristics for a N-channel enhancement-mode SIMOX MOSFET at: (a) 300K and (b) 77K with forward- and reverse-Vg scans and a grounded body contact (tsi=40nm, W/L=3/0.2).

c) Gate-Induced-Drain-Leakage

These low temperature mechanisms are also responsible for the reduction of the gate-induced drain leakage (GIDL) current in thin film SOI MOSFETs. The variations of the normalized GIDL characteristics with temperature are shown in Fig. 18 (I_{gidl}/E_{eff}=A.exp(-B/E_{eff}), E_{eff} being the effective surface electric field, and A, B constant parameters) for a p-channel thin film (80 nm) SIMOX MOSFET as a function of the inverse of the surface field [25]. A strong reduction of the leakage current, much larger than in bulk Si MOSFETs [26], is obtained at low temperature.

This strong improvement is attributed to the reduction of the PBT gain and to impurity freeze-out in the LDDs. Indeed, it has been shown that the GIDL current is enhanced in short-channel SOI MOSFETs due to the influence of PBT [27]. The GIDL current is known to originate from a high electric field in the gate/drain overlap region that causes electron tunneling from valence band to conduction band. Note that the increase in the bandgap due to energy quantization leads to a reduction in this band-to-band tunneling current in ultra-thin SOI films (<5nm) [28]. In a SOI n-MOSFET, the GIDL current is in fact the base current of the PBT, and the collector current is enhanced by the PBT gain that is exponentially dependent on temperature. This feature causes the GIDL current to substantially decrease at low temperature. Similar effects have been shown for both n and p-channel transistors. It should also be mentioned that these improvements are obtained together with a strong increase of the drain current in saturation for thin-film SOI MOSFETs.

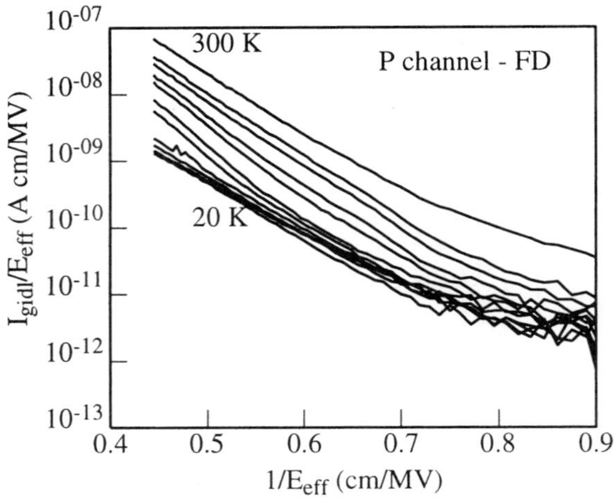

Fig. 18. Variations of the GIDL current characteristics as a function of inverse of transverse electric field for various temperatures (from the top to the bottom: 300, 250, 230, 200, 180, 160, 140, 120, 100, 80, 50, 20 K) for a p-channel fully-depleted (t_{Si}=80nm) LDD SIMOX MOSFET.

d) Self-Heating Effects

This phenomenon is enhanced at low temperature for SOI-SIMOX devices (Fig. 19) due to the increase of the thermal resistance of the buried oxide (Fig. 20) [29].

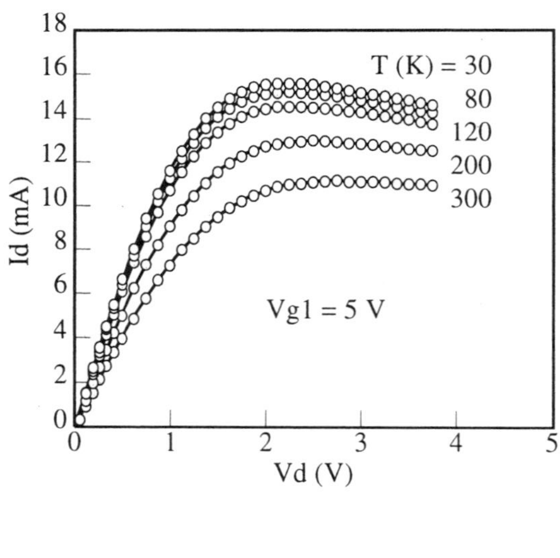

Fig. 19. Typical output characteristics for a SOI-SIMOX MOSFET showing the SH effect at various temperatures (W=40μm, L=0.8μm, 80nm Si film and 380nm buried oxide thicknesses).

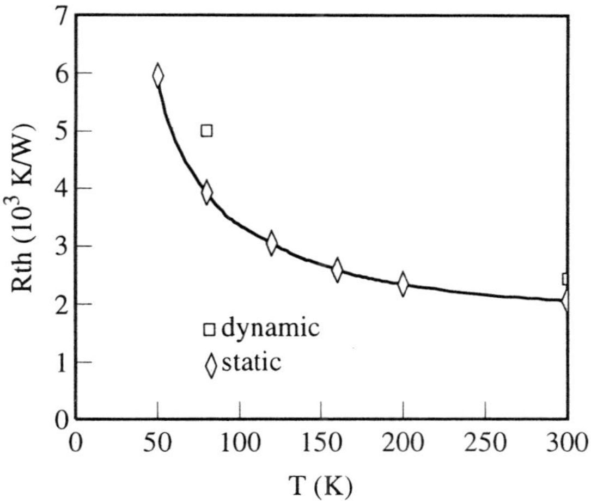

Fig. 20. Experimental variation of the thermal resistance Rth as deducted from various measurements (SIMOX MOSFET with 80nm Si film and 380nm buried oxide thicknesses).

The extracted temperature rise ΔT in the thin silicon film (80nm) is plotted versus the drain bias in Fig. 21a for various temperatures. An excess temperature of 100K can be obtained at room temperature, while this temperature rise reaches 200K at liquid nitrogen temperature.

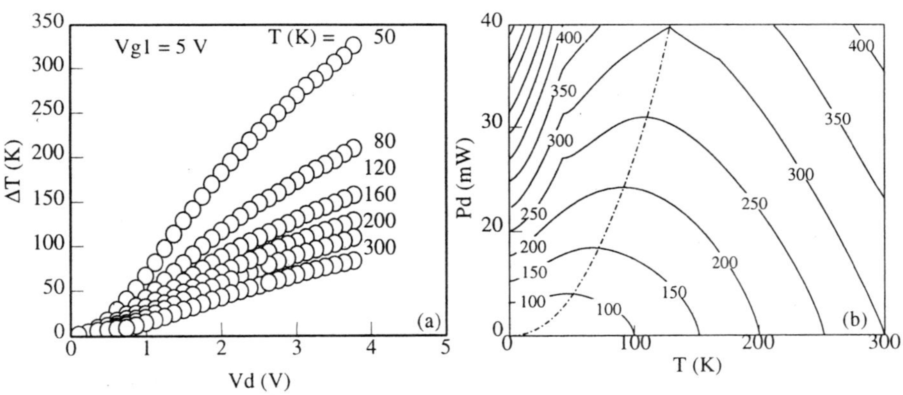

Fig. 21. (a) Variations of the device temperature rise ΔT with drain voltage for various temperatures, and (b) channel temperature versus substrate temperature T and dissipated power Pd (SIMOX MOSFET with 80nm Si film and 380nm buried oxide thicknesses).

Fig. 21b shows in a synthetic way the dependence of the local channel temperature, $T_l(K)=T+\Delta T$ ($\Delta T=Rth.Id.Vd$), with the substrate temperature T and the dissipated power Pd = Id Vd. It should be noted that the channel temperature takes two identical values for a given dissipated power, as the substrate temperature is varied. This behavior demonstrates that, in the presence of the SH effect, the reduction of the substrate temperature does not necessarily result in an increase of the device performance. Indeed, there is an optimum substrate temperature for a given dissipated power (dashed line in Fig. 21b) for which the channel temperature is minimized, and, in turn, the device performance is maximum [30]. The reduction of the buried oxide thickness, by using for instance low dose SIMOX, is an efficient method to significantly reduce the SH effects.

6.3. Influence of temperature on short-channel effects

A low temperature operation leads to a substantial reduction of short channel effects (in particular punchthrough) and allows to optimize the driving and off-state current of the devices. Indeed, Fig. 22 shows the simulated transfer characteristics at 300K and 77K for a single gate SOI MOSFET with a mid-gap gate. The mid-gap gate is used for improving the values of the threshold voltage for ultra-thin films (10nm in this study) and low doping which are needed for devices in the sub-0.1μm range. For a 0.05μm SOI MOSFET, a large leakage current is observed for Vg=0 at 300K. At 77K, a very low leakage current is obtained for gate length down to 20nm. This excellent behavior is due to the improvement of the subthreshold swing and the reduction of short channel effects.

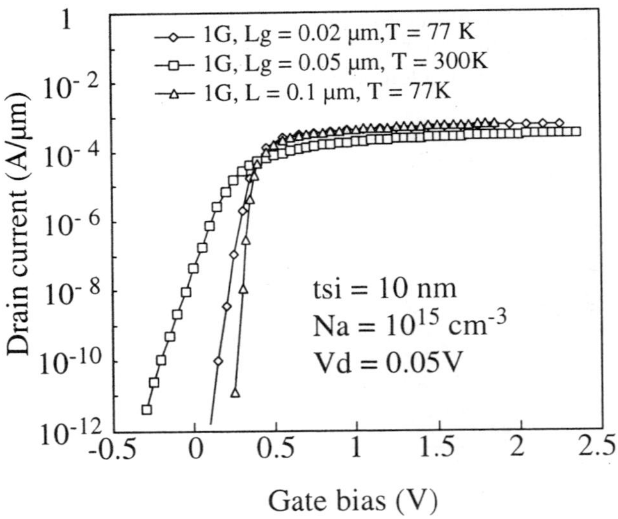

Fig. 22. 2D numerical simulation (SILVACO) of deep submicron single gate SOI MOSFETs at 300K and 77K with a mid-gap $Si_{0.3}Ge_{0.7}$ gate (tsi = 10nm, Na = 10^{15} cm^{-3},Vd = 50 mV).

For double gate MOSFETS with volume inversion, very good electrical properties with negligible leakage current are obtained at 77K for gate length down to at least 50nm, even using a traditional N+ polysilicon gate for these N-channel devices (Fig. 23).

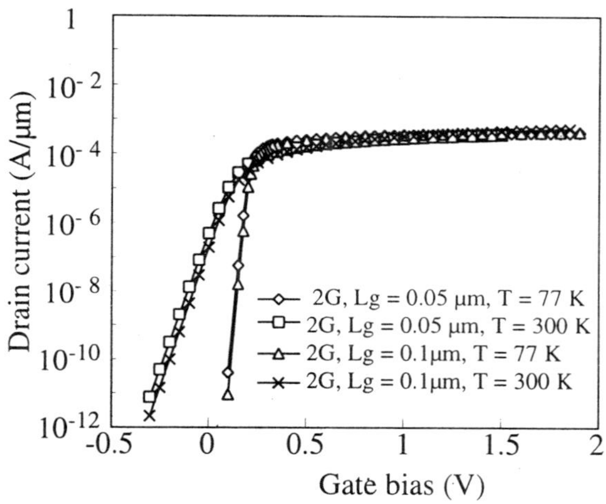

Fig. 23. 2D numerical simulation (SILVACO) of deep submicron single and double gate SOI MOSFETs with volume inversion at 300K and 77K with a N+ polysilicon gate (tsi = 10nm, Na = 10^{15} cm^{-3},Vd = 50 mV).

6.4. Influence of temperature on hot-carrier effects

For N-channel enhancement-mode SIMOX MOSFETs (Fig. 24), a strong decrease of Ibmax is obtained with reducing the temperature for a drain bias up to 4V, which extends the Vd range of reliability improvement at low temperature as compared to bulk Si devices. Two different regimes can be observed in these plots. Indeed, the second part of the curves (for Vd>3V), which is especially clear at 77 and 25K, presents a different variation with Vd. This behavior is associated with the BPT, which also leads to hot carrier effects, and induces additional impact ionization at high drain biases [31].

The impact of the temperature on hot-carrier-induced degradation in the PBT regime is now investigated in 0.1μm PD SIMOX MOSFETs. Fig. 25 shows the drain current (Id) at Vg=0 (leakage current caused by the PBT action) as a function of temperature with various drain biases for enhancement-mode devices. At 300K, an obvious decrease of Id is obtained with reducing Vd, Id reaching large values for Vd≥2.25V. In the low temperature case, the leakage current can be suppressed for Vd≤2.75V. When Vd is above 2.75V, Id is in the same order of magnitude for all the temperatures. For Vd=2.75V, a sharp Id drop is observed between T=150K and T=77K. For T≤77K, the leakage current is very small, and for T≥200K, it slightly increases with raising the temperature. A similar behavior is obtained for Vd=2.5V, but the Id drop moves to higher temperature [32].

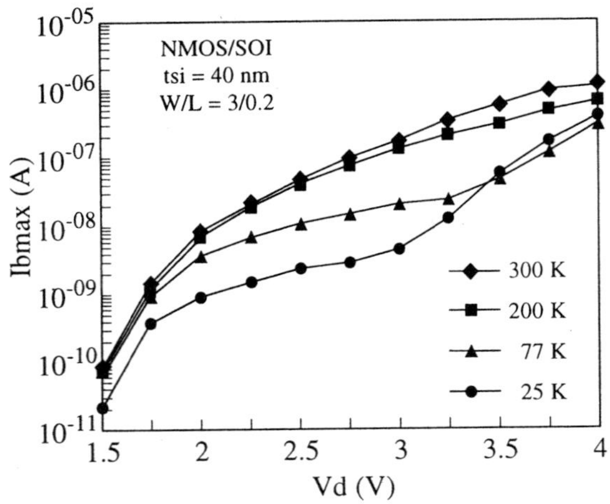

Fig. 24. Maximum substrate current versus drain bias for various temperatures in the case of a N-channel enhancement-mode SOI MOSFET (W/L=3/0.2, 40nm Si film thickness).

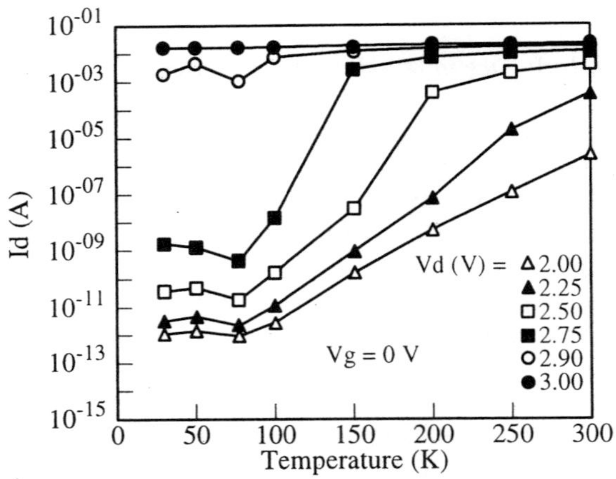

Fig. 25. Drain current (Id) at Vg=0 as a function of temperature with various drain biases for a 0.1μm N-channel SIMOX MOSFET.

A comparison of the Gmmax degradation between 300K and 77K is shown in Fig. 26. A crossing point of these two curves is observed at Vd≈2.9V. For this Vd value, Id at 300K is about 10 times that at 77K. At Vd=3V, the degradation at 77K is larger than that at 300K, with the same level of leakage current. The device aging as a function of temperature is shown in Fig. 27 (stress Vd=2.75V). The worst case is observed at T=200K for Gmmax, Vt and Id degradations. Since a sharp drop of the leakage current exists between 150K and 77K, a clear reduction of the device aging is also obtained in this range. Consequently, in the low temperature case, the reduction of the device aging is due to the strong suppression of the leakage current and the PBT action. On the other hand, with a same level of the leakage current at 300K and 77K, the higher electron energies and/or larger carrier trapping in the gate oxide at low temperature induce an enhancement of the device aging [32].

Fig. 26. Comparison of Gmmax degradation between T=300K and T=77K for 0.1μm N-channel SIMOX MOSFETs (stress Vg=0V).

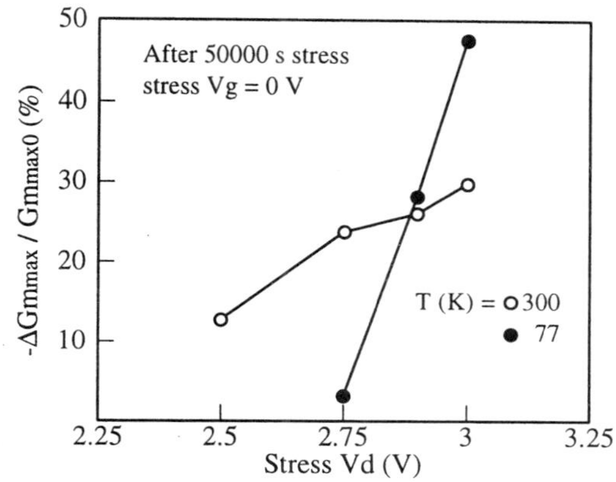

Fig. 27. Degradations of Gmmax, Id and Vt as a function of temperature for 0.1μm N-channel SIMOX MOSFETs (stress Vg=0V and Vd=2.75V).

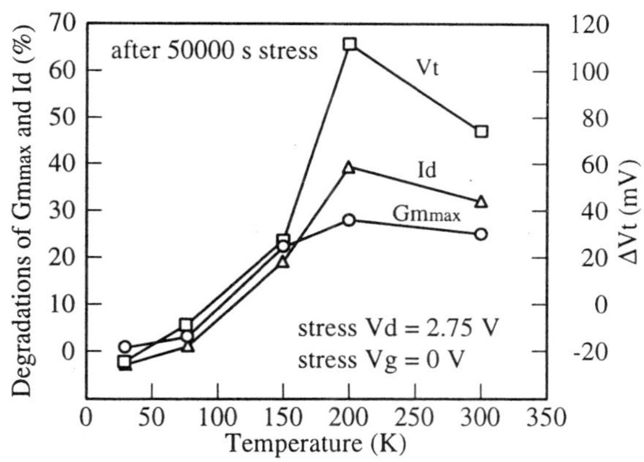

Acknowledgments

The authors would like to thank Drs. J. Jomaah, S. Cristoloveanu, S.H. Renn, E. Rauly, O. Potavin from IMEP, and C. Raynaud, J.L. Pelloie from CEA/LETI, Grenoble, for their collaboration in this work.

References

1. J.P. Colinge, *Silicon-On-Insulator technology: materials to VLSI*. Kluwer Academic Publishers, 1991.
2. S. Cristoloveanu and S.S. Li, *Electrical characterization of Silicon-On-Insulator materials and devices*. Kluwer Academic Publishers, 1995.
3. F. Balestra, *SOI devices*, Wiley Encyclopedia of Electrical and Electronics Engineering, 1999.
4. J.P. Colinge, Reduction of kink effect in thin film SOI MOSFETs, *IEEE Electron Dev. Lett.*, EDL-9, p. 97, 1988.
5. F. Balestra, T. Matsumoto, M. Tsuno, H. Nakabayashi, Y. Inoue, and M. Koyanagi, "Moderate kink effect in fully depleted thin-film SOI MOSFETs, *Electronics Letters*, 31, p. 326, 1995.
6. C.E.D. Chen, M. Matloubian, R. Sundaresan, B.Y. Mao, C.C. Wei, and G.P. Pollack, Single-transistor latch in SOI MOSFET's, *IEEE Electron Dev. Lett.*, EDL-9, p. 636, 1988.
7. M. Yoshimi, M. Takahashi, T. Wada, K. Kato, S. Kambayashi, M. Kemmoshi, and K. Natori, Analysis of the drain breakdown mechanism in ultra-thin-film SOI MOSFET's, *IEEE Trans. Electron Dev.*, ED-37, p. 2015, 1990.
8. J.Y. Choi and J.G. Fossum, Analysis and control of floating body effects in fully depleted SOI MOSFET's, *IEEE Trans. Electron Dev.*, ED-38, p. 1384, 1991.
9. J. Gautier and A.J. Auberton-Hervé, A latch phenomenon in buried N-body SOI NMOSFET's, *IEEE Electron Dev. Lett.*, EDL-12, p. 372, 1991.
10. F. Balestra, J. Jomaah, G. Ghibaudo, O. Faynot, A.J. Auberton-Hervé, and B. Giffard, Analysis of the latch and breakdown phenomena in N and P channel thin film SOI MOSFET's as a function of temperature, *IEEE Trans. Electron Dev.*, ED-41, p. 109, 1994.
11. J. Jomaah, F. Balestra, and G. Ghibaudo, Low-frequency noise in SOI MOSFET's from room to liquid helium temperature : experimental and numerical simulation results, *Proc. of ESSDERC'93*, p. 111, 1993.
12. J. Jomaah, D. Dixkens, J.L Pelloie, C. Raynaud, and F. Balestra, Impact of latch phenomenon on low frequency noise in SOI MOSFETs, *Proc. ESSDERC'96* (Les Editions Frontières), p. 87, 1996.
13. L.J. McDaid, S. Hall, P.H. Mellor, and W. Eccleston, Physical origin of negative differential resistance in SOI transistors, *Electronics Letters*, 25, p. 827, 1989.
14. F. Balestra, Performance and physical mechanisms in deep submicron SOI MOSFETs, Electron Technology, 32, pp. 50-62, 1999.
15. Y. Ohmura, An improved analytical solution of energy balance equation for short-channel SOI MOSFET's and transverse-field-induced carrier heating, *IEEE Trans. Electron Dev.*, ED-42, p. 301, 1995.
16. S.H. Renn, J.L. Pelloie, and F. Balestra, Photon emission in deep submicron N-channel SOI MOSFETs, *Electronics Lett.*, 33, p. 1093, 1997.
17. O. Potavin, S. Haendler, J. Jomaah, F. Balestra, C. Raynaud, Reliability of ultra thin film deep submicron SOI nMOSFETs, Proc. ULIS'2000, Grenoble, January 2000, pp. 43-46.

18. F. Balestra, S. Cristoloveanu, M. Benachir, J. Brini, and T. Elewa, Double-gate silicon-on-insulator transistor with volume inversion: a new device with greatly enhanced performance, *IEEE Electron Dev. Lett.*, EDL-8, p. 410, 1987.
19. K. Suzuki,T. Tanaka, T. Tosaka, H. Horie, and Y. Arimoto, Scaling theory for double-gate SOI MOSFETs, *IEEE Trans. Electron Devices*, ED-40, p. 2326, 1993.
20. D. Hisamoto, T. Kaga, Y. Kawamoto, and E. Takeda, A fully depleted lean-channel transistor (DELTA) - A novel vertical ultrathin SOI MOSFET, *IEEE Electron Dev. Lett.*, 11, p. 36, 1990.
21. J.P. Colinge, M.H. Gao, A.R. Rodriguez, and C. Claeys, Silicon-on-insulator gate-all-around device, *IEDM Tech. Dig.*, 1990, p. 595.
22. E. Rauly, O. Potavin, F. Balestra, C. Raynaud, On the subthreshold swing and short channel effects in single and double gate deep submicron SOI MOSFETs, Solid-State Electronics, vol. 43, p. 2033, 1999.
23. T. Elewa, F. Balestra, S. Cristoloveanu, I.M. Hafez, J.P. Colinge, A.J. Auberton-Hervé, and J. R. Davis, Performance and physical mechanisms in SIMOX MOS transistors operated at very low temperature, *IEEE Trans. Electron Dev.*, ED-37, p. 1007, 1990.
24. S.-H. Renn, C. Raynaud, F. Balestra, Floating body and hot carrier effects in ultra-thin film SOI MOSFETs, Proc. WOLTE-2, in J. Phy. IV, Coll. 3, Supplément au J. de Phys. III, Vol. 6, p. C3-49, 1996.
25. J. Jomaah, G. Ghibaudo, and F. Balestra, Temperature dependence of gate-induced-drain-leakage (GIDL) current in thin-film SOI MOSFETs, *The Electrochemical Society Proceedings* Volume 95-9, 1995, p. 260.
26. K. Rais, F. Balestra, and G. Ghibaudo, Temperature dependence of gate induced drain leakage current in silicon CMOS devices, *Electronics Letters*, 30, p. 32, 1994.
27. J. Chen, F. Assaderaghi, P.K. Ko, and C. Hu, The enhancement of Gate-Induced-Drain-Leakage (GIDL) current in short-channel SOI MOSFET and its application in measuring lateral bipolar current gain β, *IEEE Electron Dev. Lett.*, EDL-13, p. 572, 1992.
28. T. Ishijama and Y. Omura, Enhancement and suppression of band-to-band tunneling current in ultra-thin nMOSFETs/SIMOX: Influence of superficial Si layer thickness and it's future prospect, Proc. SSDM'96, p. 314, 1996.
29. J. Jomaah, G. Ghibaudo, F. Balestra, Analysis and modelling of self-heating effects in thin-film SOI MOSFETs as a function of temperature. Solid-State Electronics, 38, p. 615, 1995.
30. J. Jomaah, E. Rauly, G. Ghibaudo, F. Balestra, A thorough analysis of self-heating effects for SOI MOSFETs on SIMOX and UNIBOND substrates, Proc. WOLTE-3, in J. Phy. IV, France 8, p. 3-17, 1998.
31. S.-H. Renn, C. Raynaud, F. Balestra, Temperature dependence of hot carrier effects in 0.2μm N- and P-channel fully depleted Unibond MOSFETs, Proc. WOLTE-3, in J. Phy. IV, France 8, p. 3-13, 1998.
32. S.H. Renn, E. Rauly, J.L. Pelloie, F. Balestra, Impact of floating-body-induced parasitic bipolar transistor on hot-carrier effects on 0.1 μm N-channel SOI MOSFETs, ECS Meeting, Symposium on low temperature electronics and high temperature superconductivity, Montreal, Proc. p. 199, Mai 1997.

STRATEGIES FOR HIGH TEMPERATURE ELECTRONICS
A Western European Perspective

COLIN JOHNSTON AND ALISON CROSSLEY
HITEN – High Temperature Electronics Network of Excellence
AEA Technology Advanced Materials Centre, Begbroke Business and Science Park, Sandy Lane, Yarnton, Oxford, OX5 1PF, UK

1. Introduction

High Temperature Electronics (HTE) is a critical enabling technology for a range of European Union (EU) market sectors including consumer goods, aerospace, automotive, oil & gas and related power generation industries. We [1] define HTE as any electronic system which operates in an environment whose temperature is greater than 125 °C, whether this be junction or ambient. These markets represent in excess of $1,000 billion in value and employ some 12 million people in Western Europe. The current value of the HTE component of this total available market is judged to be approximately $100 million [2]. The growth in the total available market (TAM) for HTE is reproduced in Figure 1.

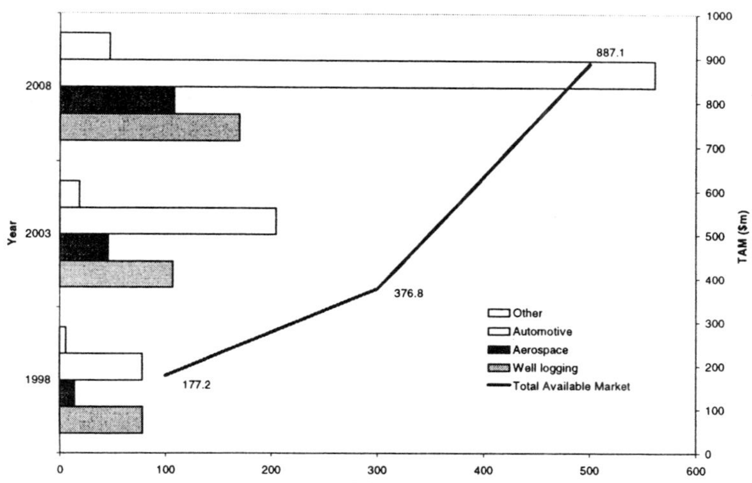

Figure 1. Growth in the market for High Temperature Electronics

F. Balestra et al. (eds.), Progress in SOI Structures and Devices Operating at Extreme Conditions, 129–137.

To realise the predicted growth a large investment is still required. A major requirement will be the availability of cost-effective commercial components. Initial applications will be the least price sensitive. The largest potential temperature range is 200 to 250 °C and will be satisfied by silicon, silicon-on-insulator (SOI) or silicon carbide semiconductor technologies [2].

The driving force for components capable of high temperature operation is the elimination of expensive and bulky cooling systems which are currently required to protect electronic systems for hostile and extreme environments as depicted in Figure 2. This would result in improved system architectures, increased safety, performance and efficiency with resultant reduced environmental impact and large operational cost savings.

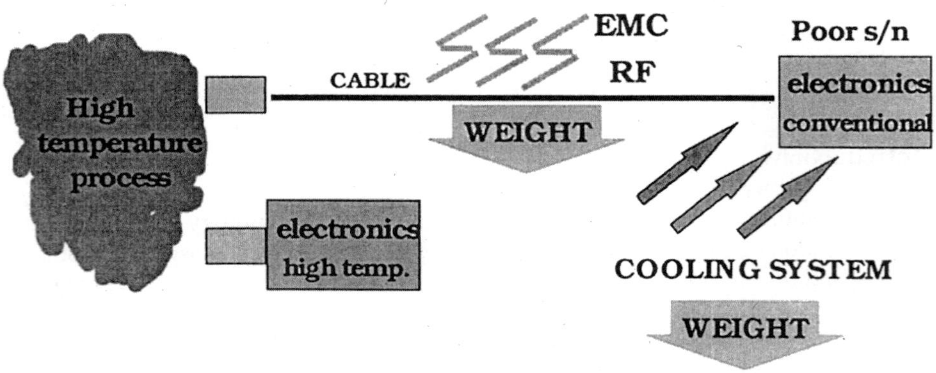

Figure 2. The benefits gained in using HTE are significant – reduced weight, increased reliability and improved safety and increased efficiency

2. Downhole applications

One of the key early market developers of commercial HTE has been well logging applications in the petroleum exploration industry [3]. Here there is a clear return in investment where high development costs can be off-set by the returns which can be accessed by more efficient working of the available reservoir. The world's largest high pressure, high temperature oil and gas field project, The Elgin/Franklin reservoir in the UK sector of the North Sea has a operational temperature of 200 °C. It has cost $30 million to develop HTE components to service this field alone. However, such investment has enabled the oil companies to access 60 million m^3 of condensate and 50 billion m^3 of gas which would otherwise have been less readily available for exploitation.

200 °C is he current limit for most down hole applications. However, as demand for petroleum and petroleum derivatives increases more extreme conditions are being

explored and the temperature range will extend to 300 °C and eventually 400 °C. It is clear that the immediate future of HTE development will be closely tied to the fate of the energy industry and particular oil and gas exploration for higher grade crudes which are often located deeper than other crudes.

Despite this it is anticipate that applications beyond 200 °C operation will account for less than 10 % of the total HTE market. The main technological drivers for well logging applications are:

- Improved accuracy and resolution of instrumentation
- Increased functionality in real time
- Decreased failure rates
- Elimination of cooling requirements
- Higher reliability
- Increased operating temperatures

The typical temperature ranges covered by well logging applications are shown in figure 3.

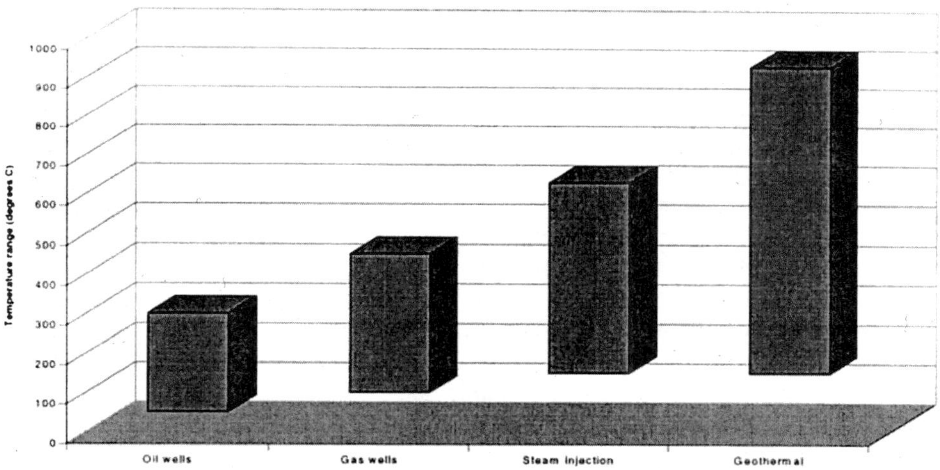

Figure 3. Typical temperature ranges encountered in down well applications

The average price for a typical system tool is $125k to $150k. Typical component requirements for well logging applications are given in table 1.

An important market emerging in "well-logging" applications is in smart completions for permanent installations. Here the actual technical requirements in terms of measurands are similar to well logging although the actual demands on the electronic systems are much more stringent with installations with a lifetime of several years and a commensurate increase in the demands on reliability of the systems.

TABLE 1. Component requirements for well logging applications

Devices	Systems
Resistors	Relays
Capacitors (low leakage)	EEPROM
Op Amps (low drift)	Voltage regulation
Zener diodes	Oscillators
Signal diodes	Transistors
A/D convertors	Line drivers (8, 10 and 16 bit)
Differential Amplifiers	8 and 16 bit microprocessors
CMOS Logic	Voltage references

3. Aerospace and Automotive Applications

Two other industrial sectors, aerospace and automotive are emerging slowly but may be of equally strategic importance to the burgeoning HTE industry [2]. At first glance the demands of the aerospace and automotive sectors may seem to be diametrically opposed. Automotive applications tend to be extremely high volume and very cost sensitive whereas aerospace applications are of lower volumes with higher value. However, the actual demands placed on the HTE components can be similar:

- High temperature operation and dynamic range
- High vibration and shock tolerance
- High reliability

3.1 AEROSPACE APPLICATIONS

For the aerospace industry HTE technology can potentially deliver enhanced functionality in a range of applications, for example in the More Electric Aircraft (MEA) programmes currently running in Europe and the US efforts are being focussed on replacing many of the mechanical systems with electric actuation. For example, replacing gear boxes in engine enclosures could significantly reduce weight and facilitate easier maintenance reducing both operating and ownership costs. However, such replacements require electronic systems capable of operating in extremely demanding environments, close to, if not in the engine combustion chamber itself. Typical application areas and temperature regimes are shown in figure 4 and Table 2.

The main drivers for this market sector is in legislation with respect to reduced emissions, pollutants and noise, necessitating improved engine control and increased engine efficiency. Another aspect is in the development of higher power dissipation engines, reduced engine footprints and reduced availability of cooling.

In an application sector where safety issue and regulations are paramount component costs are often a secondary consideration. HTE could therefore realise significant market penetration in the next decade. Present markets are small but technology uptake is accelerating making this market an attractive niche sector for future HTE providers. However, an acceleration of one or two strategic government lead programmes could see significant impact on this area and enhanced growth.

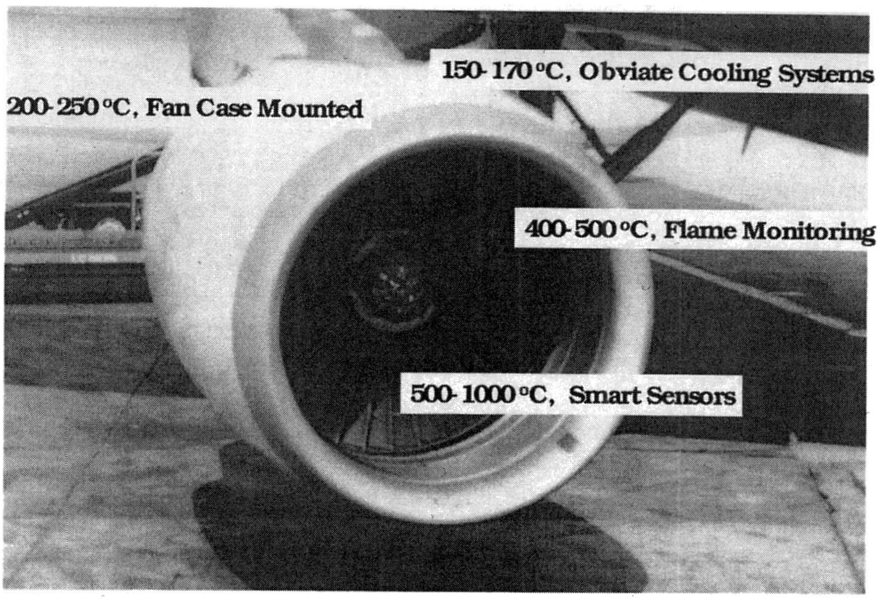

Figure 4. Typical temperature environments encountered in aerospace applications

The actual HTE market for the aerospace market could account for 10 % of the TAM by 2008. Of this almost 90% of the electronics required by the aerospace market will be demanded by systems located in or around the engines. The typical component requirements are for processors (speeds up to 4 MHz); memory; operational and power amplifiers; clocks (upto 4 MHz); power devices (10W to 100W); passive components; transformers and inductors.

TABLE 2. Application requirements for the aerospace market per temperature range

Temperature range (^{o}C)	HTE Requirement
150 to 170	Obviate cooling systems for fan case mounted unit. Or retain cooling with higher safety margins and increased reliability
200 to 250	Fan case distributed architecture, smart fuel systems, smart sensors and smart skins. Electronic braking systems
300 to 400	Distributed architectures. Smart systems
400 to 500	Core distributed systems or systems located in bypass duct areas Flame combustion monitoring
500 to 1000	Hot end distributed controlled, smart sensor/actuator systems

3.2 AUTOMOTIVE APPLICATIONS

The automotive sector represents the largest potential single market opportunity for HTE. HTE components could significantly improve system design, enhance fuel efficiency leading to overall higher performance vehicles. However, the automotive sector represents the most stringent pricing pressure of any market likely to be covered by HTE. In addition demands on the reliability of the HTE component should not be ignored – most car manufacturers demand components that last the lifetime of the vehicle which is not insignificant.

Nevertheless, since the total available automotive electronics market is so large (ca. > $14 billion/year) even a small fraction of this represents a large market opportunity (currently ca. $120m rising to ca. $550m by 2008) to HTE providers as shown in figure 5. However, much development needs to be undertaken not only in basic component architectures but also on mass production technologies for HTE before any significant penetration can be achieved.

Advanced automobiles are already heavily reliant on electronics and sensors (see figure 6). These include sensors for engine speeds, angular position, ABS, exhaust gas, power steering, engine condition monitoring and electric windows. HTE to be co-located with the sensors is currently being developed. Furthermore, high temperature mechatronic systems will become more common, offering improved safety and comfort and reduced costs.

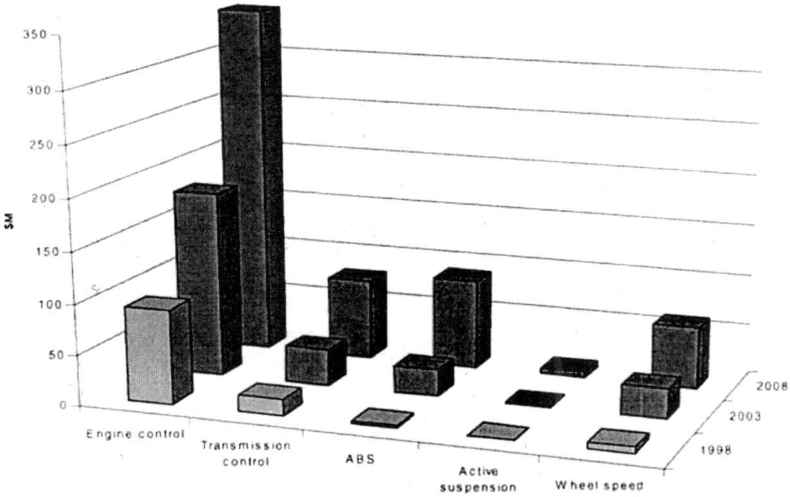

Figure 5. The automotive market for HTE

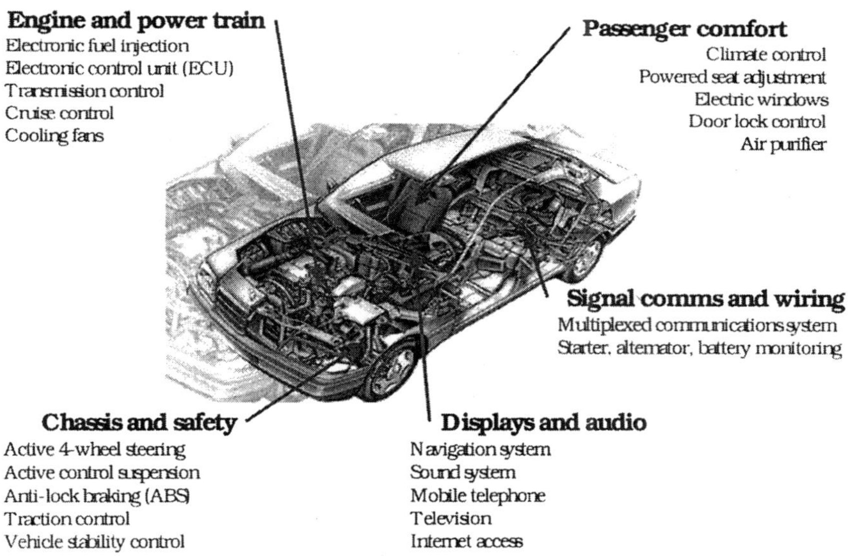

Figure 6. Typical applications for HTE in the automotive environment

The major market driver for the development of HTE for the automotive market sector is again legislation to reduce emission and improve engine efficiency and hence economy. Another driver is actually to reduce manufacturing costs for the automotive manufacturers. As a consequence of increased utilisation of HTE in automobiles there will be consumer benefits with improved reliability and increased efficiency which should work to reduce the average cost of ownership in terms of both reduced running costs and reduced maintenance requirements.

4. Conclusions

The three applications discussed generated a world-wide HTE component market of nearly $200 million in 1998. Within 10 years this could easily exceed $1,000 million. However, it will be necessary to pool limited resources and focus on key strategic objectives before the true potential of High Temperature Electronics can be realised. HITEN (The High Temperature Electronics Network of Excellence) can act as a major resource bringing together users and providers of HTE components, information and R&D, as depicted in figure 7.

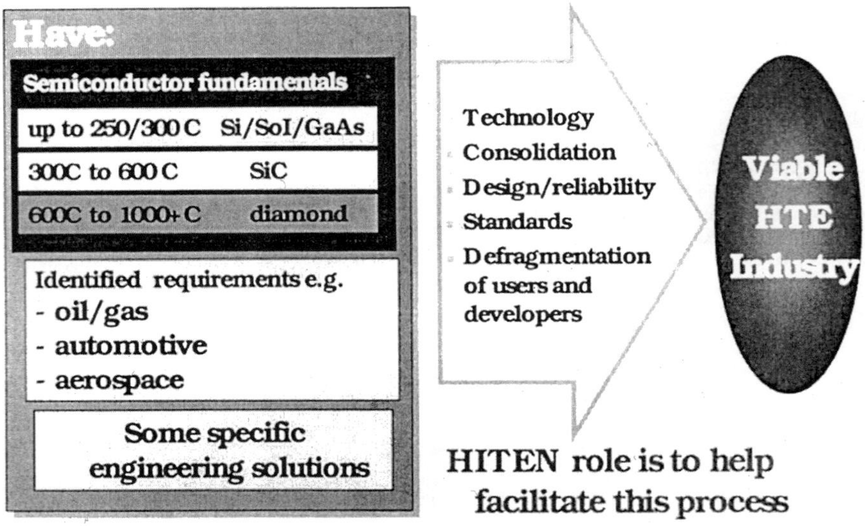

Figure 7. The role of HITEN in establishing a viable High Temperature Electronics industry

The global outlook for HTE is good. HTE is critical for the exploitation of increasingly remote oil and gas field and to facilitate improved recovery from otherwise marginal fields. HTE has become a strategic technology of modern aircraft and vehicles, where issues of safety and emissions legislation are driving increased adoption of HTE technology.

However, there are still considerable problems to be overcome before the full potential of the HTE market can be realised. In particular the question of materials integration has been largely ignored. There is a need to solve application specific problems now but more generally generic systems for high temperature joining must be sought, especially with respect to packaging for HTE.

5. Acknowledgements

This paper is supported by:

HITEN, The High Temperature Electronics Network of Excellence. www.hiten.com
The European Commission
AEA Technology plc.

6. References

1. www.hiten.com
2. HITEN (1999) *The World Market for High Temperature Electronics,* AEA Technology, Oxford.
3. J. J. den Boer (1999) *Proceedings of the HITEN Conference, Berlin 1999*, AEA Technology, Oxford.

CHARGE CARRIER INJECTION AND TRAPPING IN THE BURIED OXIDES OF SOI STRUCTURES

A.N. NAZAROV, V.I. KILCHYTSKA, I.P. BARCHUK
Institute of Semiconductor Physics, National Academy of Sciences of Ukraine
03028 Kiev 28, Prospect Nauki, 45, Kiev, Ukraine

1. Introduction

The electron injection processes in the silicon-on-insulator (SOI) devices affect strongly the reliability of device operation [1]. Usually the buried oxide (BOX)/silicon film interface shows worse structural and electrical properties than that of the gate oxide/silicon film interface [2]. This leads to enhanced charge trapping and degradation of the BOX during SOI device operation. Therefore, the promising perspectives of SOI devices for some applications (especially for high-voltage and high-temperature devices) are often limited by carrier injection processes in the BOX.

Thus, the paper is devoted to review and analysis of the charge injection processes, positive charge creation and electron trapping in the BOX during the electron injection for the main commercial SOI structures.

2. High-field conductivity in the BOX of SOI structures

Two kinds of electron high-field conductivity have been observed in the buried oxide of SOI structures. In early paper relating to single implanted (SI) and triple implanted (TI) SIMOX SOI structures [3] the Pool-Frenkel mechanism of the electron conductivity in the dielectric is supposed to be occured. This mechanism is bulk-limited one and assumes the hopping of electrons between closely spaced donor-like electron traps via the oxide conduction bond (Fig. 1a). Simple analysis, assuming a single dominant trap with energetic depth E_t below the conductivity band edge, gives for the current density, J_{PF}, following expression:

$$J_{PF} = \sigma_0 \cdot E_{ox} \cdot \exp\left[\frac{E_t - \Delta E_t(E_{ox})}{k \cdot T}\right], \qquad (1)$$

where E_{ox} is the electric field, $\Delta E_t = \sqrt{\dfrac{q \cdot E_{ox}}{\pi \cdot \varepsilon_0 \cdot \varepsilon_{ox}}}$ is the energetic depth decrease in consequence of high electric field effect, σ_0 is the electrical conductivity of the

F. Balestra et al. (eds.), Progress in SOI Structures and Devices Operating at Extreme Conditions, 139–158.

dielectric at low electric field. That is, the current is temperature activated and exhibits Ohmic law at low electric fields.

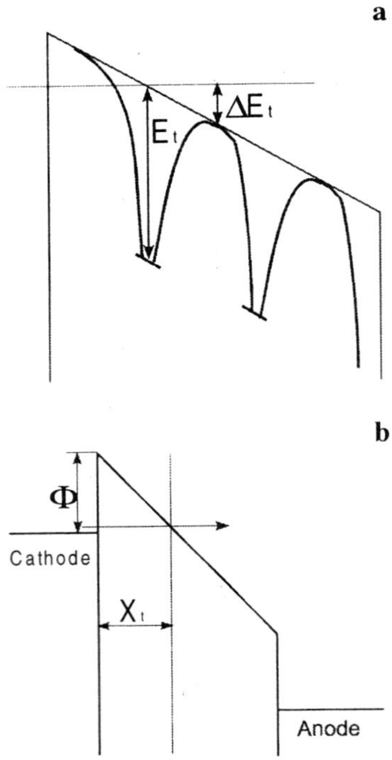

Figure 1. Schematic illustration of Pool-Frenkel (a) and Fowler-Nordheim (b) high-field electron conduction mechanisms

However it was shown by the next researches that the high-field conductivity of the BOX in SI SIMOX [4, 5] and wafer bonding [6] SOI structures depends slightly on temperature indicating that the conductivity is controlled by tunneling processes.

It is well known that high quality dioxide obeys the Fowler-Nordheim theory, which considered the electron tunneling from cathode into oxide conduction band (Fig. 1b), that is the process is interface limited. In this case a current density, J_{FN}, can be presented by classical Fowler-Nordheim expression [7]:

$$J_{FN} = C \cdot E_c^2 \cdot \exp\left(- \frac{\beta \cdot \Phi^{3/2}}{E_c} \right), \qquad (2)$$

where E_c is the electric field at cathode (in the case of small accumulated charge in the BOX $E_c \approx E_{ox}$), Φ is the barrier height, which equals to about 3.1 eV for Si-SiO$_2$ interface [8], $C = \frac{q^3 \cdot m_0}{8 \cdot \pi \cdot h \cdot m^* \cdot \Phi}$,

$\beta = \frac{8 \cdot \pi \cdot \sqrt{2 \cdot m^*}}{3 \cdot q \cdot h}$, m* is the electron effective mass in SiO$_2$ band gap ($m^* = 0.5 m_0$), other values have their conventional meanings.

In the case of SIMOX SOI structures the barrier heights for electron tunneling from silicon substrate and silicon film in the BOX is considerable lower than in the case of gate oxides and equals from 0.9 to 2.4 eV for different SIMOX techniques [9-11]. In Fig. 2a the high-field current-voltage characteristics for typical SI SIMOX and UNIBOND [12] SOI structures are presented. These characteristics can be described by Fowler-Nordheim law (Fig. 2b) with potential barriers in the case of SI SIMOX SOI structures equal to 1.20 eV and 1.35 eV for the BOX/substrate and the BOX/silicon film interface, correspondingly, and in the case of UNIBOND SOI structures, these ones equal to 2.3 eV for both interfaces.

The data obtained for the SIMOX SOI structures suggest that quality of the BOX/Si film interface is superior to that of the BOX/substrate interface. Indeed, a considerable number of researches relating to structural and defect properties of single implanted SIMOX BOX show an existence of silicon inclusions [13, 14], crystalline coesite phase of SiO$_2$ [15], dangling [16-18] and strength bonds [9,19] in the BOX in the vicinity of the BOX/substrate interface.

In the case of UNIBOND SOI structures the magnitudes of the potential barriers for electron injection from Si film and from Si substrate into the BOX are the same, but

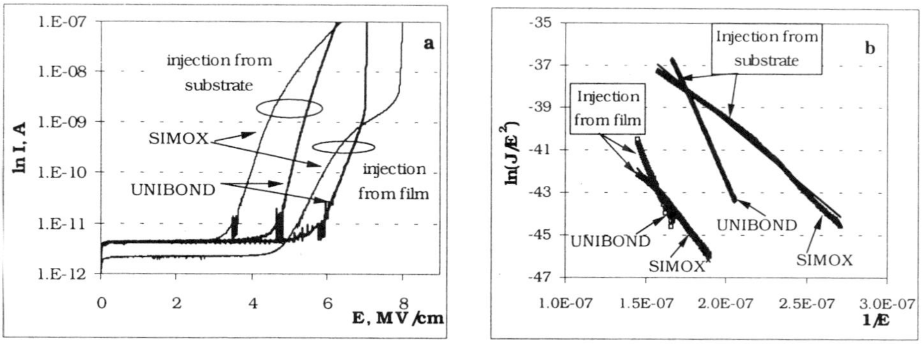

Figure 2. I-V characteristics of SIMOX and UNIBOND SOI structures in ordinary (a) and Fowler-Nordheim (b) coordinates.

the threshold voltage of FN injection for the Si film is sufficiently higher than for the substrate. This result suggests the better quality of the BOX/Si film interface than the BOX/substrate interface. Since in the studied UNIBOND SOI structures the most imperfect region is located in the BOX/substrate interface (that is the bonding place) [20,21] it is reasonable to assume that the lower threshold voltage of FN electron injection from the substrate is a result of worse quality of bonding interface in comparison with quality of the BOX/silicon film interface.

It should be noted that in both SIMOX and UNIBOND SOI structures the obtained values of potential barrier for electron injection are always smaller than that for a gate thermal oxide – silicon structures (that equals to 3.1 eV [8]). In order to explain the decrease of electron tunneling barrier for high-field injection in the papers [10,11,22] the model taking into account an appearance of interfacial asperities has been developed. Introduction of the correction coefficients into Fowler-Nordheim expression (2) allowed to achieve an agreement between theory and experimental data. Thus, the expression (2) in this model has a following view:

$$\frac{J_{FN}}{k_a} = C \cdot (k_F \cdot E_c)^2 \cdot \exp\left(-\frac{\beta \cdot \Phi^{3/2}}{k_F \cdot E_c}\right), \tag{3}$$

where k_F is the coefficient taking into account the enhancement of the cathode field, and k_a is the coefficient relating to effective area of asperities. Assuming a "circular" asperities, the asperity separation, S, can be estimated as [22]:

$$S = \frac{d}{2} \cdot \sqrt{\frac{\pi}{k_a}}, \tag{4}$$

where d is the diameter of asperities. Below this approach will be named as Hall-Wainwright (HW) one, which was carefully analyzed in the paper [22].

Estimations of asperity separation using our experimental data and HW model are presented in Table 1. It should be noted that in the case of SIMOX SOI structures, where the effective potential barrier for electron injection is low, the calculation shows unreasonably large values of asperity separation about 2 mm (S_1) for 50 nm diameter of asperities and 40 μm (S_2) for 1 nm diameter.

TABLE 1. Summary of high-field FN analysis of I-V data employing HW and RBH models

Sample	Φ_{eff} (eV)	J_{FN} (A/cm^2)	E_{ox} (MV/cm)	K_F	k_a	S_1 (cm)	S_2 (cm)	S* (cm)
SIMOX (film)	1.3	1.1x10^{-7}	5.8	3.67	6x10^{-11}	0.6	1x10^{-2}	1.1x10^{-7}
SIMOX (substrate)	1.1	1.1x10^{-7}	4.0	4.68	4x10^{-10}	0.2	4x10^{-3}	1.2x10^{-7}
UNIBOND (film)	2.3	1.1x10^{-7}	6.5	1.56	2x10^{-4}	3x10^{-4}	6x10^{-6}	2.4x10^{-7}
UNIBOND (substrate)	2.3	1.1x10^{-7}	5.5	1.56	4x10^{-2}	2x10^{-5}	4x10^{-7}	2.8x10^{-7}

S_1 and S_2 have been calculated using HW model and expression (4), where S_1 corresponds to d = 50 nm and S_2 corresponds to d = 1 nm.
S* have been calculated using RBH model and expression (9).

As it was suggested in paper [22] the interfacial asperities are associated with a flat-topped pyramidal defects observed in SIMOX BOX by XTEM study [23], which has a base diameter of about 50nm. The estimations of asperity separation for SIMOX BOX allow to conclude that reduced potential barrier and increased current flow in this case are not caused by only interfacial asperities. Reduced potential barrier for electron tunneling can be also associated with the tunneling processes between silicon clusters in the BOX near BOX/semiconductor interfaces [4]. As it was shown in paper [4] the current through dielectric in this case can be presented in following view:

$$J_T \sim E_{ox}^2 \cdot \exp\left(-\frac{\beta \cdot \Phi_{eff}^{*3/2}}{E_{ox}}\right), \tag{5}$$

where

$$\Phi_{eff}^{*3/2} = \frac{\Phi^{3/2} - (\Phi - q \cdot \Delta V)^{3/2}}{\eta} \tag{6}$$

and ΔV is the drop of voltage between clusters when tunneling takes place. The parameter η is given by following expression:

$$\eta \cdot E_{ox} = \frac{\Delta V}{S}. \tag{7}$$

Assuming that there is not voltage drop across the Si clusters:

$$\eta = 1 + \frac{d}{S}, \tag{8}$$

where d is the cluster diameter. In the paper of Revesz, Brown and Hughes (RBH) [4] it was suggested that d/S<<1 and the cluster separation can be estimated from expressions (6)-(8) as

$$S = \frac{\Phi - \left(\Phi^{3/2} - \Phi_{eff}^{3/2}\right)^{2/3}}{q \cdot E_{ox}}, \qquad (9)$$

where $\Phi = 3.1$ eV.

The results of the estimations for our data are presented in Table 1. These estimations confirm that potential barrier reducing can be caused by high density of silicon inclusions in the BOX, however for the case of studied SIMOX structures $d/S \ll 1$ approximation no valid, and for the case of UNIBOND SOI structures this estimation can not explain enhanced current during electron injection from substrate to the BOX.

Thus, we can conclude that in the case of high-quality SOI structures, such as UNIBOND ones, the reducing potential barrier for electron injection can be mainly associated with interfacial asperities; in the case of imperfect SOI structures, such as SIMOX ones, reduced potential barrier is possibly related to both interfacial aspeirties and silicon clusters in the BOX.

3. Positive charge generation and electron trapping during FN electron injection

3.1. METHODS FOR DETERMINATION OF MAIN TRAPPING PARAMETERS IN THE DIELECTRIC

It is well established for both thick and thin thermal oxides [25-27] and buried oxides in SIMOX [10,11,22] and wafer bonding (WB) [6,20,21] SOI structures that during Fowler-Nordheim electron injection the building up of positive and negative charge takes place.

Building up of the charge in a dielectric can be characterized by following main parameters: total building-up charge, $Q_{tot} = \int_0^d \rho(x)dx$, where $\rho(x)$ is the charge distribution in the dielectric (in the case of electron captured in the traps $\rho(x) \equiv n_T(x)$), d is the dielectric thickness; effective localization of the charge or charge centroid, $\overline{X_0} = \frac{\int_0^d x \cdot \rho(x)dx}{\int_0^d \rho(x)dx}$; efficiency of charge accumulation in the dielectric, $\eta(t) = \frac{Q_{tot}(t)}{Q_{inj}(t)}$, where $Q_{inj}(t)$ is a number of injected electrons per unit area, which can be calculated from measured injection current, I_{inj}, as $Q_{inj}(t) = \frac{1}{S} \cdot \int_0^t I_{inj}(t')dt$.

Important parameter for charge trap is capture cross-section, σ, that determines the charge build up dynamics. This value can be estimated from the solution of the rate equation [28]. For definiteness sake, let consider a simple electron traps with concentration of $N_{Ti}(x)$ and density of electrons trapped in the trap of $n_{Ti}(x)$. In addition, we assume that current density j_{inj} ($j_{inj}=I_{inj}/S$) is spatially independent, that is $j_{inj}(x)=j_{inj}$. In this case we can write:

$$\frac{dn_{Ti}(x,t)}{dt} = \frac{j_{inj}(t)}{q} \cdot \sigma_i \cdot (N_{Ti}(x) - n_{Ti}(x,t)) \tag{10}$$

The solution of the equation (10) with initial condition of $n_{Ti}(x, 0)=0$ is:

$$n_{Ti}(x,t) = N_{Ti}(x) \cdot \left[1 - \exp\left(-\sigma_i \cdot \int_0^t \frac{J_{inj}(t')}{q} dt' \right) \right], \tag{11}$$

$$n_{Ti}(x,t) = N_{Ti}(x) \cdot (1 - \exp(-\sigma_i \cdot Q_{inj}(t))). \tag{12}$$

Integrating equation (12) over the thickness of the SiO_2 layer we obtain :

$$Q_{tot}(t) = Q_{totmax} \cdot (1 - \exp(-\sigma_i \cdot Q_{inj}(t))) \tag{13}$$

and

$$\eta(t) = Q_{totmax} \cdot \sigma_i \cdot \exp(-\sigma_i \cdot Q_{inj}(t)). \tag{14}$$

Thus, the value of $Q_{tot\ max}$ can be obtained from saturation of the $Q_{tot} - Q_{inj}$ dependence and the value of $\sigma_i \cdot Q_{tot\ max}$ – from $\eta(0)$ and consequently we can calculate the value of σ_i. If the saturation of the $Q_{tot} - Q_{inj}$ dependence is not achieved, σ_i can be obtained from the slope of the $\ln\eta$ - Q_{inj} dependence [28].

In order to determine above mentioned parameters combination of the current and capacitance methods is employed during FN electron injection, which can be performed at constant FN voltage or at constant FN current regimes.

Constant FN current regime

This regime is convenient to control the injected charge and in order to sufficiently simply determine the total accumulated charge and its centroid from changing of the voltage needed to maintain the constant FN current [29]. Indeed, the charge building up in a dielectric leads to FN current change as consequence of cathode electric field change. In this case the expression (2) can be rewritten as following:

$$J_{FN}(t) = C \cdot (E_0 \pm \Delta E_a(t))^2 \cdot \exp\left(-\frac{\beta \cdot \Phi^{3/2}}{E_0 \pm \Delta E_a(t)} \right), \tag{15}$$

where E_0 is the average electric field determined by applied voltage ($E_0=V_g/d$), ΔE_a is the change of cathode field caused by trapping of electrons or holes in the dielectric (ΔE_a is taken as positive for electron trapping and as negative for hole trapping).

Assuming that FN electron injection occurs from silicon film into BOX and centroid of charge is calculated from the silicon film/BOX interface, from Gauss law it can be obtained:

$$\Delta E_a(t) = \frac{Q_{tot}}{\varepsilon_0 \cdot \varepsilon_d} \cdot \left(1 - \frac{\overline{X}}{d} \right). \tag{16}$$

For J_{FN}=const, $E_0 \pm \Delta E_a$ has also be constant during injection experiment. It can be easily shown, that

$$\Delta V_{FN}^{(-)} = \frac{Q_{tot} \cdot d}{\varepsilon_0 \cdot \varepsilon_d} \cdot \left(1 - \frac{\overline{X}}{d}\right) \tag{17a}$$

and

$$\Delta V_{FN}^{(+)} = -\frac{Q_{tot}}{\varepsilon_0 \cdot \varepsilon_d} \cdot \overline{X} \ , \tag{17b}$$

where index (-) and (+) corresponds to negative and positive voltage applied to silicon film of SOI structure, correspondingly.
Thus, from expressions (17a) and (17b) it can be obtained [30]:

$$Q_{tot} = \frac{\varepsilon_0 \cdot \varepsilon_d}{d} \cdot \left(\Delta V_{FN}^{(-)} - \Delta V_{FN}^{(+)}\right), \tag{18a}$$

$$\overline{X} = \frac{\Delta V_{FN}^{(+)} \cdot d}{\Delta V_{FN}^{(-)} - \Delta V_{FN}^{(+)}} . \tag{18b}$$

Main limitation of this method consists in existence of "dead" distance (equals to tunneling distance X_t) for parameters extraction from FN tunneling currents, that is FN current does not "feel" the charge located near the cathode on the distance lower than X_t (Fig. 1b) [30]. The second limitation includes the charge parameter determination from electron injection at different polarity of applied voltage. Indeed, the different direction of electron injection can result in different distribution of trapped charge in dielectric, that is expression (17a) and (17b) can be related to different Q_{tot}.

In order to avoid the second limitation the combination of capacitance-voltage (C-V) method with measuring of voltage needed to maintain the constant FN current during one-polarity injection can be employed [27]. For example, for metal-oxide-semiconductor (MOS) structures in the case of FN electron injection from metal electrode to oxide the change of FN voltage caused by charge trapping in oxide will characterize the change of potential at the metal/oxide interface and change of the midgap voltage (ΔV_{mg}) obtained of C-V characteristic corresponds to change of potential at the oxide/semiconductor interface. Thus, we can obtain:

$$Q_{tot} = \frac{\varepsilon_0 \cdot \varepsilon_d}{d} \cdot \left(\Delta V_{FN}^{(-)} - \Delta V_{mg}\right), \tag{19a}$$

$$\overline{X} = \frac{\Delta V_{mg} \cdot d}{\Delta V_{FN}^{(-)} - \Delta V_{mg}} \tag{19b}$$

<u>Constant FN voltage regime</u>

For MOS structures the change of potential at the metal/oxide interface during charge trapping in the oxide at FN electron injection from metal electrode can be obtained from the change of the FN current, and charge related to the metal/oxide interface can be estimated as [31]:

$$Q_{tot}(t)\cdot\left(1-\frac{\overline{X}(t)}{d}\right)\approx\varepsilon_0\cdot\varepsilon_d\cdot E_0^2\cdot\left(q\cdot\beta\cdot\Phi^{3/2}\right)^{-1}\cdot\ln\left[\frac{J(t)}{J(0)}\right].\qquad(20)$$

Using the SOI structures for studying the charge buildup in the oxide has an methodological preference over MOS structures because the C-V characteristic of SOI structure allows to calculate the potentials of both the BOX/film and the BOX/substrate interfaces from one measurement [32]. Since, the C-V characteristic shift is determined by total net charge (located both in interface and in bulk of oxide), the C-V method is more informative than J-t one. Thus, measurements of C-V characteristics of SOI structure during the FN electron injection into BOX allow to determine both total charge in the BOX and its centroid independently on direction of the electron injection by [33]:

$$Q_{tot}=\frac{\varepsilon_0\cdot\varepsilon_d}{d}\cdot\left(\left|\Delta V_{mg}^{film}\right|+\left|\Delta V_{mg}^{sub}\right|\right)\qquad(21a)$$

and

$$\overline{X}=\frac{\left|\Delta V_{mg}^{film}\right|\cdot d}{\left|\Delta V_{mg}^{film}\right|+\left|\Delta V_{mg}^{sub}\right|},\qquad(21b)$$

where $\Delta V_{mg}{}^{film}$ and $\Delta V_{mg}{}^{sub}$ are the midgap voltage shifts during the injection in the BOX/film and the BOX/substrate interface, correspondingly.

In order to determine the charge trapping efficiency, η, and capture cross section, σ, of the traps in the dielectric, the injection current through the dielectric as a function of injection time has to be measured during the experiment.

3.2. MECHANISMS OF POSITIVE CHARGE GENERATION IN SiO_2 DURING FN ELECTRON INJECTION

Abundant papers have been devoted to the study of positive charge creation in SiO_2 of the metal-oxide-silicon (MOS) structures during FN electron injection [26,27,29-31,34-40] and only a small number of papers considered this processes in the BOX of SOI structures [3,5,11,22,41], sometimes only mentioning on that [3,5,11]. Thus, at first, we will discuss the models of positive charge creation proposed for gate oxides of MOS structures, in order to apply then them to the BOX of SOI structures.

Nowadays, four main mechanisms of positive charge creation in SiO_2 by FN electron injection are considered. These mechanisms can be divided by dependences on applied electric field, dielectric thickness and positive charge build-up at different electric field polarity.

At low electric field ($E \leq 1.5$ MV/cm) and considerable injected electron charge ($Q \geq 10^{-3}$ C/cm^2) the ***trap creation mechanism***, depicted in Figure 3a, can be observed [30,31,34]. It was shown [34], that in thick oxides at such low electric fields a hydrogen is able to be released from anode region, located near silicon substrate, and transported to cathode. As this process takes place, positively charged sites, possibly related to the hydrogen release, were found near the anode. In addition, on the cathode side, interface states and distributed oxide electron traps have been identified. Also, the deactivation of

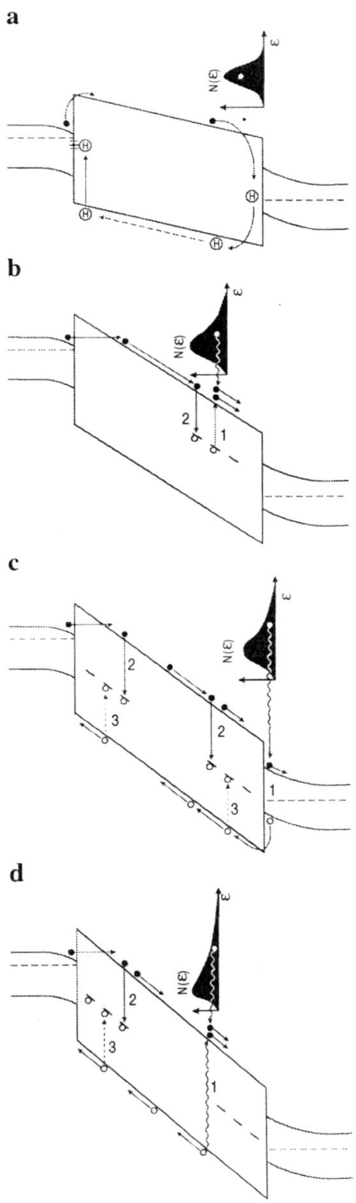

Figure 3. Schematic illustration of the positive charge generation mechanisms:
(a) trap creation mechanism;
(b) trap-to-band impact ionization;
(c) anode hole injection ;
(d) band-to-band impact ionization

boron dopants in the Si substrate have been observed [42,43], that can be initiated by mobile atomic hydrogen [44]. Thus, in the oxide near anode the positive charge associated with donor traps is created. These traps behave as "slow" states [31]. Such process of positive charge creation is thermally activated at temperature more than 200K with activation energies from 0.01 to 0.20 eV [30]. In order to create these donor traps, an average electron energy has to be more than 2.3±0.2 eV [30].

Increase of electric field gives rise of impact ionization processes. If electron energy is more than 5 eV, the ***trap-to-band impact ionization mechanism***, depicted in Figure 3b, can take place [27,35,36]. According to this mechanism, the positive oxide charge is generated by impact of hot free electrons from conduction band with the filled traps (process "1" in Fig. 3b). Together with generation of positively charged traps the electron trapping process appears (process "2" in Fig. 3b). This process can be described by the following equation [36]:

$$q \cdot \frac{dp(x,t)}{dt} = J_{inj} \cdot \beta \cdot \left(N_p(x) - p(x,t)\right) - \\ - J_{inj} \cdot \sigma_e \cdot p(x,t)\,, \qquad (22)$$

where a first component of right part describes the positive charge generation process and a second component describes the electron trapping. Here p is the positive oxide charge density; N_p is the trap density; β is the field dependent probability of a passing electron to perform an ionizing collision (or probability of positive charge generation); and σ_e is the recombination cross section of the trap.

The probability of positive charge generation, β, can be related to electric field in oxide, E_{ox}, by following expression [27, 35]:

$$\beta = \beta_0 \cdot \exp\left(-\frac{H}{E_{ox}}\right), \qquad (23)$$

where $H = \frac{E_c}{\lambda}$, E_c is the critical energy of the hot electron, which is required for ionization, λ is the mean free path of electrons.

From equation (22) it can be easily obtained:

$$\int p(t)dx = N_{\infty} \cdot \{1 - \exp[-\sigma^{+} \cdot Q_{inj}(t)]\}, \tag{24}$$

$$N_{\infty} = \frac{\int N_p dx}{1 + \frac{\sigma_e(E)}{\beta(E)}} \tag{25a}$$

$$\sigma^{+} = \beta + \sigma_e \tag{25b}$$

where N_{∞} is the maximal generated positive charge concentration, σ^{+} is the effective cross section of positive charge formation. Since, usually σ_e depends only slightly on the electric field, the electric field increase in oxide has to lead to increase of maximal generated positive charge concentration due to strong electric field dependence of β (see eq. (23)).

As a rule, for typical MOS structures the main defects are located near the SiO_2/Si interface. Therefore, in the case of trap-to-band impact ionization mechanism the electron injection from metal electrode into dioxide has to generate considerable higher positive charge concentration than that generated by electron injection form silicon substrate [36]. Thus, we can write, that

$$\frac{P_{neg}}{P_{pos}} >> 1, \tag{26}$$

where P_{neg} and P_{pos} are the positive charge concentration generated in the oxide during electron injection at negative voltage and positive voltage applied to the metal electrode, correspondingly.

In papers [29,37,38] it has been shown that at electron energy in oxide near 5 eV the positive charge generation can be related to processes of ***anode hole injection***, depicted in Figure 3c. Most brightly this process is observed during FN electron injection in thin oxides (thinner than 100 Å) [38], when the electron transport in oxides is quasiballistic. In this case when hot electrons reach the SiO_2/anode interface they can generate hole-electron pairs in anode (process 1 in Fig. 3c) through surface plasmon creation [29, 37]. Hot holes, possesed energy more than 4.5 eV (for silicon anode) are injected into oxide and trapped in defects (process 3 in Fig. 3c). Strained and dangling bonds, E'-centers (oxygen vacancy) and other structural defects of SiO_2 amorphous network can work as hole traps [45]. Simultaneously, some part of electrons can neutralize the positively charged traps (process 2 in Fig 3c). Thus, the rate equation for hole trapping can be presented in following view:

$$q \cdot \frac{dp(x,t)}{dt} = J_{inj} \cdot \sigma_p \cdot \alpha \cdot (N_p(x) - p(x,t)) - J_{inj} \cdot \sigma_e \cdot p(x,t), \tag{27}$$

where α is the hole generation probability, σ_p is the hole capture cross-section and other values are the same as in (22). The solution of equation (27) will have the same view as expression (24), where

$$N_{\infty}^{+} = \frac{\int N_p dx}{1 + \dfrac{\sigma_e}{\sigma_p \cdot \alpha}} \tag{28a}$$

and

$$\sigma^{+} = \alpha \cdot \sigma_p + \sigma_e \tag{28b}$$

In the case of hole injection from anode the holes are swept across the whole oxide film for both polarity of FN electron injection. Therefore, independently on direction of electron injection into oxide and hole trap localization in the oxide, no asymmetry is expected for the trapping of holes produced by anode hole injection. Thus, the following expression has to be valid

$$\frac{P_{neg}}{P_{pos}} \approx 1. \tag{29}$$

Some asymmetry may arise from variation in the hole injection efficiencies for various contacts [31].

The existence of anode hole injection in the thick oxides has been reported in some papers [29, 38]. In the case of anode hole injection the calculations performed in paper [38] predict a weak dependence of hole generation probability from electric field. Thus, observed increase of maximal generated positive charge concentration, N_{∞}^{+}, with increasing of electric field can be associated with a decrease of recombination cross-section for electrons, σ_e.

In the papers [38-40] it was shown, that at high electric field in oxide when the average energy of hot electrons reaches 7 eV in energetic distribution of the electrons there is an abundant number of electrons which have an energy up to 9 eV. Thus, at high electric field (near and more than 7 MV/cm) the process of ***band-to-band impact ionization*** in oxide can take place (depicted by Figure 3d). This process can be observed at high FN electric fields only in thick films because the high-energy tails in hot-electron distribution require time to form. In case of this process the holes are generated in oxide bulk near anode (process 1 in Fig. 3d). Some holes are trapped near anode, but most of holes pass to the cathode-oxide interface where some of them are trapped in deep levels (process 3 in Fig. 3d). The part of injected electrons recombine with positively charged traps (process 2 in Fig. 3d). Thus, the process can be described by rate equation (27), that has a solution, presented by eqs. (24) and (28) as in case of anode hole injection.

For band-to-band impact ionization hole generation probability have been shown to be strongly electric field dependent [39]

$$\alpha = P_0 \cdot \left[\frac{E}{E_{th}} - 1 \right]^4, \tag{30}$$

where E_{th} is the threshold electric field which for thick oxide equals to 6.4 MV/cm. This dependence is considerably different from that for anode hole injection.

Since in thick oxide during band-to-band impact ionization holes are created in bulk of oxide but not in anode/oxide interface in the case of asymmetric hole trap distribution the asymmetric hole trapping has to be observed during electron injection with different FN electric field polarity. For example, in the case of MOS structures, in which the non-uniform hole traps distribution with maximum near the oxide/silicon interface is usually observed, for the band-to-band impact ionization the following expression is valid:

$$\frac{P_{neg}}{P_{pos}} << 1. \quad (31)$$

Usually, in thick oxides the processes of band-to-band impact ionization occur simultaneously with anode hole injection [38,39] and their dividing requires additional experiments.

Thus, generation of positive charge in thick oxide during FN injection in the cases of the impact ionizing and anode hole injection mechanisms can be described by expression (24), where the maximal positive charge concentration, N_∞^+, and effective cross-section of positive charge generation, σ^+, have a different meaning in dependence on the mechanism. Existence of asymmetry in the trap location in the oxide allows to distinguish qualitatively the predominant mechanisms of the positive charge generation.

3.3. GENERATION OF THE POSITIVE CHARGE IN THE BOX OF SOI STRUCTURES

3.3.1. Single implanted SIMOX SOI structures

For single implanted SIMOX SOI structure during first seconds of FN electron injection a large concentration of positive charge has been shown to be generated in the BOX [3,10,11]. In the paper [41] it was demonstrated that this positive charge is localized near the BOX/substrate interface independently on the polarity of applied voltage (see Table 2). Such asymmetric positive charge trapping attests the asymmetric defect localization in the SIMOX BOX with maximum at the BOX/substrate interface. Actually, in a lot of papers [9,13-15,18,19] it was shown an existence of great number of silicon inclusions, strained and dangling bonds in the BOX near the BOX/substrate interface, that leads, for instance, to increased trapping of atomic hydrogen in this region after hydrogen plasma treatment [46,47].

During the following time of electron injection the right part of the C-V characteristic related to the BOX/Si film interface, begins to shift to a less positive voltages (Fig. 4a), that can be explained by both negative charge trapping near the BOX/Si film interface and positive charge centroid shifting to the BOX/substrate interface. Decrease of injection current during electron injection from the Si film into the BOX (Fig. 5) allows to conclude that the first effect is main. Thus, in the beginning of the electron injection in SI SIMOX BOX a positive charge is created very fast near the BOX/substrate interface (Fig. 4b). The following negative charge trapping in the BOX near the BOX/Si film interface on the distance more than tunneling one stabilizes the value of total positive charge (Fig. 4c).

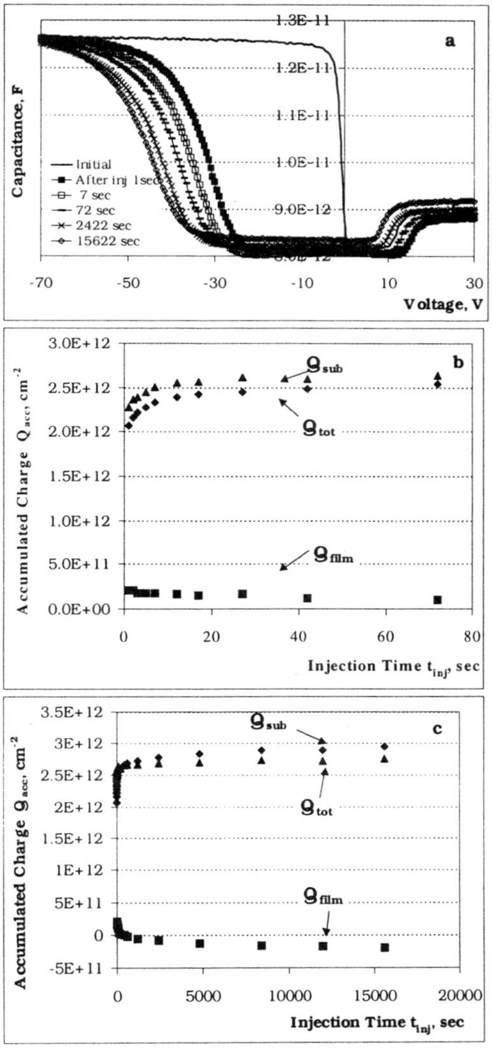

Figure 4. (a) C-V characteristics of SIMOX SOI capacitors before and after electron injection ;
(b, c) Dynamic characteristics of charge buildup into the BOX of SIMOX SOI structures.

Increase of the electric field applied to the structure during FN injection leads to increase of the maximal total positive charge, N_∞^+, in the BOX (Fig. 6), that attests the positive charge generation (see section 3.2).

It should be pointed out that the considerably larger positive charge is generated during electron injection in the BOX from the Si film than from the substrate (see Table 2). Such behavior of the positive charge generation ($P_{neg}/P_{pos}>>1$) can prove the predominance of trap-to-band impact ionization mechanism. However, the decrease of the positive charge generation efficiency with increase of applied electric field (Table 2) can not be explained by employment of only this mechanism.

Using of expression (24) for total positive charge generation allows to compare the results obtained for the BOX of SIMOX SOI structures with ones obtained for the gate oxides [29,38]. The obtained results are presented in Table 3. It can be seen, that for the same injection fields the maximal generated total positive charge in the SIMOX BOX is smaller than that for gate oxides fabricated in the middle of eighty's years [29], but considerable higher than that for the oxides produced in the middle of ninety's years [38]. Effective positive charge generation cross-section, $\sigma_f^+ \sim 1 \cdot 10^{-15}$ cm^2, in the SIMOX BOX is considerable higher than that for gate oxide ($\sim 5 \cdot 10^{-17}$ cm^2) obtained for the similar applied electric field, that can testify a different nature of the generated positive charge.

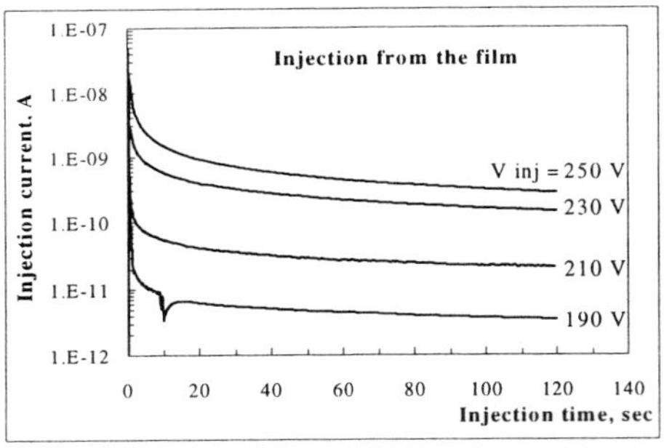

Figure 5. Current relaxation during electron injection in the BOX from the Si film performed with different applied FN fileld

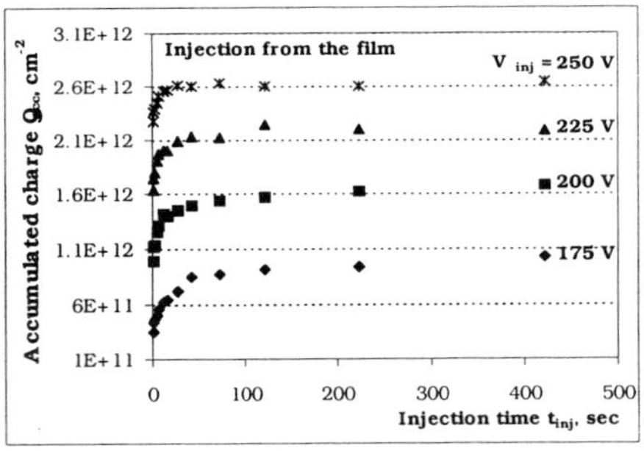

Figure 6. Dynamics of positive charge buildup into SIMOX BOX during electron injection performed with different applied FN field

TABLE 2. The generated positive charge parameters for the electron injection from the film and from the substrate

$V_{sub.}$ (V)	Q_{acc} (cm^{-2})	Q_{inj} (cm^{-2})	Q_{acc}/Q_{inj}	Xo (Å)*	Xo/d*
Injection from the film					
0				2829	0.79
190	$1.4x10^{12}$	$4x10^{13}$	$3.6x10^{-2}$	72	0.02
210	$2.0x10^{12}$	$5.9x10^{13}$	$3.5x10^{-2}$	100	0.027
230	$2.5x10^{12}$	$1.5x10^{14}$	$1.7x10^{-2}$	121	0.033
250	$2.8x10^{12}$	$3.5x10^{14}$	$8.2x10^{-3}$	72.7	0.020
Injection from the substrate					
0				2769	0.77
-130	$2.9x10^{10}$	$1.0x10^{13}$	$2.9x10^{-3}$	1800	0.5
-150	$4.1x10^{11}$	$1.2x10^{14}$	$3.4x10^{-3}$	450	0.125
-170	$7x10^{11}$	$4.8x10^{15}$	$1.5x10^{-4}$	400	0.11
-190	$9.9x10^{11}$	$8.2x10^{15}$	$1.2x10^{-4}$	540	0.15

*) - the data are given with 15% error for the injection from the film and ~25% error for the injection from the substrate

TABLE 3. Parameters of generated positive charge for different oxides ($E=6.5x10^6$V/cm; injection from film or polysilicon gate)

Parameters	SI SIMOX	UNIBOND	Gate Oxide
$\eta_{eff}(0)$	$2.6x10^{-3}$	$1x10^{-4}$	$4x10^{-3}$ [1] $2x10^{-4}$ [2]
$\sigma^+_f(cm^2)$	$1x10^{-15}$	$3x10^{-16}$	$5x10^{-17}$ [1]
$N_\infty(cm^{-2})$	$2x10^{12}$	$3x10^{11}$	$8x10^{12}$ [1] $5x10^{10}$ [2]
T_{ann} (°C)	>450	450	150-400

3.3.2. UNIBOND SOI structures

The typical measured C-V plots of UNIBOND SOI structures with p-type Si substrate and n-type Si film are depicted in Fig. 7a. From the C-V characteristics of the initial SOI structures it was obtained that the BOX/Si film interface contains a positive charge of about $7 \cdot 10^{11}$ cm^{-2}.

UNIBOND BOX exhibits a rather different behavior of charge buildup during FN electron injection performed in constant voltage regime comparing to the SI SIMOX structures. During first seconds of injection only neutralization of the positive charge localized near the BOX/Si film interface is observed (Fig. 7b). Then, the positive charge is generated near the BOX/substrate interface (Fig. 7a, b, c).

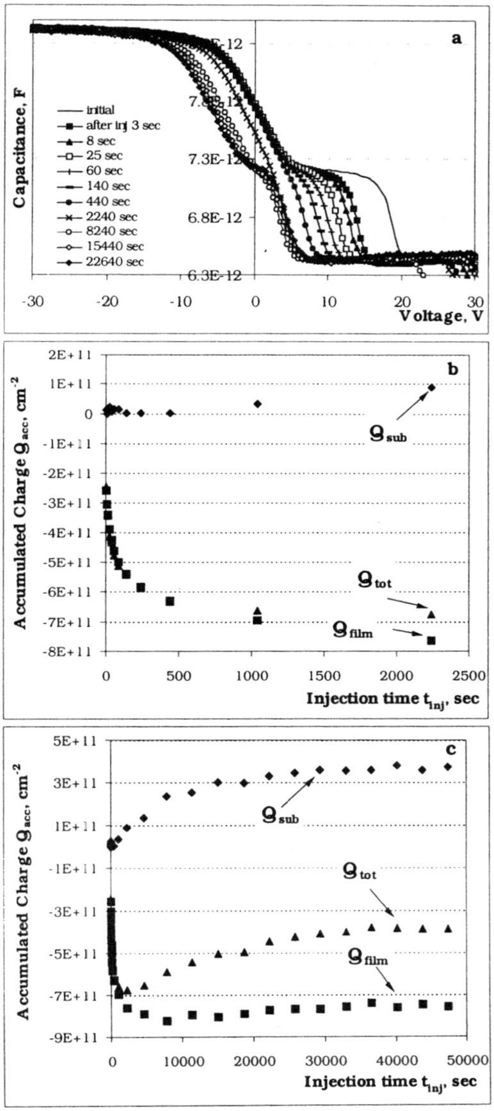

Figure 7. (a) C-V characteristics of UNIBOND SOI capacitors before and after electron injection ;
(b, c) Dynamic characteristics of charge buildup into the BOX of UNIBOND SOI structures.

Electron traps parameters have been calculated using the first order dynamics of the electron trapping (eq. (13)) and expression (14) for estimation of trapping efficiency. For electric field of 6.5 MV/cm the maximal electron traps concentration and capture cross-section for electron trapping are $6 \cdot 10^{11}$ cm^{-2} and $4 \cdot 10^{-15}$ cm^{2}, correspondingly. Electron traps with similar cross section ($\sim 2 \cdot 10^{-15}$ cm^{2}) was found in gate dioxide with considerable potassium concentration [48] and was associated with Na-related Coulomb centers.

Parameters of the generated positive charge were calculated employing of eq. (24), which is common both for impact ionizing mechanisms and anode hole injection process. The data, presented in Table 3 for electric field of 6.5 MV/cm, show considerable smaller maximal generated positive charge concentration ($\sim 3 \cdot 10^{11}$ cm^{-2}) and effective cross-section of positive charge generation ($\sim 3 \cdot 10^{-16}$ cm^{2}) in

UNIBOND SOI structures comparing with these parameters for SI SIMOX structures.

The study of electron injection at different polarity of applied electric field demonstrates that: 1) the generated positive charge is always localized near the BOX/substrate interface independently on electron injection direction (from Si film or from Si substrate); 2) the efficiency of positive charge generation is actually independent on the direction of electron injection, that is $P_{neg}/P_{pos} \approx 1$.

The first fact attests that the BOX/substrate interface in the UNIBOND SOI structure is the most imperfect place. Indeed, this interface is a bonding place of the wafers during the process of UNIBOND SOI structure formation [49].

The second fact suggests that the anode hole injection is probably a predominant mechanism of positive charge generation in the UNIBOND SOI structures.

4. Annealing of the generated positive charge in the BOX of SOI structures

4.1. THERMAL ANNEALING

Removal of the generated positive charge in the BOX of SIMOX and UNIBOND SOI structures by thermal heating was studied by annealing the samples both in isochronal and isothermal modes [41]. These annealing steps took place without electrical bias in order to avoid any tunneling effect. After each annealing step, the sample was cooled rapidly in order to record the room-temperature C-V characteristics from which the mid-gap voltage shifts were determined. The anneal-induced removing of the generated positive charge as a function of annealing temperature is depicted in the Fig. 8. It is obvious from the figure that no more than 30% of the total accumulated charge in SI SIMOX BOX is removed at anneal temperature as high as 400 °C. Furthermore, the positive charge generated in the SI SIMOX BOX at higher electric field is more thermally stable than that generated at lower electric field.

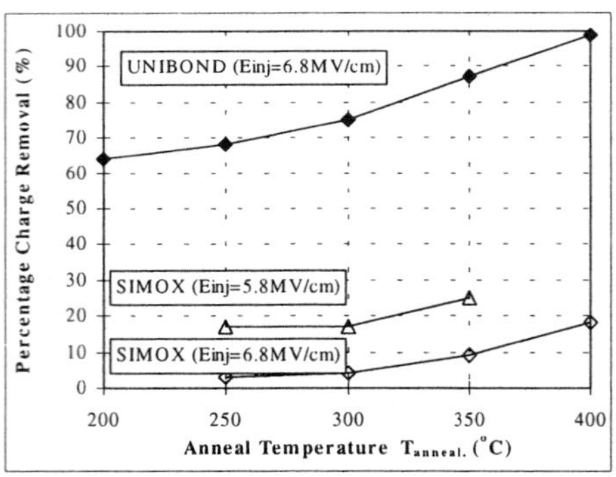

Figure 8. Temperature annealing of the injection induced postitve charge in the BOX of SIMOX and UNIBOND SOI structures.

It should be noted, that the positive charge created during injection in the UNIBOND BOX has been almost completely annealed after heating up to 400 °C during 15 minutes (see Fig. 8). The generated positive charge in the thermal gate oxides is also totally annealed at temperature range from 150 to 400 °C [48]. High thermal stability of the positively charged defects created during electron injection in SI SIMOX BOX indicates a more complicated structure of these defects in comparison with that in the thermal gate oxide and in the UNIBOND BOX.

Increased thermal stability of the positive charge generated in the SIMOX BOX at higher injection fields suggests the creation of more complicated and energetically deep defects than those at lower fields. Thus, it is reasonable to assume, that in SIMOX BOX during FN electron injection the positive charge generation occurs simultaneously with creation of new defects which have a more complicated structure and deeper energy in SiO_2 band gap.

It is worth to note that the injection-induced hole trapping defects do not related to oxygen vacancy defects (so named E'-centers) in dioxide. The detailed measurements of EPR on FN electron injected SI SIMOX SOI samples have not been shown any EPR signals [50]. Obtained results are supported by similar EPR measurements on the MOS injected structures, where E'-centers also have not been found [51]. Thus, the positively charged defects obtained during FN electron injection in SI SIMOX BOX are diamagnetic and has no relation to E'-centers.

4.2. RF HYDROGEN PLASMA ANNEALING

In order to anneal the injection-induced positive charge in the BOX of SI SIMOX structures the low-temperature RF hydrogen plasma treatment have been proposed [41]. It was found the optimal regime of the treatment, which allowed to anneal about 90% of the positive charge in the SIMOX BOX (see Fig. 9a).

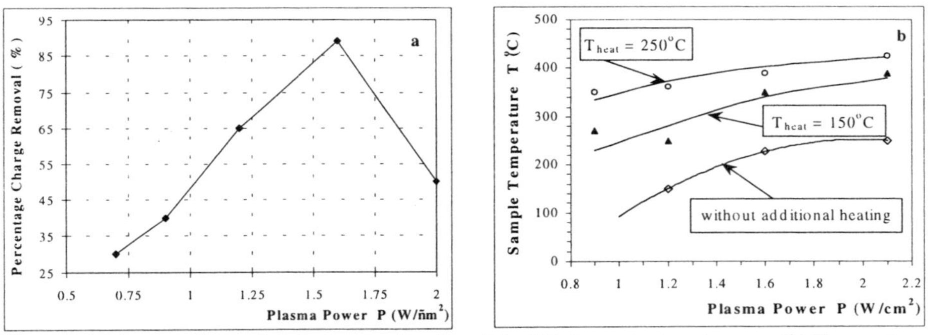

Figure 9. (a) Percent of annealed charge as a function of the RF plasma treatment power; (b) Temperature of RF plasma treated sample as a function of the RF plasma power.

The sample temperature during plasma treatment was monitored in situ employing a special thermal paints deposited on the back side of the samples. Calibration of the thermal paints was done in advance in a standard oven. Results of measurements are shown in Fig. 9b. It is evident that the sample temperature did not exceed 400 °C at maximal plasma power with additional heating. Combination of x-ray irradiation, together with electron and hole injection and trapping of atomic hydrogen in dangling and strained bonds was suggested to play a main role in decrease of annealing temperature of this positive charge.

5. Conclusions

Analysis of high-field conductivity and positive charge generation during high-field electron injection in the buried oxides of main commercial SOI structures, fabricated by SIMOX and UNIBOND techniques have been performed.

It has been shown that high-field conductivity for high-quality UNIBOND SOI structures can be described by modified Fowler-Nordheim expression, that is determined by enhomogenious properties of the BOX/semiconductor interfaces. In the case of SI SIMOX structures the tunneling processes in the BOX bulk have to be taken into account in order to describe the high-field electron conductivity.

In both types of SOI structures the injection-induced positive charge is located predominantly near the BOX/substrate interface. In the case of SI SIMOX structures the generated positive charge exhibits high thermal stability, which increases with increasing of injection electric field. Such increased thermal stability attests the more complicated structure of the defects in the SIMOX BOX in comparison with ones in the UNIBOND BOX. Application of the RF plasma treatment for annealing of the injection-generated positive charge in SIMOX BOX allows to anneal at low-temperature (up to 400 °C) up to 90% of the total injection-induced positive charge that is impossible to realize by thermal annealing in the same temperature range.

Study of the positive charge generation in the BOX of SOI structures during FN electron injection is useful for diagnostic of charge trapping in the dielectric and revealing the most unreliable regions in the structures.

Acknowledgements

The authors would like to express special thanks to Dr. M. Bruel (CEA/LETI, Grenoble, France) for providing of the UNIBOND SOI wafers. This work was partially supported by NATO Linkage Grant #SST.CLG975925 and STCU project #2332

References

1. Ioannou, D.E. (1995) Hot carrier reliability of SOI structures, in J.P. Colinge et al. (eds.), Physical and Technical Problems of SOI structures and Devices, Kluwer, Dordrecht, pp.199-210.
2. Ouisse, T., Cristoloveanu, S., and Borel, G. (1991) Hot-carrier-induced degradation of the back interface in short-channel Silicon-On-Insulator MOSFET's, IEEE Electr. Dev. Lett. 12, 290-292.
3. Mayo, S., Suchle, J.S., and Roitman, P. (1993) Breakdown mechanism in buried silicon oxide films, J. Appl. Phys. 74, 4113-4120.
4. Revesz, A.G., Brown, G.A., and Hughes, H.L. (1993) Bulk electrical conduction in buried oxide of SIMOX structures, J. Electrochem. Soc. 140, 3222-3229.
5. Wainwright, S.P., Ngwa, C., Hall, S., and Eccleston, W. A study of high field conduction and electron trapping in buried oxides produced by SIMOX technology, Microelectron. Eng. 22, 399-402.
6. Bengtsson, S., Ericsson, P., Sodervall, U., Mitani, K., and Abe, T. (1995) Charge carrier injection into the buried oxide of wafer-bonding silicon-on-insulator materials, J. Electrochem. Soc. 142, 2721-2726.
7. Weinberg, Z.A. (1977) Tunneling of electrons from Si into thermally grown SiO_2, Solid State Electron. 20, 11-18.
8. Lenzlinger, M., and Snow, E.H. (1969) Fowler-Nordheim tunneling into thermally grown SiO_2, J. Appl. Phys. 40, 278-283.
9. Revesz, A.G., and Hughes, H.L. (1995) The defect structure of buried oxide layers in SIMOX and BESOI structures, in in J.P. Colinge et al. (eds.), Physical and Technical Problems of SOI structures and Devices, Kluwer, Dordrecht, pp.133-156.
10. Wainwright, S.P., Hall, S., and Harsh, C.D. (1994) Electrical characterization of thin film, thin buried oxide SIOX SOI structures produced by low-energy, low-dose implantation, Semicond. Sci. Technol. 9, 1404-1413.

11. Ngwa, C.S., and Hall, S. (1994) Electron trapping studies in multiple- and single-implant SIMOX oxides, Semicond. Sci. Technol. 9, 1069-1079.
12. Bruel, M. (2000) Smart-Cut® technology: basic mechanisms and applications, in P.L.F. Hemment et. al (eds.), Perspectives, Science and Technology for Novel Silicon-in-Insulator Devices, Kluwer, Dordrecht, pp.1-15.
13. Stoemenos, J. (1993) Microstructure of SIMOX buried oxide, mechanisms of defect formation and related reliability issues, Microelectron. Eng. 22, 307-314.
14. Kranse, S., Anc, M., and Roitman, P. (1998) Evaluation and future trends of SIMOX material, MRS bulletin 23, 25-29.
15. Afanas'ev, V.V., Stesmans, A., and Twigg, H.E. (1996) Epitaxial Growth of SiO_2 produced in silicon by oxygen ion implantation, Phys. Rev. Lett. 77, 4206-4209.
16. Stesmans, A., Devine, R.A.B., Revesz, A.G., and Hughes, H.L. (1990) Irradiation induced ESR active defects in SIMOX structures, IEEE Trans. Nucl. Sci. NS-37, 2008-2012.
17. Warren, W.L., Shaneyfeld, M.R., Schwank, J.R., Fleetwood, D.M., and Winokur P.S. (1993) Paramagnetic defect centers in BESOI and SIMOX buried oxides, IEEE Trans. Nucl. Sci. NS-40, 1755-1764.
18. Vanhensden, K., and Stesmans, A. (1994) Similarities between separation by implanted oxygen and bonded and etch-back silicon-on-insulator material as revealed by electron spin resonance, in S. Cristoloveanu et. al. (eds), Silicon-On-Insulator Technology and Devices 94-11, ECS Inc., Pennington, 197-202.
19. Afanas'ev, V.V., Stesmans, A., Revesz, A.G., and Hughes, H.L. (1997) Structural inhomogenity and silicon enrichment of buried SiO_2 layers formed by oxygen ion implantation in silicon, J. Appl. Phys. 82, 2184-2199.
20. Bengtsson, S., Jauhiainen, A., and Engstrom, O. (1992) Oxide degradation of wafer bonded metal oxide semiconductor capacitors following Fowler-Nordheim electron injection, J. Electrochem. Soc. 139, 2302-2306.
21. Jauhiainen, A., Bengtsson, S., and Engstrom, O. (1992) Charge trapping in wafer bonded MOS structures, Microelectronic Eng. 19, 597-600.
22. Hall, S., and Wainwright, S.P. (1996) On electron conduction and trapping in SIMOX dielectric, J. Electrochem. Soc. 143, 3354-3358.
23. Margail, J., Lamure, J.M., and Papon, A.M. (1992) Defects in SIMOX structures: some process dependence, Mat. Sci. And Eng. B12, 27-36.
24. Di Maria, D.J., Wong, D.W., Falcony, C., Theis, T.N., Kirtley, J.R., Tsany, J.C., Young, D.R., and Pesavento, F.L. (1983) Charge transport and trapping phenomena in off-stoichiometric silicon dioxide films, J. Appl. Phys. 54, 5801-5827.
25. Strzalkowski, I., and Kowalski, M. (1996) Positive and negative charge creation in the SiO_2 film of a MOS transistor by high electric field stress, Appl. Phys. A63, 179-182.
26. Samanta, P., and Sarkar, C.K. (1997) Analysis of positive charge trapping in silicon dioxide of MOS capacitors during Fowler-Nordheim stress, Sol.-St. Electron. 41, 459-464.
27. Chen, C., Wilson, W.L., and Smagling, M. (1998) Tunneling induced charge generation in SiO_2 thin film, J. Appl. Phys 83, 3898-3905.
28. Ning, T.H., and Yu, H.N. (1974) Optical induced injection of hot electrons into SiO_2, J. Appl. Phys. 45, 5373-5378.
29. Fischetti, M.V. (1985) Generation of positive charge in silicon dioxide during avalanch and tunnel electron injection, J. Appl. Phys. 57, 2860-2879.
30. DiMaria, D.J., and Stasik, J.W. (1989) Trap creation in silicon dioxide produced by hot electrons, J. Appl. Phys. 65, 2342-2356.
31. DiMaria, D.J., Cartier, E., and Arnold, D. (1993) Impact ionization, trap creation, degradation, and breakdown in silicon dioxide film on silicon, J. Appl. Phys. 73, 3367-3384.
32. Nagai, K., Sekigawa, T., and Hayashi, Ya. (1985) Capacitance-voltage characteristics of SIS structures, Sol. St. Electron. 28, 789-798.
33. Nazarov, A.N., Lysenko, V.S., Gusev, V.A., and Kilchitskaya, V.I. (1994) C-V and thermally activated investigations of ZMR SOI meza structures, in S. Cristoloveanu et al. (eds), Silicon-On-Insulator Technology and Devices 94-11, ECS Publishers, NJ, pp. 236-244.
34. Bachanan, D.A., Marwick, A.D., Di Maria, D.J., and Dori, L. (1994) Hot-electron-induced hydrogen redistribution and defect generation in metal-oxide-semiconductor capacitors, J. Appl. Phys. 76, 3595-3608.

35. Klein, N., and Solomon, P. (1976) Current runaway in insulators affected by impact ionization and recombination, J. Appl. Phys. 47, 4364-4372.
36. Nissan-Cohen, Y., Shappir, J., and Frohman-Beatchkowsky, D. (1985) Dynamic model of trapping-detrapping in SiO_2, J. Appl. Phys. 58, 2252-2261.
37. Fischetti, M.V. (1985) Model for the generation of positive charge at the Si-SiO_2 interface based on hot-hole injection from the anode, Phys. Rev. B 31, 2099-2113.
38. DiMaria, D.J., Cartier, E., and Buchanan, D.A. (1996) Anode hole injection and trapping in silicon dioxide, J. Appl. Phys. 80, 304-317.
39. Arnold, D., Cartier, E., and Di Maria, D.J. (1994) Theory of high-field electron transport and impact ionization in silicon dioxide, Phys. Rev. B 49, 10278-10297.
40. DiMaria, D.J. (1999)Defect production, degradation, and breakdown of silicon dioxide films, Sol. St. Electron. 41, 957-965.
41. Nazarov, A.N., Kilchytska, V.I., Barchuk, I.P., Tkachenko, A.S., and Ashok, S. (2000) Radio frequency plasma annealing of positive charge generated by Fowler-Nordheim electron injection in buried oxides in silicon, J. Vac. Sci. Technol. B18, 1254-1261.
42. Sah, T.C., San, J. Y.-C., and Tzow, J. J.-T. (1983) Generation-annealing kinetics and atomic models of a compensating donors in the surface space charge layer of oxidized silicon, J. Appl. Phys. 54, 944-956.
43. Sah, T.C., San, J. Y.-C., and Tzow, J.J.-T. (1984) Study of the atomic models of three donorlike defects in silicon metal-oxide-semiconductor structures from their gate material and process dependencies, J. Appl. Phys. 55, 1525-1545.
44. Pearton, S.J., Corbett, J.W., and Stavola, M. (1992) Hydrogen in crystalline semiconductors, Springer-Verlag.
45. Devine, R.A.B., (1999) The structure of SiO_2, its defects and radiation hardness, IEEE Trans. Nucl. Sci. 41, 452-460.
46. Boutry-Forveille, A., Nazarov, A., and Ballutaud, D. (1998) Deuterium diffusion, trapping and stability in buried silicon dioxide layers, in N.-H. Nickle et al. (eds), Hydrogen in Semiconductors and Metals, MRS, Warrendalle, v. 513, pp.319-324.
47. Ballutaud, D., Boutry Forveille, A., and Nazarov, A. (1999) Hydrogen thermal stability in buried oxides of SOI structures, Microelectron Engineering 48, 359-362.
48. Di Maria, D.J. (1978) The properties of electorn and hole traps in thermal dioxide layers growth on silicon, in S.T. Pantelides (ed.), The Physics of SiO_2 and Its Interfaces, Pergamon Press, NY, pp.
49. Bruel, M. (198) The history, physics and applications of the smart-cut® process, MRS bulletin 23, 35-39.
50. Zaritski, I.M., Nazarov, A.N., and Kilchytska, V.I. (1995) (unpublished)
51. Conley, J., and Lenahan, P. (1996) A review of electron spin resonance spectroscopy of defects in thin film SiO_2 on silicon, in H.Z. Massond et al. (eds), The Physics and Chemistry of SiO_2 and the SiO_2 – Si interface – 3, ECS 96-1, pp.214-249.

CRYOGENIC INVESTIGATIONS OF SIMOX BURIED OXIDE PARAMETERS

V.S.LYSENKO, I.P.TYAGULSKI, I.N.OSIYUK and Y.V.GOMENIUK
Institute of Semiconductor Physics, National Academy of Sciences of Ukraine
Prospect Nauki 45, 03028 Kiev, Ukraine

1. Introduction

Substrates containing buried dielectric prepared by separation by implantation of oxygen (SIMOX) are promising for use in complementary metal oxide semiconductor (CMOS) applications where the reduced capacitance and high packing density of the MOS devices allow to reach higher operation speed and lower power consumption. The stability and long-term reliability of devices based on SIMOX technology are limited by the electrical properties of the buried oxide (BOX), especially, of silicon film–buried oxide and buried oxide–silicon substrate interfaces. It was mentioned in [1] that at low temperatures the role of hot carrier effects greatly increases, which can lead to device degradation, especially for interfaces with poor electrical properties. So, in order to take full advantage of the attractive features of this technology, the aforementioned properties require more research. Low temperature measurements make it possible to study shallow interface levels with activation energies of some meV situated in the thin non-stoichiometric transition layer. These traps are responsible for hysteresis of n-channel transistor drain characteristics at low temperatures [2]. In this paper, we present further insight into the trapping properties of the buried oxide.

2. Experimental details

The investigated SOI SIMOX structures were fabricated by single O^+ implantation with energy of 200 keV to the dose of 1.8×10^{18} cm^{-2} at temperature 600°C with following annealing at 1320 °C during 6 hours in Ar+2%O_2. The BOX thickness was 360 nm. The silicon film and substrate were p-type conductivity. For measurements the mesa-structures "Al–Si-film–BOX–Si-substrate–Al" were fabricated.

The experimental set-up allowed: (1) to fix the required temperature in the range from 6–300 K; (2) to fix or to change linearly the gate voltage in the range -30–+30 V; (3) to measure currents in the structure during temperature or gate voltage changes.

The experimental method is a combination of thermally stimulated charge release (TSCR) technique and dynamic I–V measurements using the linear voltage ramp technique. Measurements have been carried out in the structure "p-Si-film – BOX – p-

F. Balestra et al. (eds.), Progress in SOI Structures and Devices Operating at Extreme Conditions, 159–166.

Si-substrate". Charge emission can be performed in two ways. First way is the conventional TSCR, when at zero gate voltage the structure is being heated with constant heating rate and the temperature dependence of the emission current is registered. The second way consists in a linear bias sweep: $V_G(t)=V_{ch}-\alpha t$ with the rate $\alpha=0.24$ V/s at constant temperature with measurement of the dependence of the emission current on the gate voltage.

3. Results and Discussion

Figure 1 shows the dynamic I–V (quasi-static C-V) curves measured at room temperature, at 200 K and 215 K. The room-temperature curve (curve 1) is similar to those measured in [3, 4] and indicates that there is a high density of fixed positive charge in the buried SIMOX oxide, so that inversion-to-accumulation processes at each interface are separated along the voltage axis. When the voltage is varied from negative to positive values the conditions at the film/BOX and substrate/BOX interfaces change. At high negative voltages the substrate is in accumulation and the film is in inversion. When approaching zero voltage, the substrate turns into depletion and then to inversion and around V_G=+3 V both interfaces are in inversion. The further voltage sweep leads to turning the film/BOX interface from inversion to depletion and then to accumulation. Finally, at high positive gate voltages the substrate is in inversion and the film is in accumulation.

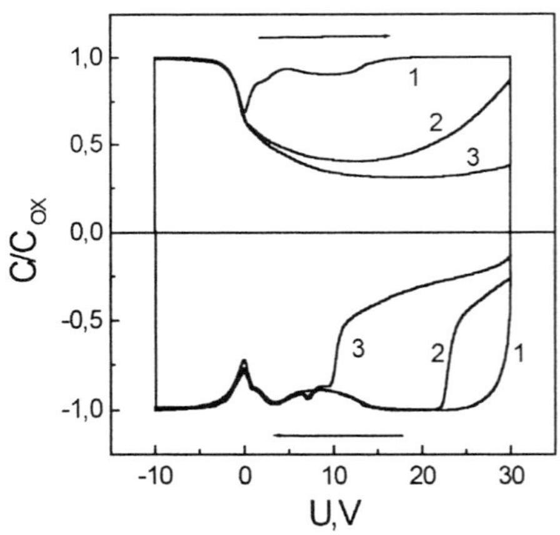

Figure 1. Quasi-static (C-V) characteristics at room temperature (1), 215 K (2) and 200 K (3).

When the measurement temperature decreases (curves 2, 3) we observe non-equilibrium effects in the I-V curves manifested by a decrease of the current magnitude below that of the quasi-static value [5]. This means that if the gate voltage is swept from zero to positive values, the generated inversion charge (electrons) in the substrate is not able to respond to varying gate voltage due to reduced thermal generation of minority carriers in the substrate. An equilibrium inversion layer can not form and the depletion layer expands deeper into the bulk of the silicon substrate.

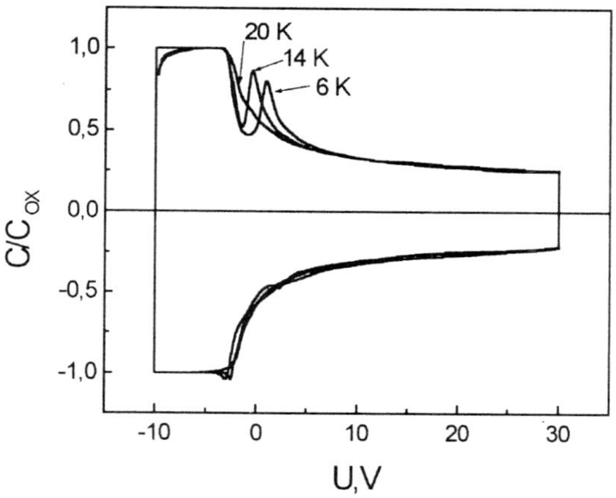

Figure 2. Quasi-static (C-V) characteristics at temperatures below 20 K.

At temperatures below 140 K the thermal generation of minority carriers in the substrate become so slow that the inversion layer in the silicon substrate/BOX interface can not be induced at positive gate voltages for the time of experiment. At positive V_G non-equilibrium deep-depleted region the bulk leads to a small capacitance of the structure. In this case only the processes related to the back SIMOX oxide interface remain to be present in the I–V curves.

At temperatures below 20 K dynamic I-V characteristics were always measured after heating the structure to 20K with –30 V at the gate. Figure 2 shows the dynamic I-V characteristics for temperatures 6, 14 and 20 K. When the gate voltage is swept from negative values to positive (corresponding to turning the substrate/BOX interface from accumulation to depletion) we observe a peak in I-V curves, at temperatures below 16 K, which is absent at the reverse direction of the voltage sweep. Such a feature can be related to the field emission of charge trapped at the interface shallow centres [6]. In order to clarify the physical nature of this peak the set of TSCR experiments was carried out.

In figure 3 thermally stimulated currents are shown measured at different conditions. It should be noted, that in these experiments all changes of the gate voltage were performed at the temperature 7 K. The initial cooling down to the temperature 7 K was made at zero gate bias. Curve 1 shows thermally stimulated current measured after the silicon substrate was turned into the accumulation conditions by changing the gate voltage from zero to –30V. In the respective I-V characteristic (see figure 4) this procedure is shown by section a-b. After TSCR measurement the structure was cooled back to 7 K, the gate voltage was set to zero (section b-c in Fig. 4) and the TSCR current was recorded at V_G=0V (curve 2). Then the structure was cooled again and TSCR measurement was repeated at zero gate voltage (curve 3). At last, the gate voltage was changed from zero to +30 V at temperature 7K, as shown by the section c-d in Fig. 4, and during heating the curve 4 was recorded.

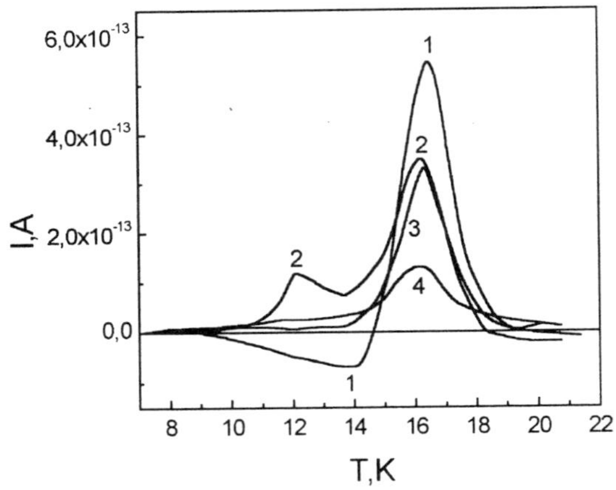

Figure 3. TSCR currents measured at different conditions (see text).

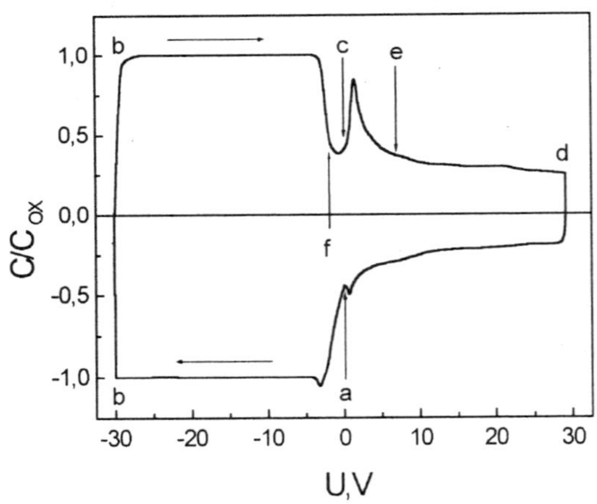

Figure 4. Quasi-static (C-V) characteristic at 7 K.

It can be seen that two current components with principally different behaviour can be found in these curves. The first current peak, with the maximum at 12 K, is similar to that observed in the "metal – thermal oxide – silicon" structures [7] and can be attributed to hole trapping at and detrapping from the shallow traps in the thin transition layer between the buried oxide and the silicon substrate. In this case the process of filling the interface traps by holes is represented by the negative part of curve 1 in Fig. 3 and the respective charge release is manifested as the first positive peak in curve 2. The repeated measurement of TSCR current (curve 3) did not show any traces of the first peak. In the case when the TSCR currents are recorded in conditions of deep-depleted substrate (curve 4) this peak is not observed. The activation energy of the charge release process determined from the slope of the initial part of the peak plotted in Arrhenius coordinates is equal to 17 meV.

It should be noted that the correlation exists between the peak observed in I-V curves at low temperatures (Fig. 2) and the first TSCR peak. This can be proved by the following experiment (Fig. 5). Firstly, the traps were filled at 20 K by application –30 V to the gate, then the sample was cooled and the I-V characteristic was recorded during the voltage sweep from –30V to the point just after the peak (in Fig. 4,5a (solid line) point e). Then the gate voltage was returned to zero and TSCR current was measured. Obtained such a way curve did not show the peak of current in the temperature range around 12 K (Fig. 5b doted line). On the contrary, if after filling the voltage sweep was stopped before the peak (in Fig. 4,5a (solid line) point f), and TSCR current was measured at this value of gate voltage (–1.3 V), then the maximum peak amplitude was observed (Fig. 5b solid line). It should be note, when the gate voltage is swept from negative values to positive (corresponding to turning the substrate/BOX interface from accumulation to depletion) once more time, we could not observe a peak in I-V curves (Fig.5a dotted line). Such correlation indicates that the same level is manifested in low-temperature I-V measurements and TSCR currents.

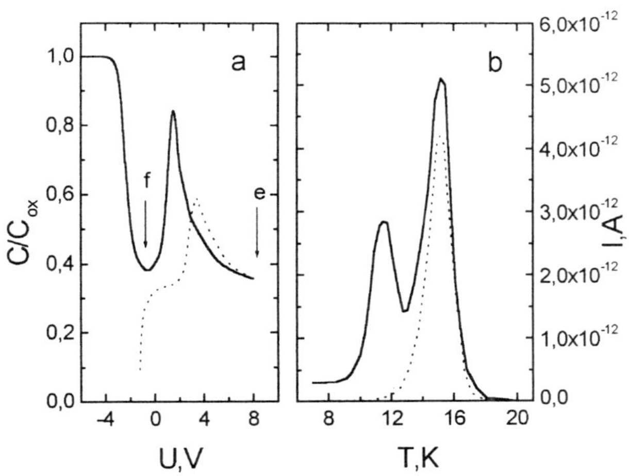

Figure 5. Field-assisted release and TSCR of shallow levels (see text).

Detailed measurements of dynamic I-V characteristics have shown that as the temperature is reduced from 16 down to 12 K the position of the peak shifts toward higher electric fields (Fig 6a). Further reduction of temperature from 12 to 6 K does not result in the shift of the peak maximum (Fig 6b). Such temperature dependence of the peak position indicates that the mechanism of trap emptying changes from thermofield to field emission of charge as the temperature decreases. Similar temperature and field

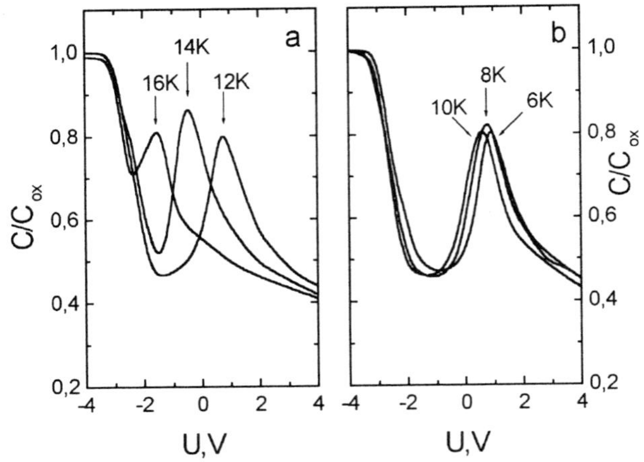

Figure 6. Regions of dynamic I-V characteristics at temperatures below 20 K.

dependences of the peak position were found in silicon p-i-n diodes at low temperatures [8, 9] in studies of field ionization from localized impurity levels in semiconductor. It was shown theoretically that this effect may be attributed to a combination of field ionization and temperature-assisted tunneling. As was shown in our previous papers [6, 7, 10], the observed thermofield effects at such low temperatures may take place with involvement of local phonons connected with defects in the transition Si-SiO_2 region. These phonons promote the carrier trapping or release at the account of carrier tunneling. Thus, emptying of the traps in the BOX–substrate interface can be performed either by heating the sample (TSCR process) or by field emission during voltage sweep turning the substrate from accumulation to depletion mode.

The second TSCR current peak, with the maximum at 15 K (Fig. 3,5b), is observed in oxide-silicon systems for the first time and has the following features. 1) Its manifestation does not depend on the voltage applied and on whether the structure was subjected to preliminary trap filling procedure at 20 K or not. 2) This peak is not observed in the dynamic I-V characteristics. 3) When the temperature is reduced from 20 to 7 K the negative peak of current is recorded, situated under the second TSCR peak. 4) Its activation energy, determined from the initial part of this peak, is 22 meV. 5) This peak is observed in the TSCR spectra when the cooling of the structure from room temperature down to 6 K was performed with different conditions at the interfaces (substrate in accumulation / film in inversion and substrate in inversion / film in accumulation), and its polarity does not depend on initial conditions of the experiment. The above-mentioned features are inherent to polarisation-type thermally stimulated current relaxation. For the low-temperature case such currents are most probably determined by the processes of relaxation of weakly-bonded charged particles (electrons or holes).

The physical nature of the observed peak seems to be determined by the clusters (silicon islands, stacking faults) in the matrix of buried oxide near the BOX/substrate interface. The presence of such clusters in SIMOX dielectrics was reported in [11]. It was also shown [12, 13] that in the range of strong fields such clusters can be responsible for the features of electron conduction and trapping in SIMOX dielectrics. Cooling and heating of the short-circuited structure lead to "freezing" or release of charge, i.e. to appearance of the current peaks of different polarity. Since the polarity of the peak observed during the sample heating is independent of the external field, the charge relaxation processes seem to be determined by the internal electric field in SIMOX dielectric. Such relaxation currents can be related to non-equilibrium initial state of the charge in the region of structural non-uniformities near the BOX-substrate interface which appears during cooling of the structure with internal electrical field [14].

Besides, since the dielectric permeabilities and conductivities of oxide matrix and of clusters are different, in the presence of electric field charges can be accumulated in interfaces (Maxwell-Wagner polarization). Such mechanism for SIMOX dielectric at higher temperatures was considered in [15].

4. Conclusions

Measurements of dynamic I-V characteristics and of thermally stimulated currents in the structures with buried SIMOX dielectric at low temperatures made it possible to study the processes of recharging of shallow traps in the BOX-silicon substrate interface. It was shown that this interface contains traps of different physical nature. The first type of traps is related to defects in the transition layer of the BOX-silicon substrate interface. The second type of traps seems to be related to processes of polarization-type relaxation due to a non-equilibrium state of the charge in the region of structural non-uniformities. Further studies are necessary to clarify the exact nature of these traps.

5. References

1. Renn, S.H., Raynaud, C., Balestra F. (1998) Temperature dependence of hot-carrier effects in 0.2 μm N- and P-channel fully-depleted Unibond MOSFETs, *J. de Physique* IV **8**, 13-16.
2. Lysenko, V.S., Tyagulski, I.P., Gomeniuk, Y.V., Osiyuk, I.N. (2000) Effect of oxide-semiconductor traps on low-temperature operation of MOSFETs, *Microelectronics Reliability* **40**, 735-738.
3. Ikraiam, F., Beck, R.B., Jakubovski, A. (1996) Ideal capacitance-voltage characteristics of SOI-MOS thick-body capacitors, *Electron Technology* **29**, 365-371.
4. Nagai, K., Segikawa, T., Hayashi, Y. (1985) Capacitance-voltage characteristics of semiconductor-insulator-semiconductor (SIS)-structure, *Solid-State Electronics* **28**, 789-798.

5. Faraone, L., Simmons, J.G., Agarwal, A., Tonner, P.D. (1981) Interpretation of non-equilibrium measurements on MOS devices using the linear voltage ramp technique, *Solid-State Electronics* **24**, 709-716.
6. Gomeniuk, Y.V., Litovski, R.N., Lysenko, V.S., Osiyuk, I.N., Tyagulski, I.P. (1992) Thermally stimulated field emission of charge from traps in the transition layer of Si-SiO_2 structures, *Applied Surface Science* **55**, 179-185.
7. Lysenko, V.S., Sytenko, T.N., Zimenko, V.I., Snitko, O.V. (1982) Investigation of traps in the transition region of Si-SiO_2 structures at cryogenic temperatures, *Phys. Stat. Sol. (a)* **71**, 619-625.
8. Banavar, J.R., Coon, D.D., Derkits G. (1978) New charge-storage effect in silicon p-i-n diodes at cryogenic temperatures, *Phys. Rev. Lett.* **41**, 576-579.
9. Banavar, J.R., Coon, D.D., Derkits G. (1979) Low-temperature field ionization of localized impurity levels in semiconductors, *Appl. Phys. Lett.* **34**, 94-96.
10. Gomeniuk, Y.V., Litovski, R.N., Lysenko, V.S., Osiyuk, I.N., Tyagulski, I.P. (1992) Current stochasticity of field emission of charge from traps in the transition layer of implanted MIS structures, *Applied Surface Science* **59**, 91-94.
11. Margail, J., Lamure, J.M., Papon, A.M. (1992) *Mater. Sci. Eng.* **B12**, 27-32.
12. Waiwright, S.P., Hall, S., Marsh, C.D. (1994) Electrical characterization of thin film, thin buried oxide SIMOX SOI substrates produced by low-energy, low-dose implantation, *Semicond. Sci. Technol.* **9**, 1404-1413.
13. Hall, S., Waiwright, S.P. (1996) On electron conduction and trapping in SIMOX dielectric, *J. Electrochem. Soc.* **143**, 3354-3358.
14. Gorokhovatski, Yu.A., Gubkin, A.N., Malinovskii, V.K., Sturman, B.I. (1982) On the theory of electret effect, *Fiz. Tverd. Tela* **24**, 2483-2485 (in Russian).
15. Ioannou-Sougleridis, V., Papaioannou, G.J., Dimitrakis, P., Cristoloveanu, S. (1993) Characterization of the buried oxide in SOI structures by a rate window method, *J. Appl. Phys.* **74**, 3298-3302.

GATE-ALL-AROUND TECHNOLOGY FOR HARSH ENVIRONMENT APPLICATIONS

J.P. COLINGE
Dept of Electrical and Computer Engineering, University of California, Davis, USA

Abstract

Gate-All-Around (GAA) transistors are thin, fully depleted SOI MOSFETs with a double gate structure. When used at high temperature GAA devices present low leakage current, minimal threshold voltage shift and, in general, better characteristics than bulk or even SOI MOSFETs. The radiation hardness of GAA devices is reported as well, and the dose evolution of parameters such as threshold voltage, subthreshold slope and output conductance is analyzed.

1. Introduction

It is well known that the double-gate (top and bottom gate) silicon-on Insulator (SOI) MOSFET is the most suitable device structure for deep-submicron applications [1]. Several techniques have been proposed to fabricate such devices [2,3,4,5] but, to-date, the Gate-All-Around (GAA) device [6] is the only double-gate MOSFET that has been used to fabricate functional CMOS circuit. GAA devices offer interesting properties, such as increased tansconductance due to volume inversion [7]. Two-dimensional confinement quantum effects have been observed in GAA MOSFETs at low temperature.[8,9] This paper describes the performance of GAA devices operating in a harsh environment. GAA devices and circuits can operate at high temperature. They show minimal threshold voltage variation with temperature and leakage currents smaller than any other silicon device. Since GAA devices are made in a thin silicon film that is in contact only with good-quality gate oxide (as opposed to a buried oxide with lower quality) they are quite hardened against radiation effects (total dose and single-event upset).

2. Fabrication process

The Gate-All-Around devices are fabricated using a simple 3μm process and commercially available SIMOX or Unibond material with a 400nm buried oxide. The final active silicon thickness is adjusted to 80nm by successive oxidation and strip down. A thin pad oxide is then grown, and silicon nitride is deposited. Using a photolithography step the nitride and the pad oxide are patterned to define the active areas. Masking p-channel devices, a rather heavy dose of boron is then implanted to increase the threshold voltage of the lateral n-channel transistors such as to avoid edge transistor turn-on (Fig.1A). Using the nitride pattern as a mask, the silicon film is now patterned by plasma etching to define the active areas (MESA process). A wet oxidation

F. Balestra et al. (eds.), Progress in SOI Structures and Devices Operating at Extreme Conditions, 167–188.

step is used to round the edges of the silicon islands (Fig.1B), after which the pad oxide and the nitride are stripped. This has been reported to improve the gate oxide breakdown properties [[10]].

A mask step is then used to cover the entire wafer with resist except the areas which correspond to an oversize of the intersection between the active area and the poly gate layers. The wafers are then immersed in buffered HF. At this step the oxide on the sidewalls of the silicon islands as well as the buried oxide are etched, and a cavity is created underneath the center part of the silicon islands (Fig.2), resulting in the formation of a free-standing silicon bridge whose both ends are supported by the remaining buried oxide. Gate oxidation is then carried out. A 30-nm-thick gate oxide is grown over the exposed silicon (Fig. 1C). Boron is implanted to adjust the threshold voltage, and the polysilicon gate material is deposited and doped n-type. Because of the extremely good conformality of the LPCVD polysilicon, the gate oxide over the cavity is completely coated with polysilicon and a gate is formed on the top, the sides and the bottom of the channel area. The polysilicon is patterned using conventional lithography and anisotropic plasma etch. Source and drain are formed using arsenic implantation followed by an annealing step. A CVD oxide layer is deposited and contact holes are opened. An aluminum metallization step is finally used to contact the devices. A complete view of the Gate-All-Around MOS transistor finally obtained is represented in Fig.3. A SEM view of the device is illustrated in Fig.4. Figure 5 presents a bird's eye view of the device.

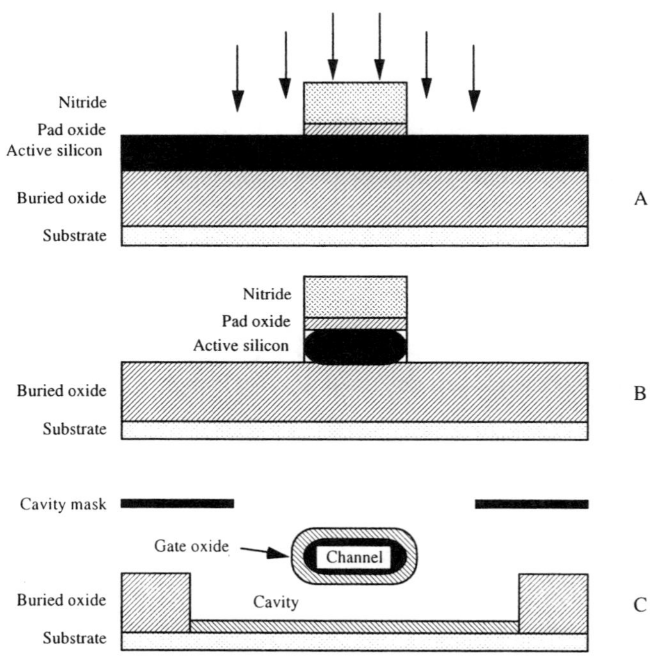

Figure 1: Successive steps of the GAA fabrication process: A. Boron implantation to increase the threshold voltage of lateral n-channel transistor to avoid edge transistor turn-on. B. Wet oxidation performed to smooth the corners of the silicon island. C. Cavity etching and gate oxide growth.

One can observe that only two steps are added to a standard SOI process in order to create a cavity underneath the central part of the silicon islands. No special oxide hardening technique is used. However, a drawback of the technique is that the length and width of all the individual transistors have to be equal to the minimal device dimensions in order to avoid too long Si islands which could break when underetched, as well as too wide Si islands which may not be completely underetched. Furthermore, the isotropic cavity etching results in significant lateral overetching which contributes to an increase of the parasitic gate to source/drain capacitance.

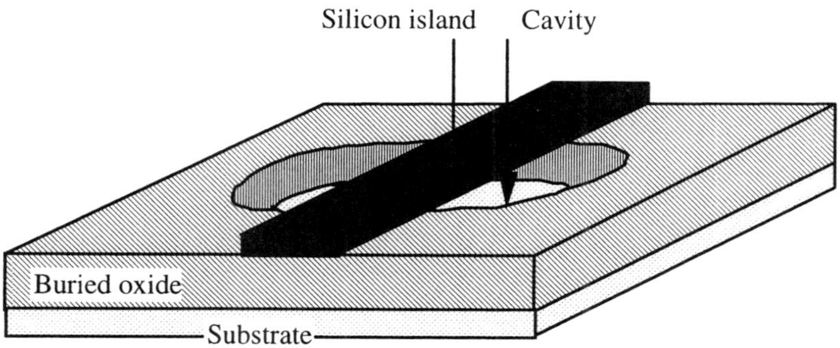

Figure 2: 3D view of the GAA device after etching the cavity underneath the silicon island.

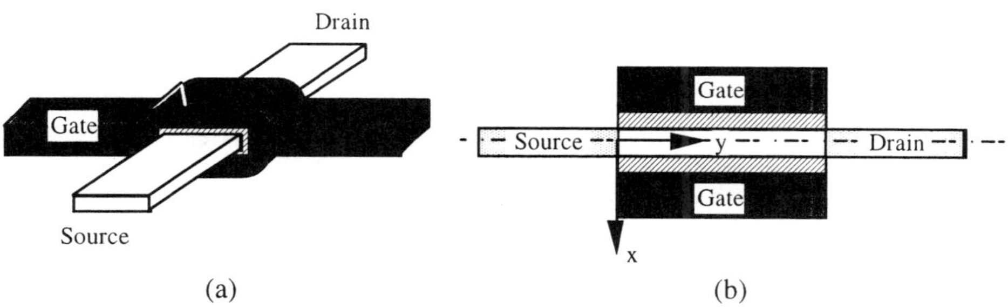

Figure 3: (a) 3D view and (b) longitudinal cross section of the GAA transistor.

The narrow channel of the GAA device results in small drain current drive. The transistors are therefore usually arranged in parallel but the density of the transistors is limited by the lithography resolution. Recently, a new technique, based on silicon anisotropic etching, was proposed to increase the density of the wire channel and current drive [[11]] in GAA devices. By using anisotropic etching and selective oxidation, two silicon wires can be formed from one mesa, increasing the density by a factor two.

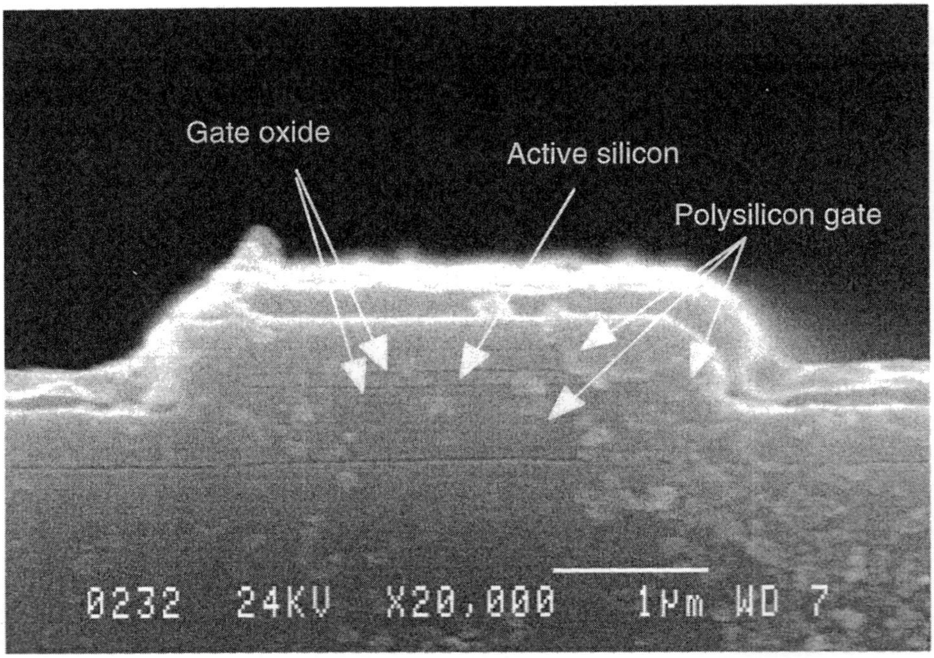

Figure 4: SEM transverse cross-section of the GAA transistor.

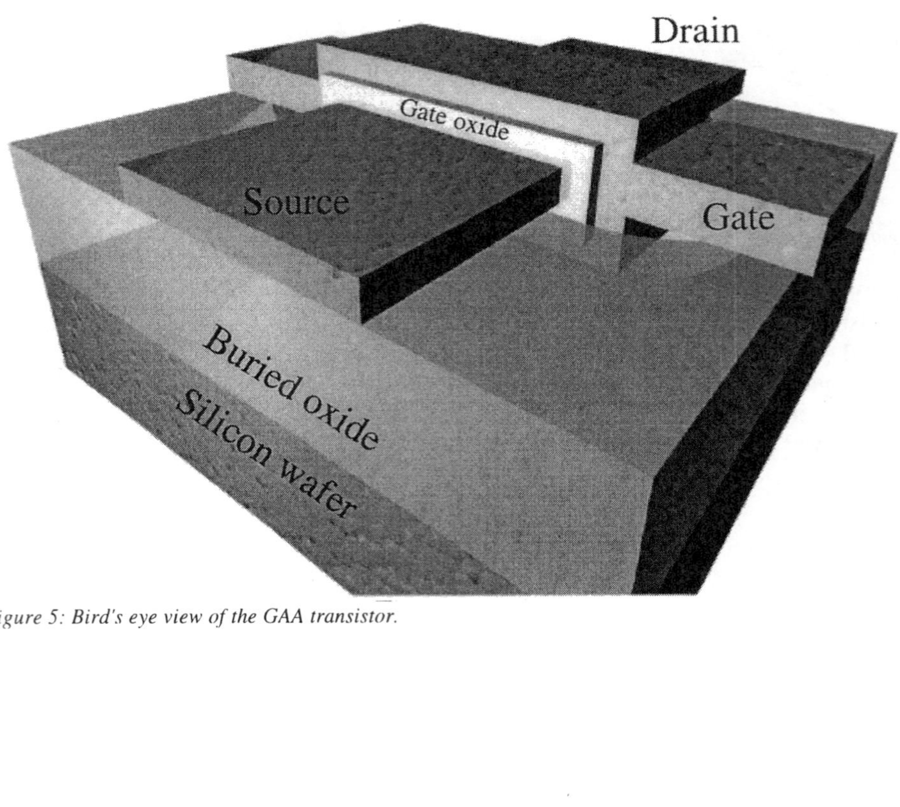

Figure 5: Bird's eye view of the GAA transistor.

3. High-temperature characteristics

Operation at elevated temperature is required for many applications. While bulk MOSFETs usually cease to function in the range of 150°C to 200°C, SOI FD MOSFETs can operate up to 350°C. Elevated temperature induces larger junction leakage currents and lowers the threshold voltage. SOI FD MOSFETs exhibit 3 advantages over their bulk counterparts [12]:

i) absence of thermally activated latchup (no parasitic thyristor structure),
ii) reduced leakage currents due to smaller junctions,
iii) smaller variation of the threshold voltage.

Gate-All-Around devices are also fully-depleted SOI and therefore exhibit the same advantages [13]. The high-temperature behavior of SOI device parameters is well understood and can be modeled [14]. The evolution of the GAA device parameters are first illustrated at high temperature

GAA transistors have been wafer-probed on a temperature-regulated hot chuck at a temperature varying between room temperature and 350C. This paragraph briefly illustrates the evolution with temperature of the threshold voltage, mobility, subthreshold slope and leakage current parameters. The evolution of these basic parameters easily explains the scaled transconductance, g_m/I_d, dependence on temperature using the same procedure as the one described in chapter V for total dose results. The evolution of the Early voltage versus temperature is also illustrated. The drain current versus gate voltage evolution in ohmic region is illustrated in Fig. 6, for a n-channel GAA transistor. Figure 7 presents similar characteristics for an n-channel (inversion mode) and a p-channel (accumulation mode) GAA transistors.

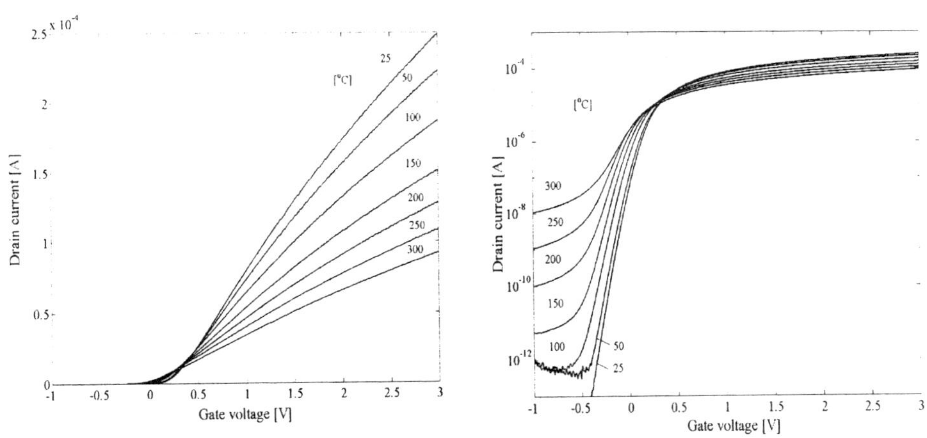

Figure 6: I_d versus V_g characteristics from room temperature to 300°C, in a n-channel GAA MOSFET.

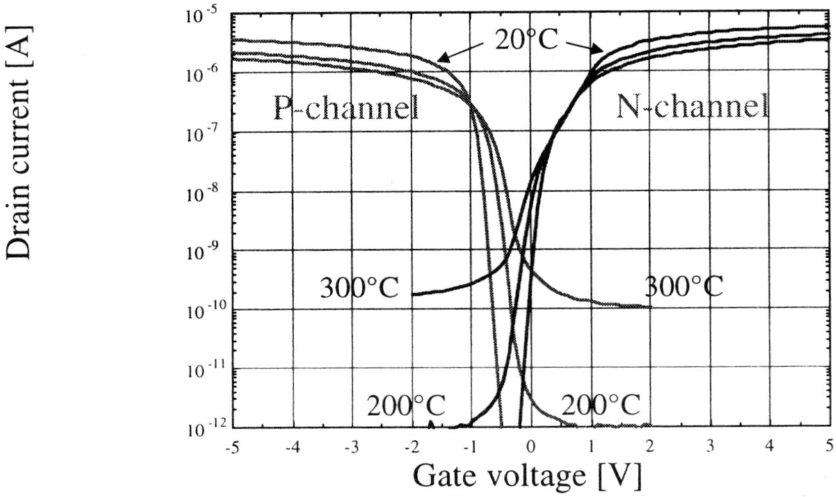

Figure 7: I_d versus V_g characteristics from room temperature to 300^oC, in an n-channel and a p-channel GAA MOSFET.

3.1. MOBILITY

The Gate-All-Around device exhibits the same kind of high temperature behavior as the fully-depleted SOI MOSFET. The mobility in a GAA device varies as that of an SOI transistor according to a standard dependence with temperature given by [15]:

$$\mu_o = \mu_{300\,K}\left[\frac{T}{T_o}\right]^{-m}$$

where μ_0 is the nominal mobility and T_0 is room temperature (300K). The coefficient m ranges from 1.6 to 1.8,[16] being governed by acoustic phonon scattering. The nominal mobility dependence versus temperature is illustrated in Fig.8 for both n-channel and p-channel GAA MOSFETs.

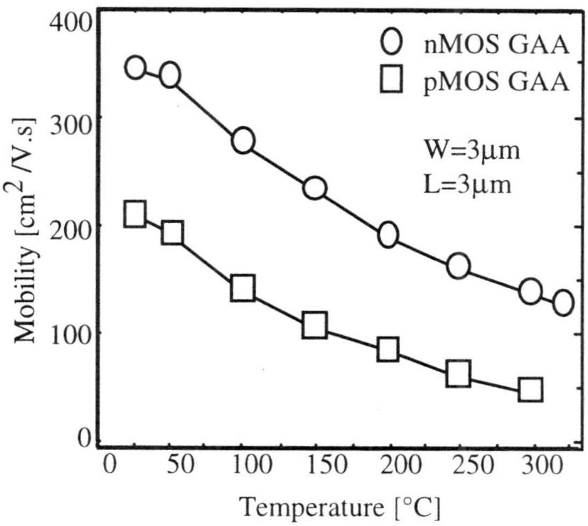

Figure 8: Nominal mobility evolution versus temperature in GAA n-channel and p-channel transistors [17].

3.2. THRESHOLD VOLTAGE

The threshold voltage variation with temperature for both nMOS and pMOS GAA transistors is limited to about 0.3V between room temperature and 350°C and is illustrated in Fig.9. When neglecting the presence of interface traps and volume inversion phenomenon as well as the variation of bandgap with temperature, the temperature dependence of V_T in GAA devices operating in full depletion is equal to the variation of the Fermi potential with temperature:

$$\frac{\partial V_T}{\partial T} = \frac{\delta \phi_F}{\delta T}$$

The threshold voltage dependence varies from –0.6 to –0.8mV/K in the temperature range between room temperature and 350°C for FD SOI/GAA devices. As long as they operate in full depletion, this variation is about 2 to 3 times smaller than in bulk transistors. As the temperature increases, the extension of the depletion region decreases. When the depletion region becomes smaller than half the silicon film thickness, the GAA device enters partial depletion operation. The threshold voltage evolution with temperature follows then the same temperature law as in bulk devices, dependent on the Fermi potential variation and the change in depletion charge with temperature, according to:

$$\frac{\delta V_T}{\delta T} = \frac{\delta \phi_F}{\delta T} + \frac{\delta}{\delta T}\left(\frac{\sqrt{4qN_a \varepsilon_{si} \phi_F}}{C_{ox}}\right)$$

The breaking point between full depletion and partial depletion operation occurs at a critical temperature, T_K, dependent on the doping concentration in the channel. For low doping concentrations, T_K is close to 250°C, while at higher doping concentrations ($10^{17}cm^{-3}$), T_K can reach the 400°C range for an 80 nm-film thickness. As the silicon film thickness decreases, GAA devices can operate in full depletion at even higher temperature, reaching 460°C for a silicon film thickness of 10 nm and a channel doping concentration of $10^{17}cm^{-3}$ [18].

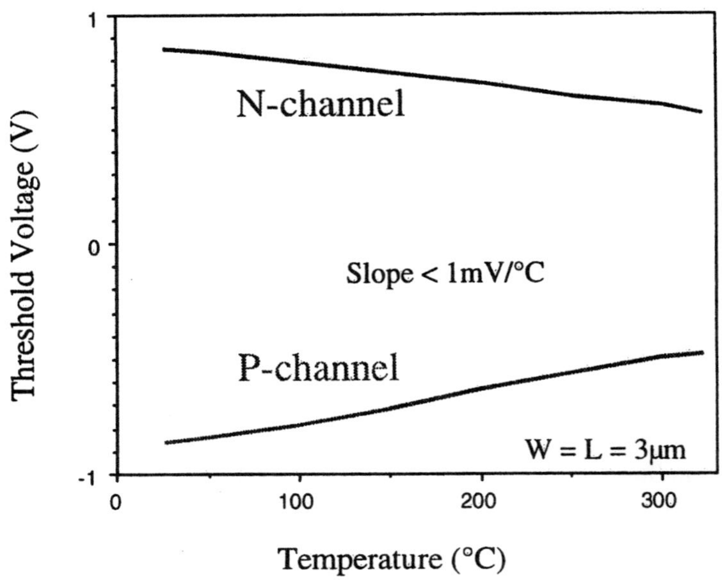

Figure 9: Threshold voltage variation with temperature in GAA n-channel and p-channel transistors.

In the case of accumulation-mode p-channel GAA transistors, two conduction components are present. The current conduction first occurs due to the onset of body conduction in the central quasi-neutral part of the film. Surface accumulation turn-on occurs later on, when the gate voltage decreases down to the flatband voltage. The turn on of accumulation-mode single-gate SOI devices has been studied by D. Flandre et al. [19] and a model of the threshold voltage sensitivity was derived in [20]. We have adapted this model to symmetrical structures and the threshold voltage component due to body conduction is given by:

$$V_{th,body} = V_{FB} + \frac{qNat_{si}^2}{8\varepsilon_{si}} + \frac{qNat_{si}}{2C_{ox}}$$

Both $V_{th,body}$ and V_{FB} exhibit the same temperature dependence. Due to the mechanism of volume accumulation, the definition of the body threshold voltage may be inadequate and may not fully reproduce the threshold voltage behavior with temperature in GAA devices.

3.3. SUBTHRESHOLD SLOPE

The subthreshold slope evolution with temperature is presented in Fig.10. Neglecting the presence of interface traps, the subthreshold slope increases linearly with temperature. The overlinear dependence observed in Fig.10 above 200°C, is an artifact due to the difficulty to extract the slope when the leakage currents become important.

3.4. LEAKAGE CURRENT

The junction leakage current is plotted versus temperature in Fig.11. GAA devices, as well as single-gate SOI devices, exhibit smaller junction leakage currents than their bulk counterpart. This advantage is conserved at high temperature as well. The reverse-biased PN junction leakage current is given by the following expression [21]:

$$I_{leak} = qA\sqrt{\frac{D_n}{\tau_n}\frac{n_i^2}{Na}} + qA\frac{n_i W}{\tau_e}$$

with D_n the electron diffusivity, τ_n the electron recombination lifetime and τ_e the effective generation lifetime. The first term is a diffusion component in the neutral region and is proportional to n_i^2. The second term is a generation component in the depletion region and is dependent on n_i. In FD SOI, the quasi-neutral region is totally suppressed. As a result, at low temperature, the generation current dominates and the leakage current varies with temperature as n_i (I_{leak} varies as $\exp(-E_g/2kT)$). Above a certain temperature, the diffusion component becomes preponderant and the leakage current increases as n_i^2 (I_{leak} varies as $\exp(-Eg/kT)$). This change in mechanism can be explained by the fact that the diffusion component increases more rapidly and the devices do not remain fully depleted above a certain temperature.

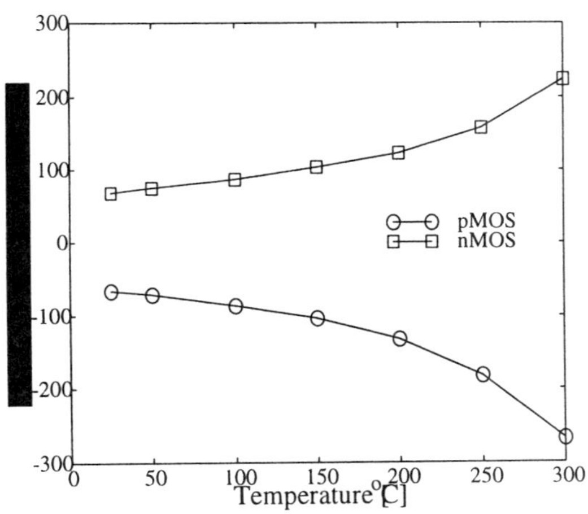

Figure 10: Subthreshold slope variation with temperature in GAA n-channel and p-channel MOSFETs.

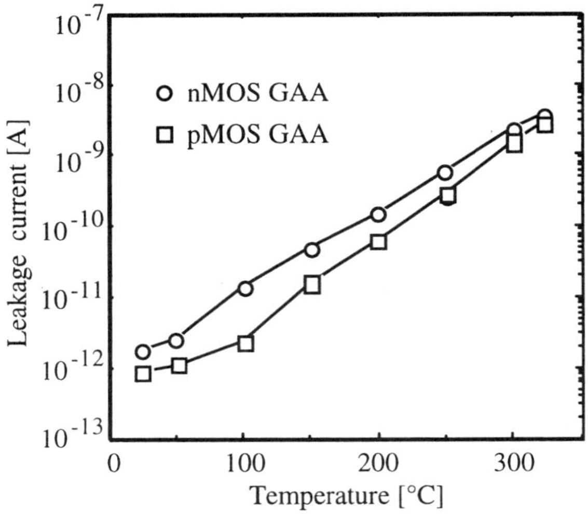

Figure 11: Junction leakage current variation with temperature in GAA n-channel MOSFETs.

3.5. SCALED TRANSCONDUCTANCE

Figure 12 presents the scaled transconductance, g_m/I_d, as a function of the scaled drain current at different temperatures in nMOS GAA transistors. The subthreshold slope factor degradation with temperature explains the lowering of the maximum plateau level. The junction leakage current increase induces the g_m/I_d degradation at low current values (at the left of the plateau), whereas the mobility reduction lowers the g_m/I_d values with temperature in the strong inversion region.

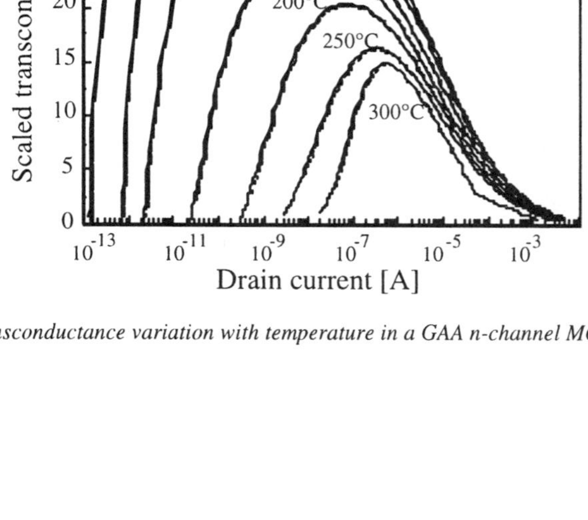

Figure 12: Scaled transconductance variation with temperature in a GAA n-channel MOSFET.

3.6. EARLY VOLTAGE

The Early voltage evolution (Fig.13) is difficult to interpret because is neither stable nor identical for pMOS and nMOS. The nMOS transistors show a steady decrease of V_{ea} with temperature while pMOS transistors present more fluctuations but remains approximately constant. The fluctuation of the Early voltage is most likely due to limitations in the extraction method used. Indeed, the Early voltage is extracted by taking the first derivative of the I_{ds} versus V_{ds} curve. The variation of the drain current in the saturation region with drain voltage is very small and is even smaller in GAA transistors where the control of the gate on the active area is better. As a result, a numeric noise (inaccuracy in the current measurement due to limited precision of the equipment) appears. This renders the extraction of the Early voltage difficult. One way to improve the accuracy of the method is to increase the drain voltage step.

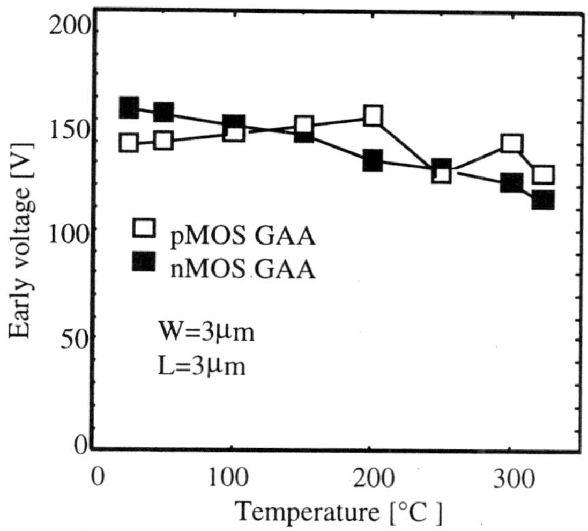

Figure 13: Early voltage variation with temperature in GAA n-channel and p-channel MOSFETs.

4. Dose hardness

^{60}Co gamma irradiation was performed on GAA transistors with a 30 nm gate oxide thickness. Both 3μm x 3μm n- and p-channel transistors were irradiated with source and drain grounded and with the gate voltage at 0V, 1.5V or 3V. Linear measurements on n- and p-channel devices biased with the gate voltage at 3V during irradiation are presented in Figures 14 and 15, respectively.

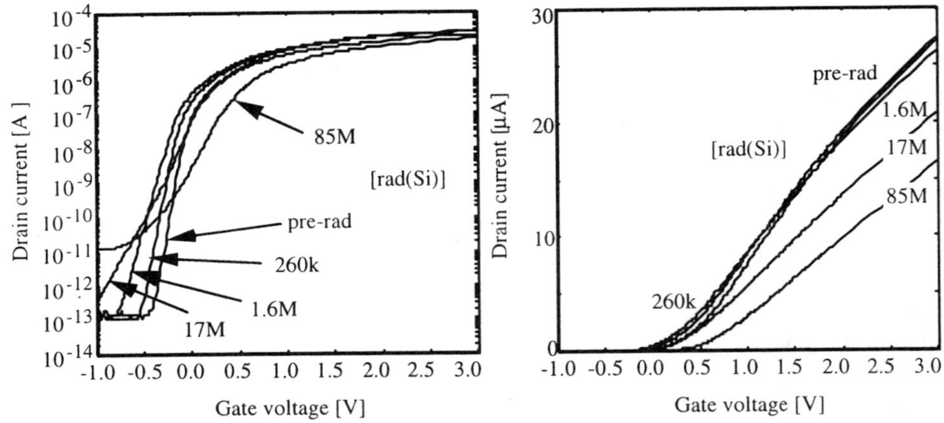

Figure 14: Drain current vs. gate voltage curves of a 3mm x 3mm nMOS/GAA device exposed to different irradiation doses. $V_S = V_D = 0V$ and $V_G = 3V$ during irradiation.

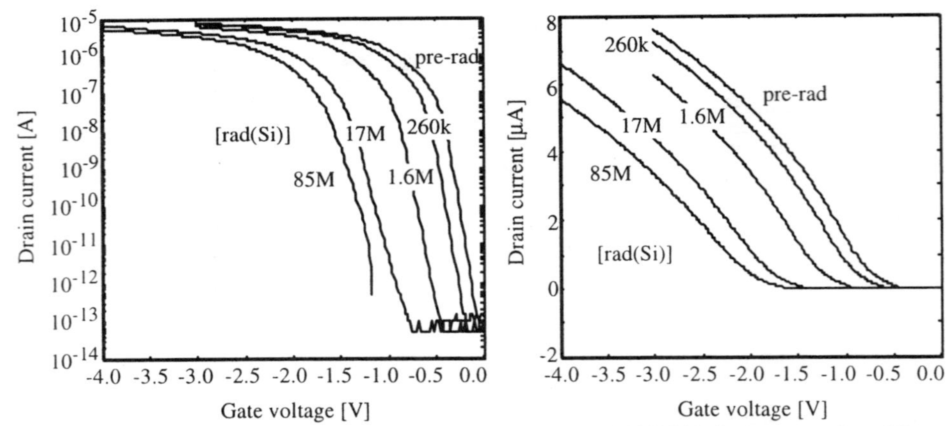

Figure 15: Drain current vs. gate voltage curves of a 3µm x 3µm pMOS/GAA device exposed to different irradiation doses. $V_S = V_D = 0V$ and $V_G = 3V$ during irradiation.

I_D-V_D characteristics were also recorded using V_D and V_G steps of 0.1V and 0.25V, respectively. Some of the results are shown in Figure 16 and indicate that both mobility and output conductance decrease under irradiation. The rebound of n-channel device threshold voltage can also be observed.

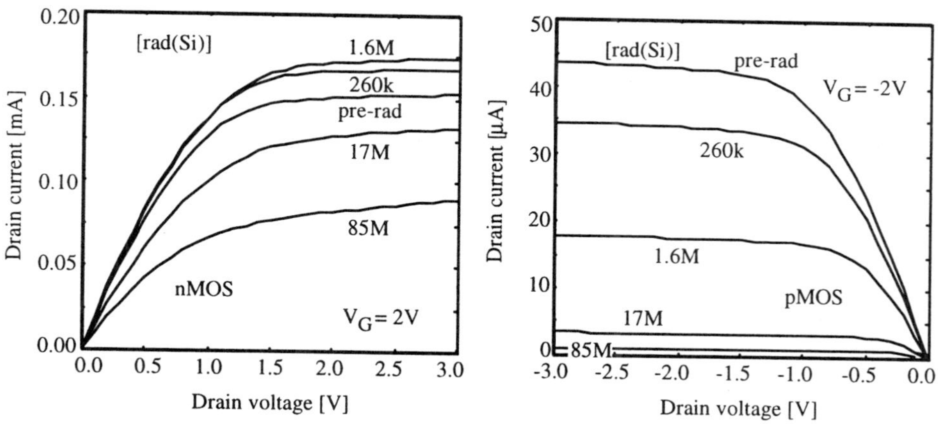

Figure 16: Drain current vs. drain voltage curves of 3•m x•3•m n- and pMOS/GAA devices exposed to different irradiation doses. $V_S = V_D = 0V$ and $V_G = 3V$ during irradiation, $V_G = \pm 2V$ during parameter extraction.

4.1. THRESHOLD VOLTAGE

Figure 17 presents the threshold voltage (V_{thn} and V_{thp}) in n- and p-channel GAA transistors as a function of dose when the gate is maintained at 0V and 3V during irradiation with all other terminals grounded (static bias). Figure 17 also presents results with devices mounted in an inverter configuration during irradiation, the common gate being continuously switched between 0V and 3V at the frequency of 200Hz. The maximum amplitude of ΔV_{thn} is only 0.3V up 85Mrad(Si) irradiation. P-channel devices with $V_G = 3V$ and $V_{DS} = 0V$ during irradiation suffer from a more important threshold voltage shift that reaches -1.2V at the highest dose. But they are rarely biased in this way when embedded in digital as well as analog circuits. Figure 17 shows that with $V_{GS} = V_{GD} = 0V$, ΔV_{thp} is much less dramatic and only reaches -0.6V at 85Mrad(Si) irradiation. Furthermore, threshold voltage shifts are also reduced with the gate continuously switched in the inverter configuration.

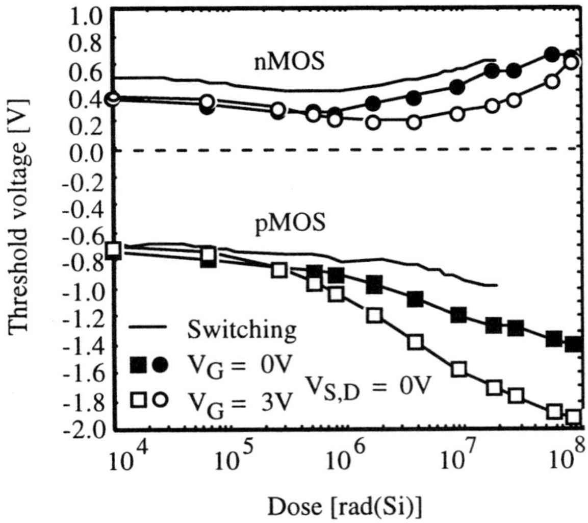

Figure 17: Threshold voltage as a function of dose in 3μm x 3μm n- and p-channel GAA devices irradiated with V_G = 0V or 3V and source/drain grounded, or placed in the inverter configuration with the gate continuously switched between 0V and 3V.

4.2. MOBILITY

In linear operation, the mobility μ is defined as the extrapolation to $V_G = V_{th}$ of the best linear fit to the $g_m(V_g)$ curve above threshold, divided by $2C_{ox}V_{DS}W/L$. In saturation regime, $\mu = 2m^2L/(C_{ox}2W)$ with *m* being the slope of the linear least-square fit to the $\sqrt{I_D}$ -V_G curve. The device width is doubled owing to conduction at the front and back interfaces. Figure 18 shows the mobility (left graph) and the mobility degradation μ/μ_o (right graph) extracted from saturation measurements as a function of dose with the gate bias as parameter. The mobility degrades upon irradiation due to the creation of interface states. The mobility obtained from measurements in linear operation is about 35% larger than from saturation measurements but the degradation with dose is also slightly larger such that less difference between the two definitions is observed at high dose (only 10%).

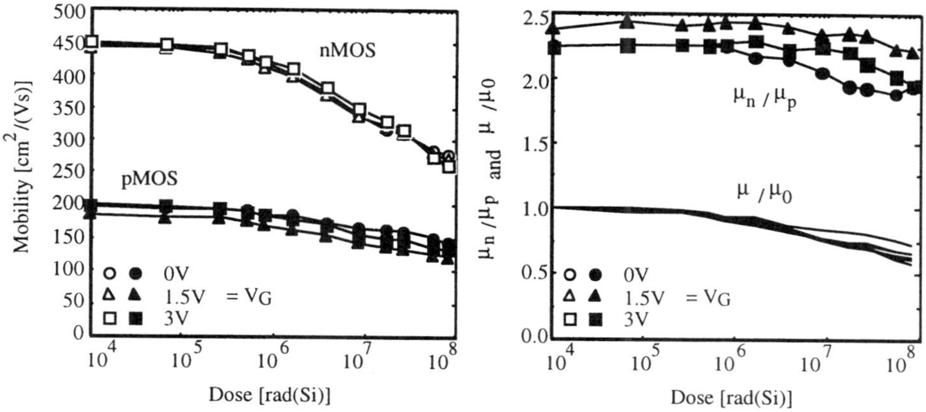

Figure 18: Mobility (left) and mobility degradation (right) as a function of dose with gate bias during irradiation as parameter in 3μm x 3μm n- and p- channel GAA devices.

4.3. SUBTHRESHOLD SLOPE

The evolution of the subthreshold slope in GAA devices is presented as a function of total-dose in Figure 19. The initial value of S is close to the theoretical limit for both n- and p-channel devices (62mV/dec) owing to the strong control of the surrounding gate on the body potential. It is well-known that S drops under irradiation due to the formation of interface states. The degradation after 85 Mrad(Si) irradiation is larger in n-type transistors (S = 210mV/dec) than in p-channel devices (S = 110mV/dec) because of the creation of larger interface trap densities in N-channel devices. S clearly depends on the gate bias in n-channel devices. The irradiation produces less degradation with the gate grounded than under high electric field (S = 217mV/dec with V_G = 1.5V and S = 189mV/dec with V_G = 0V after 85Mrad(Si) irradiation). In pMOS devices, on the other hand, the influence of the gate electric field is remarkably small.

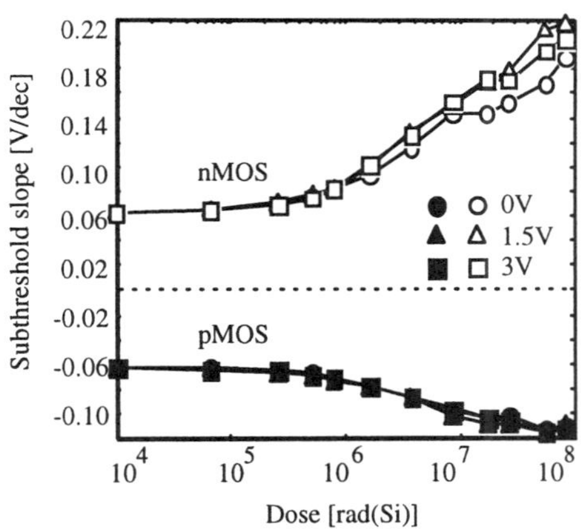

Figure 19: Subthreshold slope as a function of dose with gate bias during irradiation as parameter in 3μm x 3μm n- and p- channel GAA devices.

4.4. OUTPUT CONDUCTANCE

The output conductance g_d is one of the key parameters in analog circuit design. Until recently no studies have been carried out concerning the evolution and the origin of the degradation of this specific parameter with total dose. The predominant effect of the output conductance g_d on analog circuits such as operational transconductance amplifiers has already been analyzed [22]. The g_d degradation is most likely caused by radiation induced interface states built-up at the Si-SiO_2 interface along the channel. This degradation is explained by the combined effect of the variation of the surface potential at pinch-off with drain to source voltage V_{ds} and the increase of the interface state density along the channel. The proposed model makes use of the EKV model for the g_d computation based on the pinch-off voltage parameter V_p and its dependence with V_{ds} and the increase of interface states.[23]

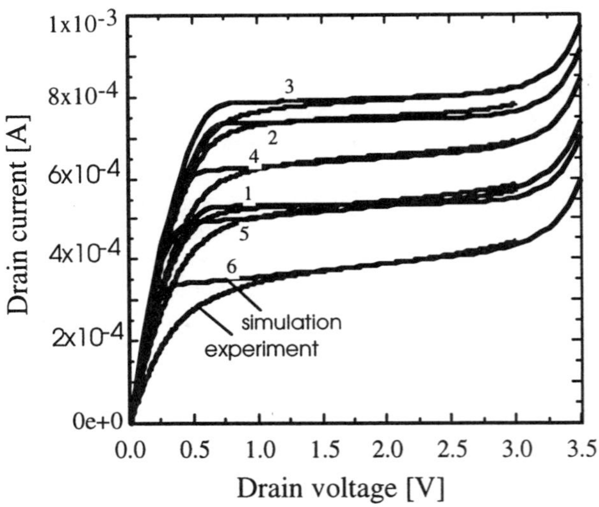

Figure 20: Experimental and simulated output characteristics with increasing dose (from '1'=pre-rad to '6'=14.6Mrad(Si)) (Vgs=2V).

Irradiation tests were performed on GAA devices using a ^{60}Co source with a dose rate of 9 krad(Si)/h, progressively increased up to 24 krad(Si)/h. The maximum total dose reached is 24Mrad(Si). The devices were kept biased during irradiation at Vg=0V or Vg=3V, Vs=Vd=0V. The output characteristics $I_d(V_{ds})$ and the output conductance are represented as a function of dose in Figs 20 and 21 for Vg=1V. Beside a clear variation of the DC drain current due to threshold voltage variation, a clear degradation of the slope can also be observed. Both of these effects contribute to the variation of the output conductance related by the general expression of $g_d=I_d/V_{ea}$. The extracted Early voltage, V_{ea}, shows a degradation by a factor 10 after a total dose up to 24Mrad(Si).

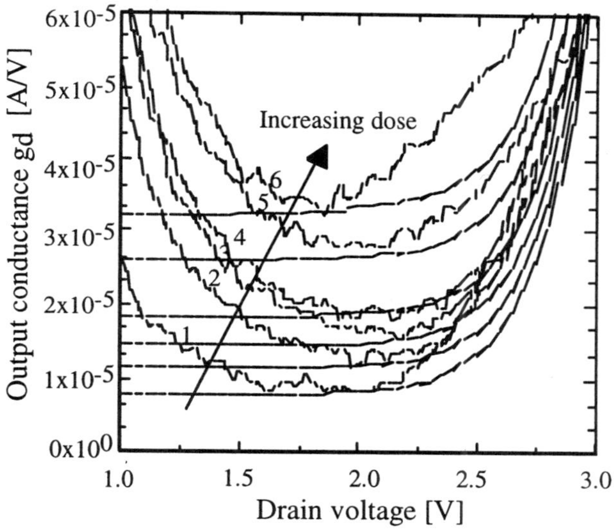

Figure 21: Experimental and simulated evolution of the output conductance as a function of drain voltage with increasing dose (from '1'=pre-rad to '6'=14.6Mrad(Si) (Vgs=2V).

The dependence of the surface potential at pinch-off with V_{ds} has been extracted from SILVACO simulations and presents a linear behavior almost independent of the gate voltage (Fig. 22). The average extracted value of the slope α is equal to 0.021. As V_{ds} is increased, the surface potential at the location of the pinch-off decreases causing more interface traps to be filled with electrons. The quantity of electrons trapped at the interface is also proportional to the effective density of interface traps which increases with total dose. Hence, the variation of V_{ds} allows more interface energy states to be filled at the location of the pinch-off while the dose increase the density of interface states. This double dependence was introduced in the EKV (Enz-Krummenacher-Vittoz) model [[24]] through an additional term in the definition of the pinch-off voltage V_P:

$$V_P = V_g - V_t + \frac{qN_{it}\alpha V_{ds}}{C_{ox}}$$

Using the above simple model for V_p and accounting for the threshold voltage variation and the increase of interface trap density with dose, the output conductance g_d profile was simulated and is represented in Figs 20 and 21.

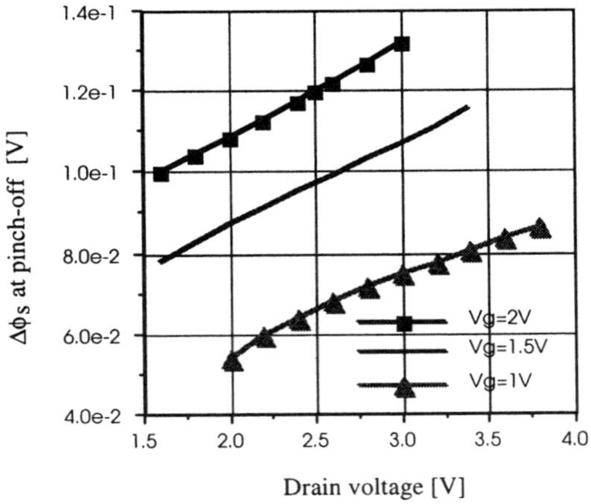

Figure 22: Variation of the surface potential at pinch-off versus drain voltage for different values of gate voltage.

Both the subthreshold slope and the output conductance show a quasi-linear dependence on interface trap density. As a result, there should be a linear relationship between the inverse of the Early voltage ($1/V_{ea}$) and the subthreshold slope, as the device is exposed to different radiation doses. This behavior can readily be observed in Figure 23.

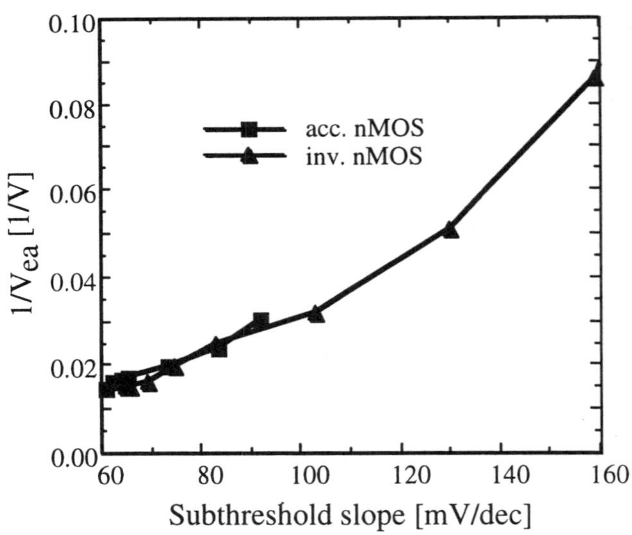

Figure 23: Evolution of the inverse of the Early voltage versus subthreshold slope in accumulation-mode and inversion-mode GAA nMOSFETs.

5. Proton irradiation

GAA devices were tested with high-energy proton (189MeV) irradiation up to a total dose of 300krad(Si). The threshold voltage and subthreshold slope behavior versus total dose is illustrated in Figs. 24 and 25, for both proton irradiation and gamma rays under the same bias condition during exposure, respectively. A very tight correlation exists between both experiments. This is not surprising given that both ^{60}Co and high-energy protons follow the same geminate recombination process with identical yield, close to unity. The early gate oxide breakdown reported for the devices may be due to single-event gate rupture phenomena.

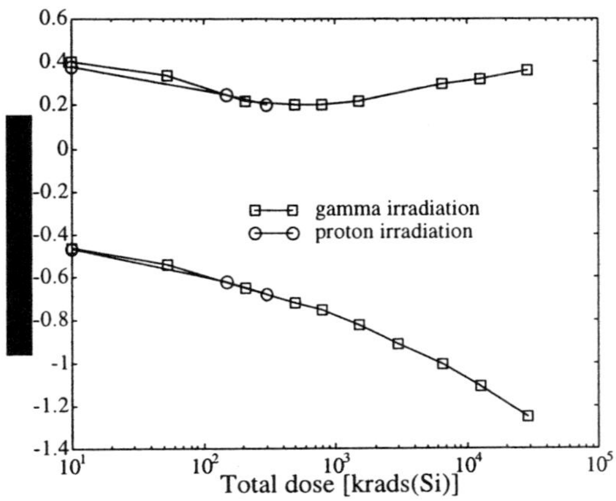

Figure 24: Threshold voltage versus total dose in nMOS and pMOS GAA devices with high-energy protons and gamma irradiation.

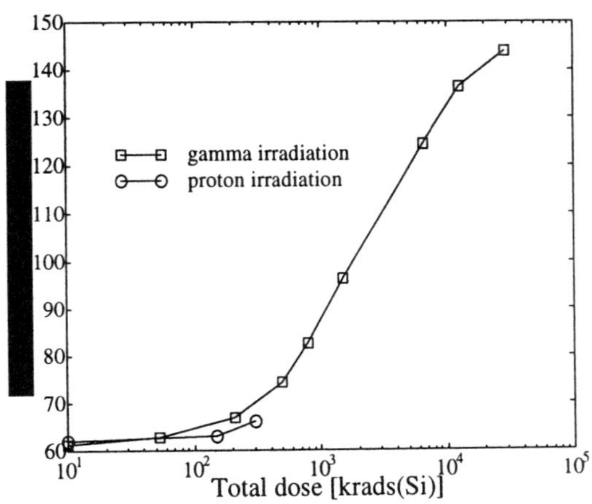

Figure 25: Subthreshold slope versus total dose in nMOS GAA devices with high-energy protons and gamma irradiation.

Publications about proton induced radiation damage on MOS transistors are rare. The authors in [25], however, illustrate the shift of threshold voltage in CMOS bulk technology after pulses of proton, which show good agreement with the results reported here.[26]

6. Single-event upset

GAA devices also offer excellent Single-Event Upset (SEU) radiation hardness. Classical SEU hardening techniques involve either adding capacitors to the SRAM cells (to absorb current pulses) or using Silicon-on-Insulator (SOI) technology. Indeed, SOI transistors are made in a thin silicon film, and heavy ions generate less electrical charges in a thin-film device than in a bulk transistor. Figure 26 present the SEU hardness of a 1-kbit GAA (Gate-all-Around) SRAM and that of other (bulk and SOI) SRAM chips.

The devices were irradiated with 459 MeV-Xenon ions using a cyclotron. The values of the LET were varied by tilting the samples in order to increase the length of the ion track into the devices. Not a single bit flip were observed when the devices were operated with a standard supply voltage of 3 V. The supply voltage had to be reduced below 2.5 V to reduce the noise margin of the transistors in order to observe some upsets. It was necessary to further reduce the supply voltage to 1.9 V to observe cross sections comparable to data found in the literature.

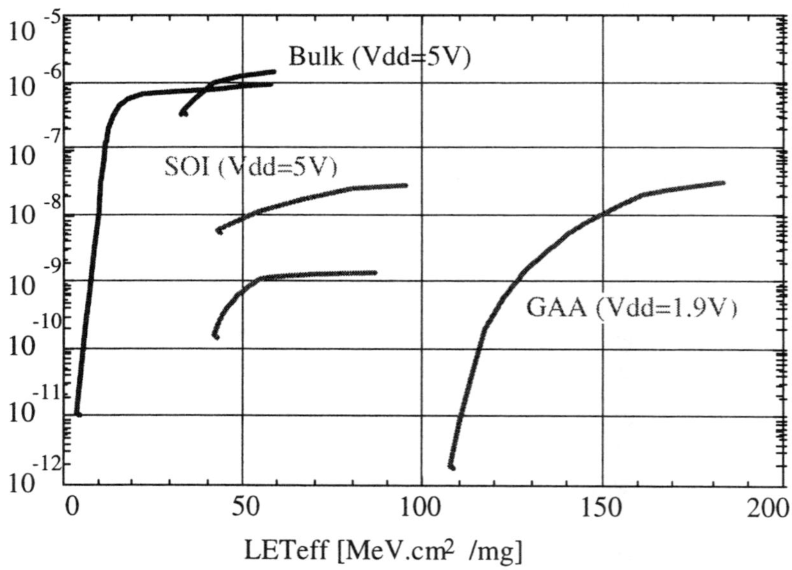

Figure 26: Upset cross section vs. LET. Curves A and B represent bulk SRAMs with V_{DD}=5V, curves C and D represent SOI SRAMs for V_{DD}=5V, curve E represents a GAA SRAM for V_{DD}=1.9V. [27]

Conclusion

Gate-All-Around (GAA) transistors are thin, fully depleted SOI MOSFETs with a double gate structure. When used at high temperature GAA devices present low leakage current, minimal threshold voltage shift and, in general, better characteristics than bulk or even SOI MOSFETs. The radiation hardness of GAA devices is reported as well, and the dose evolution of parameters such as threshold voltage, subthreshold slope and output conductance is analyzed.

Acknowledgment

The author would like to thank Drs. X. Baie, P. Francis and A. Vandooren for the invaluable research work they carried out on GAA devices.

References

1 H.S.P. Wong, D.J. Frank, P.M. Salomon, "Device design considerations for double-gate, ground-plane, and single-gated ultra-thin SOI MOSFETs at the 25 nm channel length generation", *Tech. Digest of IEDM*, p. 98, 1998

2 H. Takato, K. Sunouchi, N. Okabe, A. Nitayama, K. Hieda, F. Horiguchi, F. Masuoka, "High performance CMOS Surrounding Gate Transistor (SGT) for ultra high density LSIs", *Tech. Digest IEDM*, pp. 222-225, San Francisco, 1988

3 D. Hisamoto, T. Kaga, Y. Kawamoto, E. Takeda, "A fully depleted Lean-channel Transistor (DELTA) - A novel vertical ultrathin SOI MOSFET", *Tech. Digest of IEDM*, pp. 833-836, Washington, 1989.

4 H.S.P. Wong, K.K. Chan, Y. Taur, "Self-aligned (top and bottom) double-gate MOSFET with a 25 nm thick silicon channel", *Tech. Digest IEDM*, p. 427, 1997

5 J.P. Denton and G.W. Neudeck, "Fully depleted dual-gated thin-film SOI P-MOSFETs fabricated in SOI islands with an isolated buried polysilicon backgate", IEEE Electron Device Letters, Vol. 17-11, p.509, 1996

6 J.P. Colinge, M.H. Gao, A. Romano-Rodriguez, H. Maes, C. Claeys, "Silicon-On-Insulator Gate-All-Around Device"," *Tech. Digest IEDM*, pp. 595-598, 1990

7 F. Balestra, S. Cristoloveanu, M. Benachir, J. Brini, T. Elewa, "Double-Gate Silicon-on-Insulator Transistor with volume inversion: A new device with greatly enhanced performance", *IEEE Electron Device Letters*, vol.8, no.9, pp. 410-412, September 1987

8 X. Baie, J.P. Colinge, "Two-dimensional confinement effects in gate-all-around (GAA) MOSFETs", Solid-State Electronics, Vol. 42, No. 4, pp. 499-504, 1998

9 B. Majkusiak, T. Janik, J. Walczak, "Semiconductor Thickness Effects in Double-Gate SOI MOSFET", *IEEE Trans. on Electron Dev.*, vol. 45, no. 5, May 1998

10 M. Haond, O. Le Neel, G. Mascarin, J.P. Gonchond, "Gate oxide breakdown in an SOI CMOS process using MESA isolation", *Proc. ESSDERC*, Springer-Verlag, pp.893-896, Berlin, 1989

11 T. Mukaiyama, K.I Saito, H. Ishikuro, M. Takamiya, T.Araya and T. Hiramoto, " Fabrication of Gate-All-Around MOSFET by silicon anisotropic etching technique", *Solid-State Electronics*, vol.42, n.7, pp.1623-1626, 1998.

12 J.P. Colinge, "*Silicon-On-Insulator Technology: Materials to VLSI*", 2nd edition, 1997

[13] P. Francis, A. Terao, B. Gentinne, D. Flandre, J.P. Colinge, "SOI technology for high-temperature applications", *Tech. Digest IEDM*, p. 353, 1992

[14] D.S. Jeon, D.E. Burk, "A temperature-dependent SOI MOSFET model for high-temperature application (27oC-300oC)", IEEE Trans. on Electron Devices, vol. 38, no. 9, September 1991

[15] C. Lombardi, S. Manzini, A. Saporito, M. Vanzi, "A physically based mobility model for numerical simulations of nonplanar devices", *IEEE Trans.* CAD 7(11), p. 1164, 1988

[16] D.S. Jeon, D.E. Burk, "A temperature-dependent SOI MOSFET model for high-temperature application (27oC-300oC)", IEEE Trans. on Electron Devices, vol. 38, no. 9, September 1991

[17] P. Francis, A. Terao, B. Gentinne, D. Flandre, J.P. Colinge, "SOI technology for high-temperature applications", *Tech. Digest of the IEDM*, p. 353, 1992

[18] P. Francis, "Double-gate SOI/MOS devices and circuits in hostile environments", thesis UCL, 1996.

[19] D. Flandre, A. Terao, "Extended theoretical analysis of the steady-state linear behavior of accumulation-mode, long-channel p-MOSFETs on SOI substrates", *Solid-State Electronics*, vol. 35, no. 8, pp. 1085-1092, 1992

[20] D. Flandre, A. Terao, P. Francis, B. Gentinne, J.P. Colinge, "Demonstration of the potential of accumulation-mode MOS transistors on SOI substrates for high-temperature operation (150-300C)", *IEEE Electron Device Letters*, vol. 14, no. 1, pp. 10-12, 1993

[21] S.M. Sze, *Physics of semiconductor Devices*, 2nd ed., New York, Wiley, p. 91, 1981

[22] A. Vandooren, P. Francis, D. Flandre, J.P. Colinge, "Performance of gamma-irradiated gate-all-around SOI MOS OTA amplifiers", Proc. IEEE International SOI Conference, pp. 62-63, 1997

[23] "Drain conductance degradation in gate-all-around inversion-mode nMOSFETs with total dose", A. Vandooren and J.P. Colinge, in "Silicon-on-Insulator Technology and Devices IX", Ed. by. P.L. Hemment, Electrochemical Society Proceedings Volume 99.3, pp. 323-328, 1999

[24] C.C. Enz, in *Low-power HF microelectronics: a unified approach*, edited by G.A.S. Machado, IEE circuits and systems series 8, the Institution of Electrical Engineers, p. 247, 1996

[25] H.J. Ziock et al., "Measurement of proton induced radiation damage to CMOS transistors and PIN diodes", IEEE Trans. Nucl. Sci., vol. 37, no. 3, June, 1990

[26] A. Vandooren, "Advanced characterization of double-gate/gate-all-around SOI devices and circuits", Doctoral Dissertation, U.C. Davis, Sept. 2000

[27] P. Francis, J.P. Colinge, G. Berger, "Temporal analysis of SEU in SOI/GAA SRAMs", IEEE Transaction on Nuclear Science, Vol. 42-6, pp. 2127-2137, 1995

LOW-NOISE HIGH-TEMPERATURE SOI ANALOG CIRCUITS

V. DESSARD, B. INIGUEZ, S. ADRIAENSEN AND D. FLANDRE
Université catholique de Louvain
Place du Levant, 3
1348 Louvain-la-Neuve, Belgium

Abstract

This paper deals with SOI n-MOS low frequency noise measurements, analysis and modeling from room temperature up to 250°C. Based on this, general considerations about low-noise high-temperature analog circuits are discussed. In the second part of the paper, we introduce a new low-noise high-temperature fully-differential instrumentation preamplifier implemented on a thin-film fully-depleted SOI process.

1. Introduction

It is well known that use of SOI technology presents several advantages over bulk-Si technologies concerning parasitic capacitances, power consumption, leakage current, latch-up, radiation hardness, compactness [1]. These advantages give outstanding performances for very low power or high frequency circuits [2] [3]. Another niche application for SOI is the high temperature market thanks to reduced leakage currents [4] [5]. In such applications, a typical measurement chain would consist of a sensor, amplifier, filter and analog to digital converter. Most sensors are based on a resistor variation depending on the physical measured parameter. This resistor value is usually converted to a differential voltage by way of a Wheatstone bridge. This small voltage is then amplified and filtered before its conversion to digital world. To achieve a high signal to noise ratio, the amplifier noise has to be as small as possible compared to the bridge noise. Section 2 presents low-frequency noise measurements, analysis and modeling up to *250°C* on n-channel SOI MOS devices. It is followed by a short section presenting an original low-noise fully-differential preamplifier for high-temperature applications.

2. Low-frequency noise study on n-MOS SOI

The aim of this section is to understand, extract and model the n-MOS SOI low-frequency (LF) noise component over a large range of bias and temperature conditions. The study of noise is important because it represents a lower limit to the value of the electrical signal that can be amplified by a circuit without significant deterioration in signal quality. The main objective of such analysis is to give general guidelines on using SOI n-MOSFETs in analog circuits where noise level may be critical. A lot of research works have dealt with the low frequency noise in MOS devices. The originality of this section is relative to the noise measurement and study at high temperature. As will be seen, most usual low-frequency noise considerations remain valid at high

F. Balestra et al. (eds.), Progress in SOI Structures and Devices Operating at Extreme Conditions, 189–209.

temperature while some usually neglected noise components can appear to be dominant in some special conditions.

Section 2.1 gives an overview of the main low frequency noise components. Section 2.2 presents the measurement setup and a summary of most significant obtained measurements. Section 2.3 analyses measurements, derives a noise model and concludes on some analog circuits considerations.

2.1 MAIN COMPONENTS OF MOSFET LOW-FREQUENCY NOISE

2.1.1 GENERALITIES

The electrical noise that will be considered hereafter originates from small current and voltage fluctuations generated within the devices themselves. External noise sources (crosstalk, glitches, ...) are not considered here. The existence of noise is basically due to the fact that electrical charge is carried in discrete amounts equal to the electron charge [6, Chapter 11]. There are several ways to represent an electrical noise. For low-frequency (LF) applications, the noise is usually described by its power spectral density. Voltage (current) noise unit is therefore V^2/Hz (resp. A^2/Hz). These units will be used further in this chapter. In order to obtain the actual root mean square (RMS) voltage or current noise level of a circuit, the power spectral density has to be integrated over the circuit bandwidth. The square root of this integration gives the RMS noise value that would be observed on an oscilloscope for example. In some cases, the noise can be expressed by the noise figure (F) or the noise temperature (T_n). This is mainly the case for telecom applications. These units will not be considered in our work.

2.1.2 1/f NOISE (FLICKER NOISE)

The 1/f source of noise represents one of the main limitations of MOSFETs when compared to bipolar devices. The theoretical basis of this noise component characterized by a 1/f-like spectrum, is still a matter of long-lasting debate. In the McWhorter model, the 1/f noise is primarily generated by fluctuation in the number of carriers and is therefore also called the Δn model [7]. The Hooge theory, on the other hands, assumes that fluctuations in mobility cause the LF noise (Δμ model) [8]. In MOSFETs, the 1/f noise magnitude is usually related to the interface quality (i.e. number of traps) between the silicon film and the gate oxide [9]. In strong inversion regime, the drain current 1/f noise spectral density can be expressed as [10] [11] [12]

$$S_{I_{d,1/f}} = \frac{K_F . gm^2}{C_{ox}^2 . W . L . f} \tag{1}$$

where C_{ox} is the front gate capacitance per unit area, W and L are the device width and length respectively, gm is the front channel transconductance and K_F [C^2/m^2] is a characteristic constant for a given process. Relation 1 has been validated on all our measured SOI enhancement mode n-MOSFETs. This means that $\frac{S_{I_{d,1/f}}}{gm^2}$ is basically independent of applied bias voltages and temperature. This confirms the fact that the *1/f* noise originates from the fluctuation of carrier concentration in such devices. Typical

K_F value for a bulk n-MOSFET device is about 5.10^{-9} $fC^2/\mu m^2$ (this value is about 10 times lower for p-MOSFETs). It is interesting to note that this noise is inversely proportional to the MOSFET active area *(W.L)*.

2.1.3 WHITE (THERMAL) NOISE

The white noise component originates from the thermal random motion of carriers in the channel. For a resistor with a value *R*, the white noise is given by

$$S_{I_{White}} = \frac{4.k.T}{R} \tag{2}$$

where *k* is the Boltzman constant and *T* the temperature expressed in degree Kelvin. In the case of a long channel MOSFET device, the white noise associated with the channel current is given by

$$S_{I_{d,White}} = 4.k.T.gm.\gamma \tag{3}$$

where γ slightly depends on the biasing conditions. In linear mode, γ is close to one whatever the inversion level, while γ tends to 2/3 in strong inversion (s.i.) [6] and to 1/2 in weak inversion (w.i) [13] in saturation mode.

2.1.4 LORENTZIAN-LIKE NOISE

Depending on the MOS technology and on bias conditions, some MOSFET present low-frequency excess noise called Lorentzian-like noise (or "burst noise" [6]). Such noise is characterized by a plateau value in the power spectral density, followed by a $1/f^2$ decrease above a characteristic frequency f_c. Such spectrum can have different origins. Usually, it is associated with generation-recombination (GR) or trapping phenomena in the silicon volume or surface [14]. It has also been demonstrated that traps distributed uniformly in the oxide from the interface to a certain distance can create 1/f noise spectrum (i.e. summation of Lorentzian with different characteristic frequency and plateau value leads to a 1/f noise spectrum) [15]. Voltage Lorentzian like noise spectrum is also obtained when a white noise current is filtered across a low pass R-C circuit [16] [17].

2.1.5 SHOT NOISE

The shot noise originates from the discrete flow of carriers across a junction. The spectral power density of this current noise is given by

$$S_{I_{diode}} = 2.q.I_D \tag{4}$$

where *q* represents the electron charge and I_D is the current flowing across the junction. This noise component is obviously present in diodes and bipolar devices, but also in MOSFET due to junctions associated with source and drain areas.

2.2 MEASUREMENTS

2.2.1 EXPERIMENTAL DATA

The following section exhibits noise measurements performed on devices from various SOI process. The first part is relative to our FD-SOI UCL process (*t_{ox_f} =30nm, t_{si}=80nm, t_{ox_b}=390nm).* The second part is relative to LETI FD & Nearly-Fully Depleted (NFD) SOI processes *(t_{ox_f} =4.5nm, t_{si}=30 and 40nm, t_{ox_b}=380nm).* The third part is relative to the X-fab PD-SOI process (*t_{si}=260nm, t_{ox_b}=1μm*). These data will be interpreted in section 2.3.2 and will validate the physical model of the Lorentzian like noise of section 2.3.1. A typical observed input referred noise spectrum is represented in figure 1. In some cases, the Lorentzian noise component can be hidden by the 1/f noise component. For the UCL FD-SOI process, measurements bias correspond to devices working at their zero temperature coefficient (ZTC) [18].

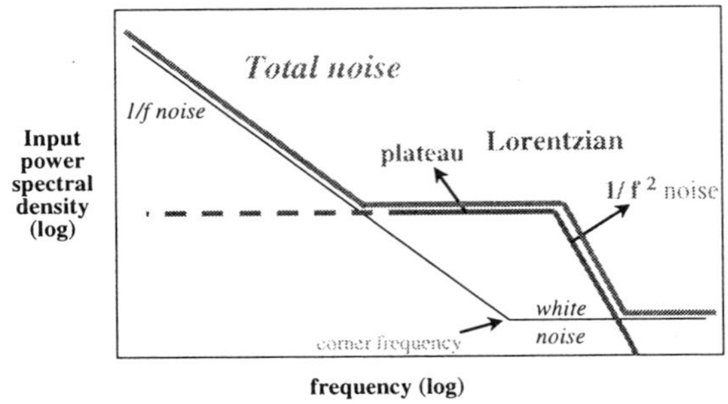

Figure 1: Typical input refered low-frequency noise voltage spectrum.

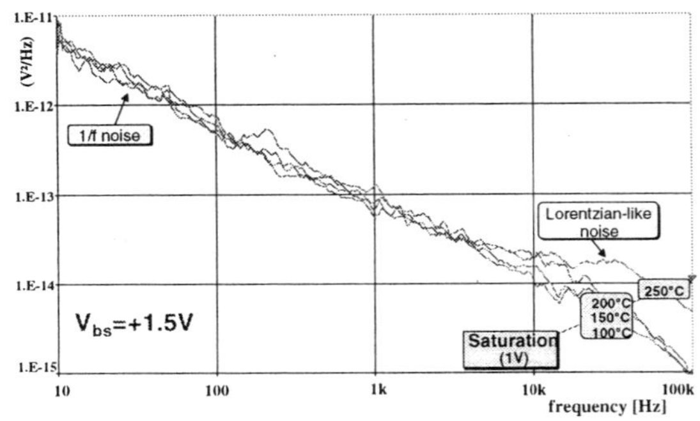

Figure 2: Input referred noise power spectral density. UCL process. W/L=20/2. n-MOS, V_{ds}=1V, V_{bs}=1.5V, V_{gs}~ZTC

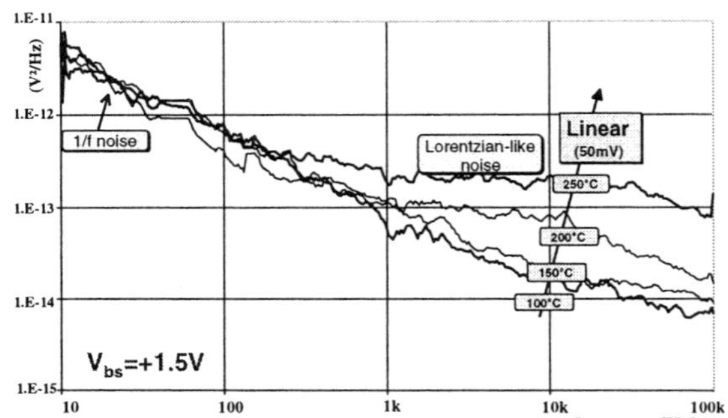

Figure 3: Input referred noise power spectral density. UCL process. W/L=20/2. n-MOS, V_{ds}=50mV, V_{bs}=1.5V, V_{gs}~ZTC

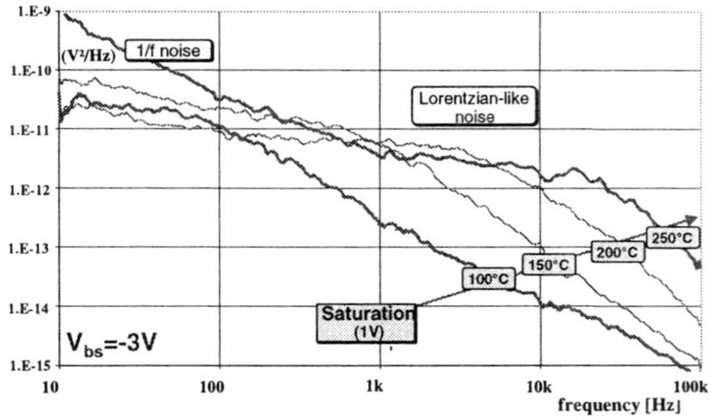

Figure 4: Input referred noise power spectral density. UCL process. W/L=20/2. n-MOS, V_{ds}=1V, V_{bs}=-3V, V_{gs}~ZTC

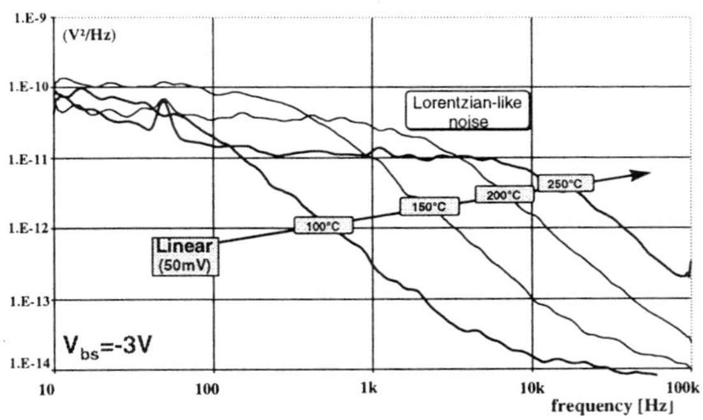

Figure 5: Input referred noise power spectral density. UCL process. W/L=20/2. n-MOS, V_{ds}=50mV, V_{bs}=-3V, V_{gs}~ZTC

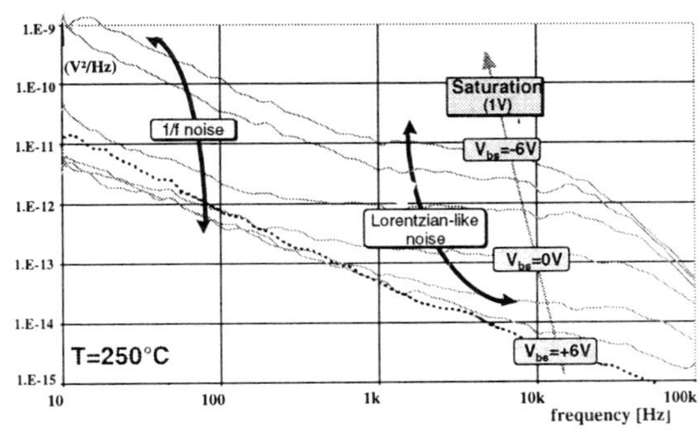

Figure 6: Input referred noise power spectral density. UCL process. W/L=20/2. n-MOS, T=250°C, V_{ds}=1V, V_{gs}~ZTC

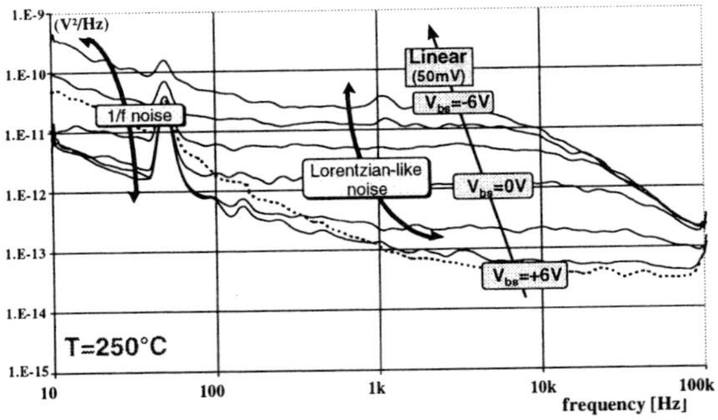

Figure 7: Input referred noise power spectral density. UCL process. W/L=20/2. n-MOS, T=250°C, V_{ds}=50mV, V_{gs}~ZTC

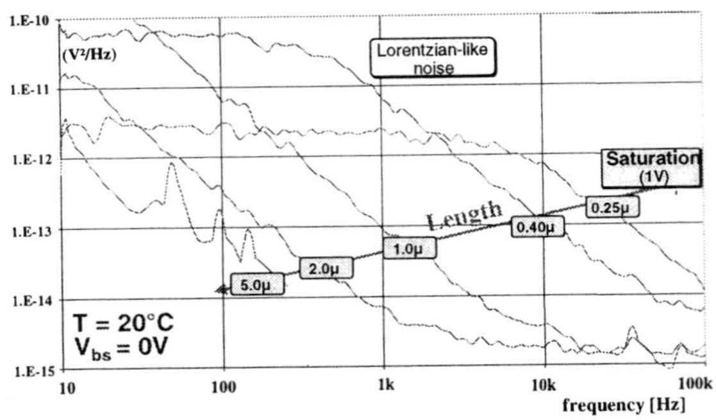

Figure 8: Input referred noise power spectral density. LETI Nearly FD (NFD)-SOI process (t_{si}=40nm). $W=25\mu m$. n-MOS, $T=20°C$, $V_{ds}=1V$, $V_{gs}=0.5V$, $V_{bs}=0V$

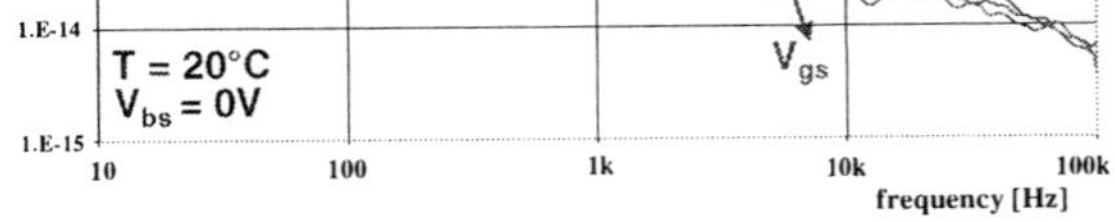

Figure 9: Input referred noise power spectral density. LETI FD-SOI process (t_{si}=30nm). $W=25\mu m$, $L=0.25\mu m$,. n-MOS, $T=20°C$, $V_{ds}=1V$, $V_{bs}=0V$

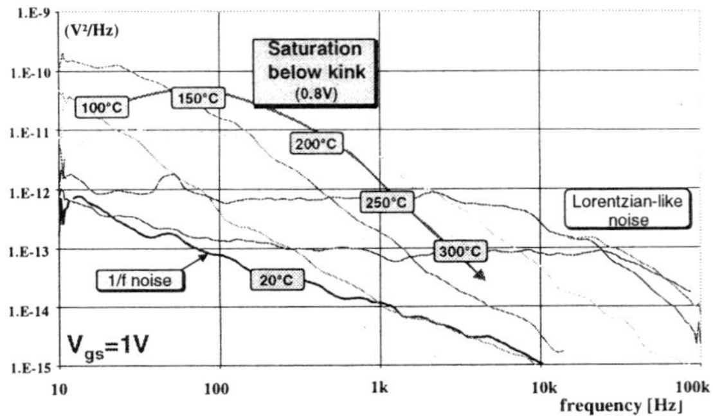

Figure 10: Input referred noise power spectral density. X-fab PD-SOI process. W=80µm. L=2µm. n-MOS, T=20°C, V_{ds}=0.8V (Below kink), V_{bs}=0V, V_{gs}=1V.

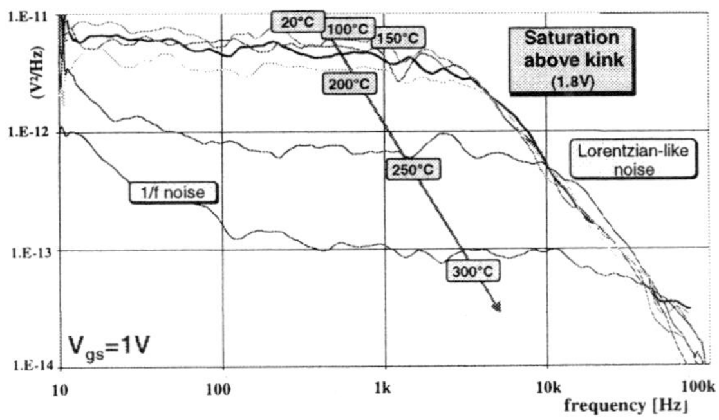

Figure 11: Input referred noise power spectral density. X-fab PD-SOI process. W=80µm. L=2µm. n-MOS, T=20°C, V_{ds}=1.8V (Above kink), V_{bs}=0V, V_{gs}=1V.

2.3 MEASUREMENTS ANALYSIS AND PHYSICAL NOISE MODEL

As can be seen on figures 2 to 11, the Lorentzian-like spectrum is very sensitive to the temperature, the bias voltages and the used technology (FD or PD-SOI). For a better comprehension of this dependence, this section presents an engineering model that will be further validated on noise measurements.

2.3.1 PHYSICAL LORENTZIAN-LIKE NOISE MODEL

The following model is related to an enhancement mode SOI n-MOSFET. Let consider a PD-SOI n-MOSFET as depicted on figure 12. This figure is relative to a device biased in strong inversion mode at low drain to source voltage. This is represented by the almost uniform front electron channel without pinch-off near the drain. Below this front channel appears the depletion area which can roughly be splitted in three parts. Firstly, the central part, controlled by the front gate voltage. Secondly the left part, corresponding to a forward biased PN (body-source) junction [19]. This area gives rise to recombination current (I_{Rs}) Finally, the right part of the depletion region is relative to a reverse biased PN (body-drain) junction. This reverse biased diode gives rise to generation current (I_{Gd}). As the device is PD-SOI, a quasi neutral region remains close to the back interface. Complex modeling can be found in the literature, taking into account all electrons and holes currents components [17].

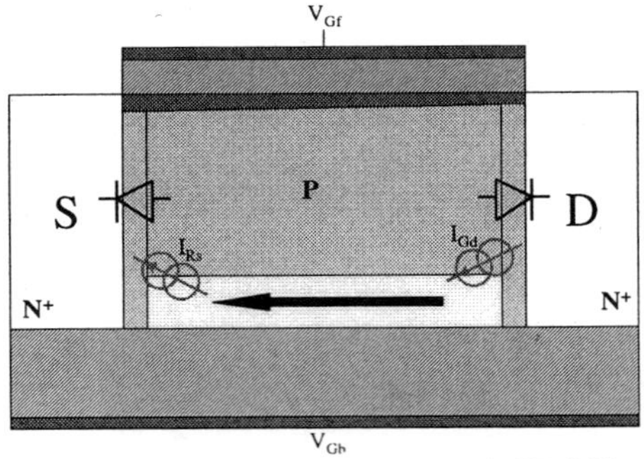

Figure 12: Representation of an enhancement mode PD-SOI n-MOSFET

For the purpose of our Lorentzian physical model, all these currents sources can be summarized in an electron front channel current and a hole body current. This hole current is the sum of a generation current (as depicted on figure 12) and a diffusion current [20]. In a more general point of view, this hole current can also include all other hole sources, as electron-hole generation near the drain at high drain voltage biases. The physical Lorentzian noise model consists in a discretisation of the physical device into an equivalent RC circuit. In the case of a PD-SOI MOSFET, the bottom quasi-neutral region can be considered as an equipotential node, because of the low value of the body hole current. This equipotential will correspond to the "body" node of the equivalent discrete circuit. Note that this body is equivalent to the body of a bulk device as long as we consider PD-SOI MOSFET. The linearized body effect associated with this node will be noted by "***n***", with a typical value around 1.4 (at least at room temperature) for bulk MOSFET (depending on process parameters). This body effect has not to be confounded with the FD-SOI substrate effect, usually represented by "λ" with a typical value around 1.08.

Figure 13 represents the "large signal" discrete modelling of a PD-SOI MOSFET with floating body.

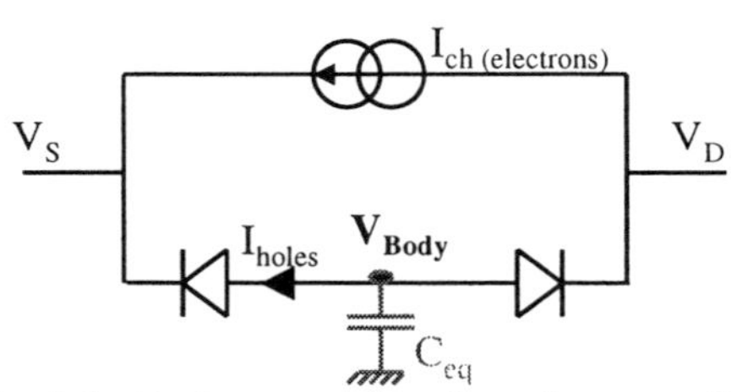

Figure 13: Large signal electrical representation of an enhancement PD SOI n-MOSFET.

The capacitor C_{eq} represents the total capacitance seen by the body (i.e. capacitances to the front and back gates as well as source and drain junction capacitances). In terms of non-noisy small signal analysis, figure 13 can be simplified to figure 14.

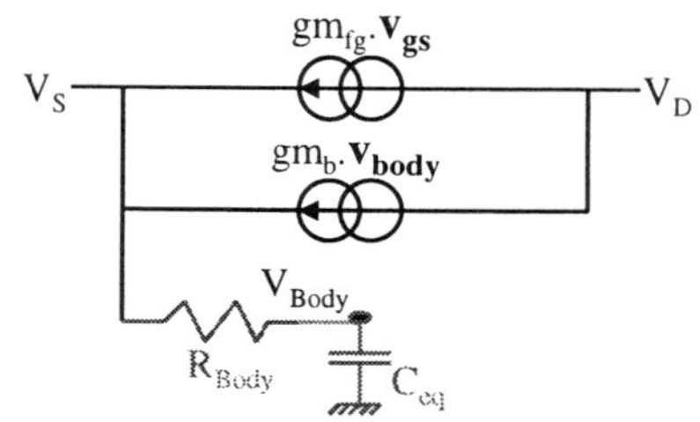

Figure 14: Noiseless equivalent small signal model of a PD-SOI n-MOSFET

R_{Body} represents the impedance of the forward biased body-source PN junction. Its value is given by $m\frac{kT/q}{I_{holes}}$, where kT/q is the thermal voltage ($=U_T$), m is the junction non-ideality factor (~1.4) and I_{holes} correspond to the total leakage current (including holes generated at high drain voltage). As the leakage current is very small, R_{Body} typical value is in the order of giga ohms. Note that associated with the equivalent body capacitance C_{eq}, the body cut of frequency $1/(2\pi R_{Body}. C_{eq})$ occurs in the range of kHz. gm_{fg} represents the front channel small signal transconductance (i.e. $\delta I_{DS}/\delta V_{GS}$). gm_b represents the body transconductance (i.e. $\delta I_{DS}/\delta V_{body}$). Going one step further in our model, figure 15 depicts the noisy equivalent small signal model.

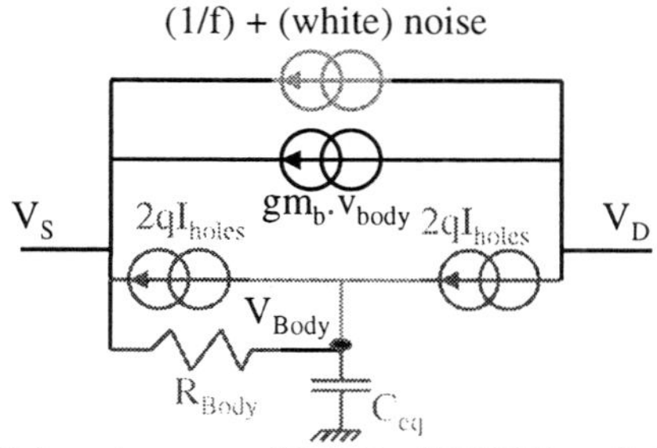

Figure 15: Noisy enhancement PD-SOI n-MOSFET small signal model

As presented in section 2.1.5, a diode generates a shot noise given by $2qI_{Diode}$, with $I_{Diode}=I_{holes}$ in our case. The upper current source represents the 1/f and white noise associated with the front channel current. Figure 16 gives another equivalent representation of figure 15. A Lorentzian-like voltage noise on the body now clearly appears, generated by a white noise current filtered by the body RC equivalent circuit. The Lorentzian body voltage noise generates a Lorentzian-like current noise between drain and source due to the body transconductance gm_b. Figure 16 introduces a parameter "α", defined as gm_b/gm_{fg}. In case of PD-SOI, this α parameter corresponds to the body effect parameter (n-1) previously introduced in this section. For the FD-SOI case, we can make the hypothesis that our noisy small signal electrical model is still valid, even if it is intuitively difficult to define an equipotential silicon area corresponding to the node "body" (V_{body}). In case of FD-SOI, α value goes to zero, due to the strong interaction between front and back gates [17]. In order to validate the model and the upper hypothesis, the α parameter will be extracted from measurements in the following section.

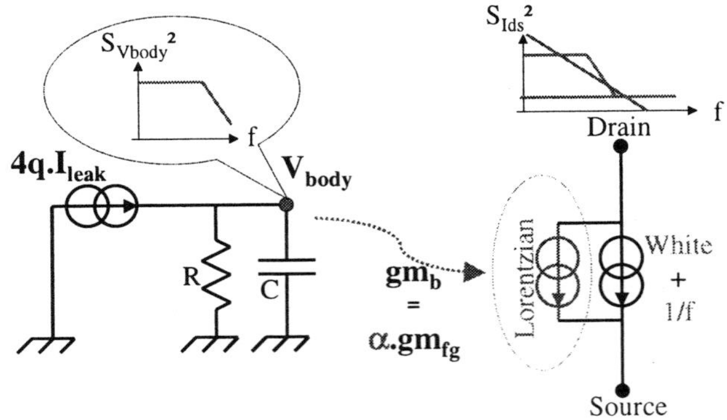

Figure 16: Lorentzian physical noise model

2.3.1.a LORENTZIAN MODEL VALIDATION BASED ON MEASUREMENTS

Based on the model of figure 16, the theoretical input referred Lorentzian plateau value and cut-off frequency (f_c) can be expressed as follows:

$$plateau_{theory} = \frac{4(m.k.T.\alpha)^2}{q.I_{leak}} \quad [V^2/Hz] \tag{5}$$

$$f_c = \frac{I_{leak}}{2\pi.m.(kT/q).C_{eq}} \tag{6}$$

Based on equation 5, the α value can be expressed as

$$\alpha = \frac{\sqrt{q.I_{leak}.plateau_{theory}}}{2.m.k.T} \tag{7}$$

In order to extract α dependence on bias voltages, temperature and technology (FD or PD-SOI), the theoretical plateau value of equation 5 is replaced by the measured plateau. It is also required to know the leakage current value. This leakage current has been measured on the UCL process at negative front gate voltages. Sufficient precision on the leakage current value is obtained for temperature above 150°C only. Table 1 (linear mode) and 2 (saturation mode) summarize the extrapolated leakage current values corresponding to noise measurement conditions. Table 3 (linear mode) and 4 (saturation mode) summarize the measured input referred plateau levels. Based on tables 1 to 4 and on equation 7, figures 17 (linear mode) and 18 (saturation mode) represent the extracted body effect (α) dependence versus temperature and back gate bias.

T \ Vback	1.5	0	-1.5	-3	-6
200	1E-11	1E-11	1E-11	1E-11	1E-11
250	5E-10	4E-10	2E-10	2E-10	2E-10

Table 1: Measured leakage current values [A] in linear mode (V_{ds}=50mV) and negative V_{gs} as function of temperature (°C) and back gate to source voltage. UCL SOI process.

T \ Vback	1.5	0	-1.5	-3	-6
150	2E-12	1.8E-12	1.8E-12	1.8E-12	1.8E-12
200	6.6E-11	5.94E-11	5.35E-11	4.81E-11	4.33E-11
250	9E-10	8.1E-10	7.29E-10	6.56E-10	5.9E-10

Table 2: Measured leakage current values [A] in saturation mode (V_{ds}=1V) and negative V_{gs} as function of temperature (°C) and back gate to source voltage. UCL SOI process.

T \ Vback	1.5	0	-1.5	-3	-6
100			3.4E-12	3.4E-11	2.2E-09
150		4.2E-13	7.74E-12	7.74E-11	8.15E-10
200		5.7E-13	7.74E-12	3.4E-11	1E-10
250	1E-13	1E-12	7.74E-12	1.2E-11	1.7E-11

Table 3: Extracted input referred plateau value [V^2/Hz] *in linear* mode (V_{ds}=50mV) and strong inversion (V_{gs} at ZTC) as function of temperature (°C) and back gate to source voltage. UCL SOI process.

T \ Vback	1.5	0	-1.5	-3	-6
100			1.6E-12	2.5E-11	1E-09
150		1.5E-13	1.6E-12	2.5E-11	2.5E-10
200	1.7E-14	1.5E-13	1.6E-12	5.7E-12	4.6E-11
250	1.7E-14	1.5E-13	9E-13	2.8E-12	9E-12

Table 4: Extracted input referred plateau value [V^2/Hz] *in saturation* mode (V_{ds}=1V) and strong inversion (V_{gs} at ZTC) as function of temperature (°C) and back gate to source voltage. UCL SOI process.

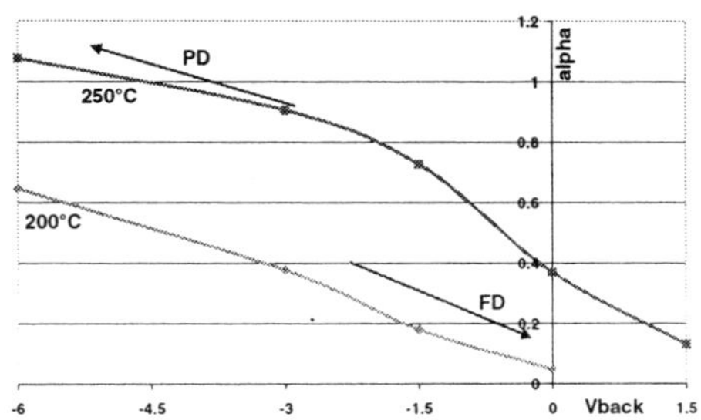

Figure 17: Extracted body effect (α) in linear mode (V_{ds}=50mV) as function of temperature and back gate to source bias voltage V_{gs}~ ZTC. UCL SOI process

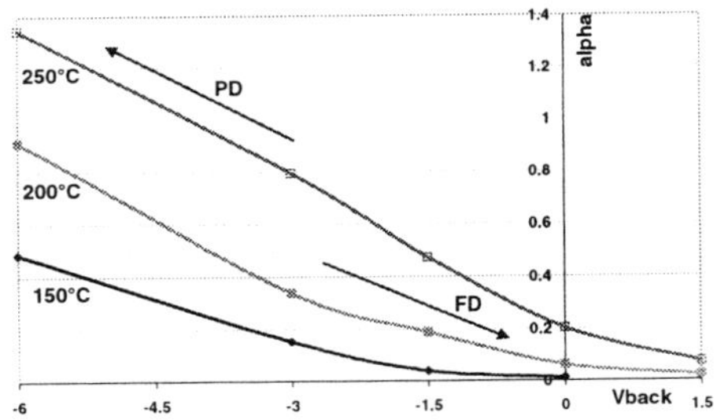

Figure 18: Extracted body effect (α) in saturation mode (V_{ds}=1V) as function of temperature and back gate to source bias voltage V_{gs}~ ZTC. UCL SOI process

As expected, α goes to zero when the MOSFET device goes into deep depletion (which is favored by lower temperature and higher back gate voltage). For high temperature and negative back gate voltages, UCL SOI n-MOSFETs reach partial depletion progressively. Note that the transition between FD and PD SOI is very progressive in our case (figures 17 and 18) due to the strong inversion regime of the device. In this case, the front gate mainly controls the film depletion while the back gate voltage has less control on it. Similar measurements in weak inversion should lead to sharper transition between FD ($\alpha \approx 0$) and PD ($\alpha = n$-1 $\approx$ 0.4 at room temperature). Such measurement is not available due to the low sensitivity of our measurement setup. Based on equation 6, on leakage current measurements and on measured Lorentzian's f_c, values of C_{eq} close to $2.C_{ox\_front}$ whatever the MOSFET regime are obtained, which is a good order of magnitude with regards to its definition given in §2.3.1. This gives further confidence in our model.

2.3.1.b LORENTZIAN MODEL VALIDATION BASED ON POISSON EQUATION

In order to physically confirm the previously presented empirical model, the α parameter has been simulated as a function of temperature and applied back gate voltage conditions. The simulation is based on a numerical resolution of the complete Poisson equation using Matlab. In order to get a continuous value for α parameter from full depletion to partial depletion, the standard hypothesis consisting on neglecting carriers in the depletion region cannot be considered anymore. Supposing our enhancement mode SOI n-MOSFET in strong inversion and with a low drain to source voltage, the Poisson equation can be written as

$$\frac{\partial^2 \phi(x,T)}{\partial x^2} = -\frac{q}{\varepsilon_{si}}\left(p(x,T) - n(x,T) - N_A^-\right) \tag{8}$$

where $\phi(x)$ is the potential over the active silicon film thickness (from x=0 (front interface) to $x=t_{si}$ (back interface)), ε_{si} is the silicon permittivity, $p(x)$, $n(x)$ and N_A^- are respectively the hole, the electron and the ionized acceptor levels in the silicon film. For boron doping, N_A^- can be considered equal to N_A for standard doping levels (i.e. for Fermi levels not too close from the valence energy band, which would degenerate the semiconductor) and above temperatures of about 100 K. The device being in strong inversion at low drain to source voltage, the electron quasi-Fermi level can be fixed to the source potential, i.e. zero. On the other hand, the hole quasi-Fermi level would be the potential applied on a lateral body contact of our n-MOSFET. If no lateral contact is used and for very low body current, the hole quasi-Fermi level in regime is also fixed by the source voltage, i.e. zero. However, as we are interested by the influence of a noisy hole body current, the hole quasi-Fermi level will not be tied to the source voltage but rather will have an average value equal to this source voltage. The body noise voltage is then added to this source voltage level. If v_{body} is defined as the hole quasi-Fermi level, it has a zero value in static, but presents some instantaneous value due to the random body noise generated by the body leakage current across drain and source junction diodes. Using Boltzmann equations [21], *p(x)* and *n(x)* can therefore be expressed as

$$n(x,T) = \frac{n_i^2(T)}{N_A}.e^{\frac{\phi(x,T)}{U_T}} \qquad p(x,T) = N_A.e^{-\left(\frac{\phi(x,T)-v_{body}}{U_T}\right)} \tag{9}$$

where $n_i(T)$ is the silicon intrinsic carrier density over temperature and U_T is the thermal voltage kT/q. In order to solve the Poisson equation 8, two boundary conditions are required. The first one is based on a physical threshold voltage definition, which corresponds to a surface potential of two times the electron quasi-Fermi level $(2.\phi_F)$, defined as

$$\phi_F = U_T.\ln\left(\frac{N_A}{n_i}\right) \tag{10}$$

The second is fixed by the back gate surface potential. This surface potential is scanned from 0 to $2.\phi_F$. Front and back gate voltages are then computed thanks to their respective surface potential and interface electric field. Using Kirchoff voltage law and Gauss equation over the oxides, the front gate voltage corresponding to a surface potential of $2.\phi_F$ is given by

$$V_{G_{front}} = 2\phi_F - \frac{\varepsilon_{si}.t_{ox_{front}}}{\varepsilon_{ox}}.\frac{\partial\phi}{\partial x_{(x=0)}} + V_{FB_{front}} \tag{11}$$

where t_{ox_front} is the front oxide thickness and V_{FB_front} is the front oxide flat-band voltage, including oxide charges as well as the work functions difference between the gate material and the doped silicon film. Similarly, for the back gate voltage, one obtains

$$V_{G_{back}} = \phi_{S_{back}} - \frac{\varepsilon_{si}.t_{ox_{back}}}{\varepsilon_{ox}}.\frac{\partial\phi}{\partial x_{(x=t_{si})}} + V_{FB_{back}} \tag{12}$$

where t_{ox_back} is the back gate oxide thickness, V_{FB_back} is the back gate related flat-band voltage and ϕ_{S_back} is the scanned back gate surface potential. Figure 19 corresponds to a Matlab simulation of a UCL FD-SOI floating body n-MOSFET. It presents the potential evolution across the device from the front interface to the back interface.

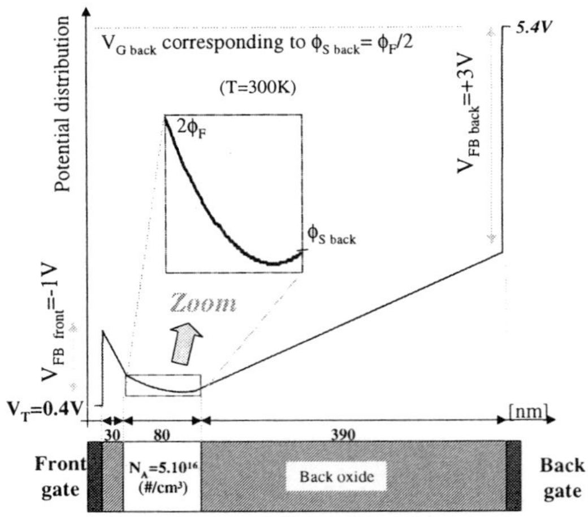

Figure 19: Matlab simulation based on Poisson equation. T_{si}=80nm, t_{ox_front}=30nm, t_{ox_back}=390nm, $N_A=5.10^{16}[cm^{-3}]$, T=300K, $\phi_{S\_front}=2\phi_F$, $\phi_{S\_back}=\phi_F/2$

Next step is to extract our α parameter from such simulation. This is obtained thanks to the relation

$$\alpha = \frac{\partial I_D / \partial V_{bs}}{\partial I_D / \partial V_{gs}} = \frac{(\partial I_D / \partial V_{T_{front}}).(\partial V_{T_{front}} / \partial V_{bs})}{\partial I_D / \partial V_{gs}} = -\frac{\partial V_{T_{front}}}{\partial V_{bs}} \qquad (13)$$

where I_D is the front channel electron current, V_{T_front}, V_{gs} and V_{bs} are respectively the front channel threshold voltage, the front gate to source voltage and the body to source voltage. As exposed previously, V_{T_front} can be simulated as function of the back gate voltage and as function of the temperature. In order to extract our α parameter, a simulation over back gate voltage and over temperature is performed for hole quasi-Fermi level equal to zero as well as for a slightly positive hole quasi-Fermi level (in order to get a δV_{bs}). Figure 20 is the result of such simulation.

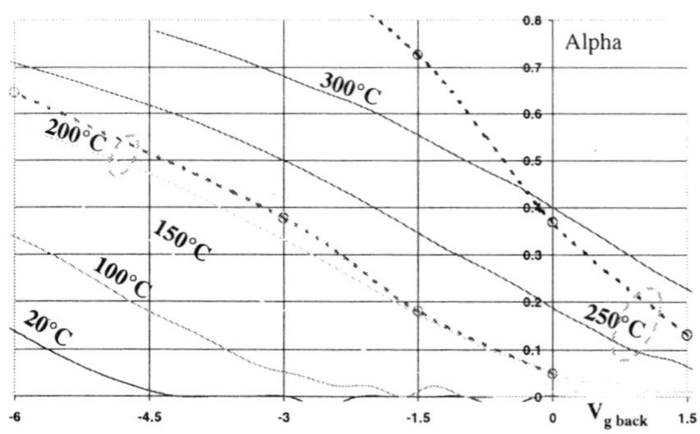

Figure 20: Alpha parameter Matlab simulation over temperature as function of the back gate voltage. The two dashed curves are measured alpha values extracted from noise and leakage current measurements at *200* and *250°C*. t_{si}=80nm, t_{ox_front}=30nm, t_{ox_back}=390nm, $N_A=5.10^{16}[cm^{-3}]$, $\phi_{S\_front}=2\phi_F$

The two dashed curves are extracted from noise and leakage current measurements. For sake of comparison, the simulated and extracted alpha values at *200°C* are in very good agreement. This is not as perfect for the *250°C* curves. This could be attributed to the low precision of both Lorentzian plateau level and leakage current measurements or to the precision of the diode ideality factor *m*. It could also be attributed to back oxide charges movement at such temperature. However, this simulation clearly validates the proposed model, continuously from partial depletion to full depletion.

2.3.2 MEASUREMENTS INTERPRETATION

Thanks to this model, figures 2 to 11 can now be explained.

- figure 2: A Lorentzian appears only at *250°C* with a low plateau and a high cut-off frequency due to the high leakage current associated with high temperature and positive back gate voltage. For temperature lower than *250°C* due to the very low value of α^2 for a positive back gate voltage, Lorentzians are hidden by the 1/f noise source (see equation 5). Note that the input referred 1/f noise is always constant.

- figure 3: Similar to previous case. However, as the drain to source voltage is only 50mV, the leakage current is smaller [20], which means that the plateau value is higher and the cut-off frequency f_c is lower than for the previous saturated case. Plateau values are also higher than the previous case due to the much higher value of α^2 in nearly fully depleted region when the drain voltage is low. Indeed, a high drain voltage increases the depletion area near the drain, which reduces α value.

- figure 4: Around *100°C*, the leakage current is very low (even more with a negative back gate voltage). This means a theoretical high plateau level and a very low cut-off frequency. At *100°C* with negative back gate voltage, the MOSFET is nearly fully depleted (NFD) (i.e. $0<\alpha<n-1$). When the temperature increases, the device goes toward more partial depletion, i.e. increasing the value of α, which should increase the plateau level. However, a temperature increase is also associated with a leakage current increase, which tends to decrease the plateau level and increase the cut-off frequency. The opposite trends between the effect of α and the effect of I_{leak} leads to a merely constant plateau value while the f_c increases. It is interesting to note an excess 1/f noise over *250°C*. This excess noise probably originates from the back interface. Indeed, the back interface presents a lot of traps at some energy levels. When the drain to source voltage is high, the surface potential along the back interface can activate some trap levels. These traps, due to their spatial distribution, can generate a 1/f noise spectrum in the body [15]. This back related 1/f noise is then transferred to the front channel by way of the body effect α. At lower temperature, no 1/f excess noise is observed due to the low value of α^2. Similar measurement using a good body contact remove the Lorentzian and the observed 1/f excess noise. This tends to prove that this excess noise is related to the back oxide.

- figure 5: As V_{ds} is very small, the leakage current is lower than the previous case. Plateau levels are therefore higher. Low V_{ds} value also means more PD than for similar conditions with high V_{ds} value. Compared to previous case, the Lorentzian plateau decreases with temperature due to the leakage current increase and due to the "saturation" of α value close to $n-1$. Some excess 1/f noise still appears at *250°C*.

This excess noise level is however one order of magnitude lower than the previous case in saturation. This is probably due to the fact that the potential evolution along the back gate interface is weak and does not activate a lot of back interface traps.

- figures 6 and 7: Similar justification than previous cases.
- figure 8: Lorentzian curves are observed. The very low plateau level of the $0.25\mu m$ MOSFET compared to other curves is due to the leakage current increase in deep sub-micron MOSFET [22] and due to the kink effect at high drain to source voltage. In addition, due to short channel effect, the device is close to the full depletion (i.e. low α). For lengths above $1\mu m$, the leakage and kink effect related currents remain approximately constant. In this case, the shift to the left of observed Lorentzian with MOSFET length increase is associated with the C_{eq} increase, approximately proportional to the device length.
- figure 9: The 1/f noise is weakly dependent on the front gate bias.
- figure 10: We observe a typical Lorentzian shift to lower plateau and higher cut-off frequency due to the leakage current increase. At *20°C*, only 1/f noise is observed. In fact, there is still a Lorentzian in this case but shifted to lower frequency, below our measurement frequency window.
- figure 11: The Lorentzian is approximately constant up to *200°C*. This is due to the front gate electron-hole generation near the drain. Generated holes are equivalent to an excess leakage current. Up to *200°C*, this current contribution dominates the standard leakage current. Above *200°C* , the leakage current is dominated by the drain to body junction generation current, which explains the decrease of the plateau and the increase of f_c. Note that the 1/f noise remains constant over temperature in this PD-SOI MOSFET. This could be associated with the zero back gate voltage applied during this measurement. The back gate surface potential is probably far from hole traps energy. This is similar to figure 6 for zero back gate voltage, while a negative back gate voltage bend the valance band such that hole traps close to the valance band can be easily activated. In the UCL process, this comes in parallel with the increase of α with back gate voltage decrease.

2.3.3 NOISE CONSIDERATIONS FOR ANALOG CIRCUITS

Up to now, we have observed and modeled the low-frequency noise. We have to remember that the aim of this analysis was to give practical considerations for analog designers. Designers are usually interested by root means square (RMS) noise value in their circuits. This RMS value will fix their signal to noise ratio in a given bandwidth. In order to get the RMS value, it is required to integrate over the circuit bandwidth the input referred noise spectrum.

Generalized integrated noise model

In order to be as general as possible, an external body contact is considered in the model of figure 21.

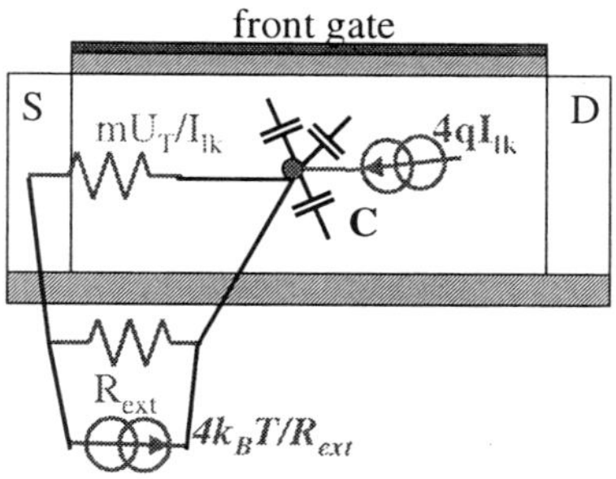

Figure 21: Generalized equivalent noise model for a SOI MOSFET

For the general case of an external resistor R_{ext} connected between the body and the source, the gate referred voltage noise spectrum is given by

$$4\left(q.I_{leak}+\frac{k.T}{R_{ext}}\right)\left|\frac{R_{eq}^2}{1+(2\pi.R_{eq}C.f)^2}\right|.\alpha^2 \tag{14}$$

with R_{eq} corresponding to R_{ext} in parallel with the body-source impedance mU_T/I_{leak} (In fact, when R_{ext} is very small, I_{leak} mainly flows towards this resistor and no more in the body-source junction. The shot noise of this junction should therefore disappear. This can however be neglected as in this case, the dominant noise will be created by R_{ext} as in [12]). The integration over frequency of this spectrum from zero to infinity yields

$$\begin{aligned}&\int_0^\infty 4\left(q.I_{leak}+\frac{k.T}{R_{ext}}\right)\left|\frac{R_{eq}^2}{1+(2\pi.R_{eq}C.f)^2}\right|.\alpha^2.df\\ &=\frac{m.k.T}{C}\left(\frac{U_T+R_{ext}.I_{leak}}{m.U_T+R_{ext}.I_{leak}}\right)\alpha^2\\ &\cong\frac{m.k.T}{C}.\alpha^2\end{aligned} \tag{15}$$

It is interesting to note that this integrate noise is merely independent on R_{ext}. However, this property has to be interpreted correctly. In practice, in PD-SOI, the use of a weakly resistive body contact will strongly reduce the Lorentzian plateau value and will increase its cut-off frequency f_c (equation 14). If the analog application where this n-MOSFET is used involves somewhere a low-pass filter, the noise spectrum has to be integrated only on this frequency range. In this case, the use of a low resistive body contact is interesting as the Lorentzian noise contribution is strongly reduced in such limited bandwidth application. For very large bandwidth application however, the use of a body contact will not reduce the integrated noise. In this case anyway, the main noise contribution is due to the white noise and it is not relevant to take into account the Lorentzian contribution. However, if the low frequency noise spectrum can affect the phase noise of a high frequency oscillator for example [23, p535], it would be relevant to use a body contact for PD SOI.

Based on equation 15, it can be seen that FD-SOI MOSFETs, with their α close to zero, exhibit no contribution of the body noise. In PD-SOI however, the body related integrated Lorentzian noise cannot be neglected, compared to the integrated *1/f* noise. In order to give an order of magnitude of the Lorentzian noise contribution on the total noise, lets compare the noise of the UCL $2\mu m$ process in partial depletion (at $250°C$ with V_{Back}=-6V) with the spectrum in full depletion (at $250°C$ with V_{Back}=+3V) (figure 6). Considering only the PD-SOI Lorentzian contribution (not its 1/f noise), one obtains about $400\mu V_{RMS}$. Considering now only the *1/f* noise of the FD-SOI MOSFET on a bandwidth of *1MHz*, one obtains about $40\mu V_{RMS}$. In this case, the Lorentzian noise reduces the signal to noise ratio (S/N) by about *20dB*. If we would considèr the excess *1/f* noise created by the back interface, this S/N ratio would be even more reduced. Still based on equation 15, it is possible to reduce the integrated noise by way of an increase of the body to ground capacitance. This can be achieved either by increasing the MOSFET size (figure 8) or by an external capacitance. Note that such capacitance will low-pass filter the body noise, but will not affect the front channel speed.

3 Preamplifier for high-temperature applications

In order to keep a good signal to noise ratio after a Wheatstone bridge, a low noise fully-differential instrumentation preamplifier is required. Usual implementation of such amplifier (figure 22) is based on two conventional high gain single-ended amplifiers and three matched resistors to fix the gain by way of a feedback [24] [25].

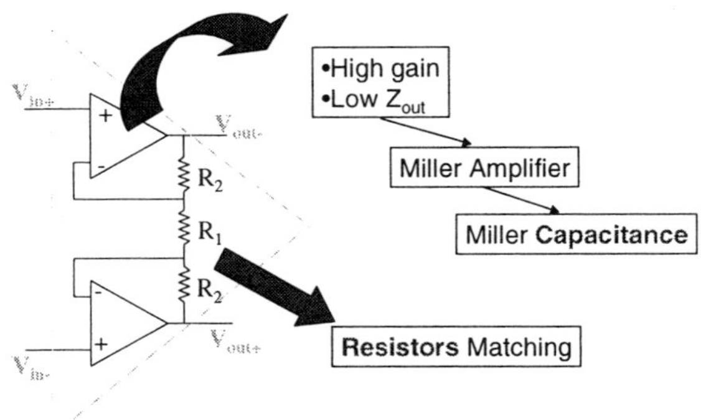

Figure 22: Standard instrumentation fully-differential amplifier architecture

Such architecture presents very good characteristics but can't intrinsically exhibit a very low noise. Low frequency noise level is mainly fixed by the area of both single-ended amplifier input transistors. For stability consideration, these transistors have an upper size limit, which limits the noise performance of the overall fully-differential amplifier. Such architecture also requires the use of two Miller capacitances with the operational amplifiers to ensure stability.

Figure 23 presents an original architecture where the gain is only fixed by way of transconductance (*gm*) ratios.

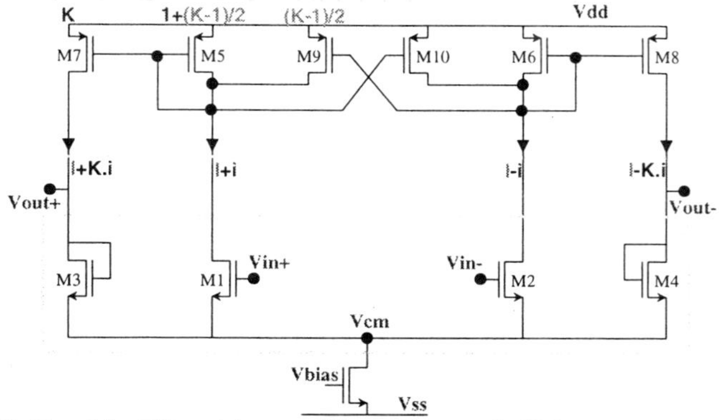

Figure 23: New fully-differential preamplifier architecture for high temperature application

The main advantages are the absence of feedback in the circuit and the absence of integrated resistors which are heavy silicon die area consumers. Low noise performances can therefore be reached using large transistors area for critical transistors without limitations due to stability problems. An optimized sizing of input outputs transistors (M1 to M4) could also provide some gain dependence over temperature. This could for example be useful to compensate at first order the sensitivity decrease of most sensors over temperature. On the other hand, absence of feedback means poor distortion performances. This means that the new architecture can only handle a low differential voltage (around *1mV* at the input) while keeping acceptable distortion. This, however, is perfectly compatible with the fact that the amplifier is very low-noise and therefore will mainly handle small signals. If transistors M1 to M4 of figure 23 are identical, the gain is fixed by the active current mirror (M5 to M10). For the common mode current, this active current mirror appears like two regular unity current mirrors, which means that the total bias current is divided into four identical currents throughout transistors M1 to M4. For the differential current, the positive transconductance of transistor M5 (*gm5*) appears in parallel with the (negative) transconductance of M9 (*gm9*). As *gm5* is slightly larger (size='*(K+1)/2*' in figure 23) than the negative gm9 (size='*(K-1)/2*'), the equivalent observed transconductance is $|gm_5|-|gm_9|$ (equivalent to a size of '1'). Note that *K* should not be too large (~10) in order to keep some precision on the gain. As transistor M7 has a size of *K*, this means that the active current mirror presents a differential gain current of *K*. If transistors M1 to M4 are identical, the voltage gain is *K*. The gain precision is limited by matching considerations and by transistors output conductance. The conductance effect on the gain precision is reduced if long channel transistors are used (which is the case for low noise consideration). The gain being fixed by transconductance ratios and all transistors working at same current densities, the gain is therefore not very sensitive to the temperature.

The architecture has been successfully validated on two different SOI processes. Figure 24 summarizes measurements as a function of the bias current at room temperature.

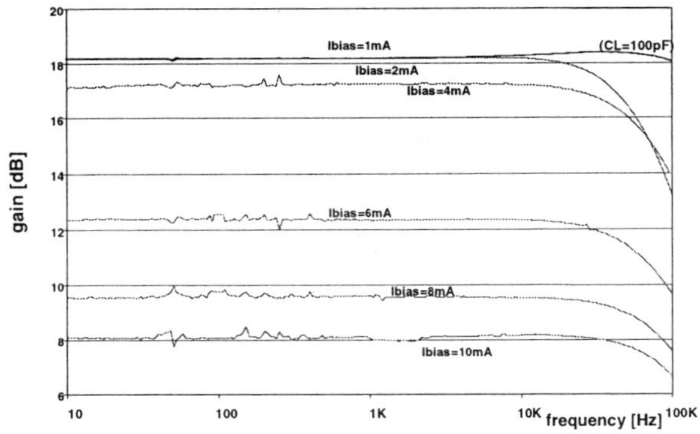

Figure 24: Preamplifier gain measurement vs bias current. $K=10$, $C_L=10nF$ if not specified, $T=20°C$, FD UCL SOI process, $L=3\mu m$.

K has been fixed to 10, i.e. a theoretical gain of $20dB$. In the measured circuit, the output conductance has not been taken into account during its design. The observed lower gain come from the effect of this conductance appearing in parallel with transconductances. An interesting application for such amplifier would be for example a wide bandwidth integrated accelerometer. Indeed, a high bandwidth accelerometer is mechanically more rigid and therefore presents a lower sensitivity. The sensitivity of such sensor also slightly decreases with temperature, which can be roughly compensated with the presented architecture.

4 Conclusion

Our work demonstrates that SOI-CMOS technology exhibits constant input referred noise over temperature as long as the device remains fully-depleted or if a low resistive body contact is used. In these conditions, analog low noise circuits can be designed at high temperature using same methodology and parameters values as for room temperature. We also demonstrate a new compact low-noise fully-differential preamplifier architecture suited for sensors with poor sensitivity.

References

[1] Jean-Pierre Colinge, silicon-on-insulator technology: Materials To VLSI, Kluwer, 2nd edition, 1997

[2] D. Flandre, L.F. Ferreira, P.G.A. Jespers, and J.P. Colinge, "Modelling and application of fully depleted SOI MOSFETs for low-voltage, low-power analogue CMOS circuits", Solid State Electron., vol. 39, no. 4, pp. 455–460, 1996.

[3] C. Raynaud, O. Faynot, J-L Pelloie, C. Tabone, A. Grouillet, F. Martin, G. Dambrine, M. Van-mackelberg, L. Picheta, E. Mackowiak, H. Brut, P. Linares, J. Sevenhans, E. Compagne, G. Fletcher, D. Flandre, V. Dessard, D. Vanhoenacker, and J-P Raskin, "70 GHz fmax fully-depleted SOI MOSFET's for low-power wireless applications", in GAAS, Paris, 2000, pp. 268–271.

[4] B. Gentinne, J-P Eggermont, D. Flandre, and J-P Colinge, "Fully depleted SOI-CMOS technology for high temperature IC appliçations", Materials Science and Engineering (B),pp. 1–7, 1997.

[5] Denis Flandre, "Silicon-on-insulator technology for high temperature metal oxide semiconductor devices and circuits", Materials Science and Engineering (B), pp. 7–12, 1995.

[6] P. R. Gray and R. G. Meyer, Analysis and Design of analog integrated circuits, John Wiley and Sons, third edition, 1992.

[7] L.K.J. Vandamme, X. Li, and D. Rigaud, "1/f noise in MOS devices, mobility or number fluctuations",

IEEE trans. on electron devices, vol. 41, no. 11, pp. 1936–1945, Nov. 1994.

[8] F.N. Hooge, "1/f noise sources", IEEE trans. on electron devices, vol. 41, no. 11, pp. 1926–1935, Nov. 1994.

[9] R. Jayaraman and C.G. Sodini, "A 1/f noise technique to extract the oxide trap density near the conduction band edge of silicon.", IEEE trans. on electron devices, vol. 36, no. 9, pp. 1773–1782, Sept. 1989.

[10] Y. P. Tsividis, Operation and Modeling of the MOS Transistor, McGraw-Hill Book Company, 1987.

[11] J.A. Badcock, D.K. Schroder, and Y-C Tseng, "Low-frequency noise in near-fully-depleted TFSOI MOSFET's", IEEE Electron device letters, vol. 19, no. 2, pp. 40–43, Feb. 1998.

[12] Frederico Faccio, Etude Des Transistors MOS Avancés sur Silicium sur Isolant (SOI): Bruit, Dégradation En Environnement Radiatif et Applications, PhD thesis, INPG, CERN, France, Novembre 1997.

[13] Christian Enz, High Precision CMOS Micropower Amplifiers, PhD thesis, EPFL, Lausanne, 1989.

[14] L.D. Yau and C-T. Sah, "Theory and experiments of low-frequency generation-recombination noise in MOS transistors.", IEEE Trans on electron devices, vol. 16, no. 2, pp. 170–177, Feb.1969.

[15] Z-H Fang, "Low-frequency pseudogeneration-recombination noise of MOSFET's stressed by channel hot electrons in weak inversion.", IEEE Trans. on electron devices, vol. 33, no. 4, pp. 516–519, Apr. 1986.

[16] W. Jin, P.C.H. Chan, S.K.H. Fung, and P.K. Ko, "Shot-noise-induced excess low-frequency noise in floating-body partially depleted SOI MOSFET's", IEEE Trans on electron devices, vol. 46, no. 6, pp. 1180–1185, June 1999.

[17] G.O. Workman and J.G. Fossum, "Physical noise modeling of SOI MOSFET's with analysis of the lorentzian component in the low-frequency noise spectrum.", IEEE Trans. on electron devices, vol. 47, no. 6, pp. 1192–1200, June 2000.

[18] J-P Eggermont, V. Dessard, A. Vandooren, D. Flandre, and J-P Colinge, "SOI current and voltage reference sources for applications up to 300c",in Hiten'97, Manchester, England, Sept. 1997, pp. 14–17.

[19] M. Bellodi, B. Iniguez, D. Flandre, and J.A. Martino, "Diodes model for the leakage drain current in enhancement-mode SOI n-MOSFETs at 300 ",in SBMICRO 2000, XV Int. Conference on Microelectronics and Packaging, Sept 18-24 2000, pp. 203–207.

[20] T.E.Rudenko, V.S.Lysenko, V.I.Kilchytska, and A.N. Rudenko, "A comprehensive analysis of the high-temperature off-state and subthreshold characteristics of SOI MOSFETs", in Perspectives, Science and Technologies for Novel Silicon on Insulator Devices, P.L.F. Hemment et Al., Ed., pp. 281–293. Kluwer Academics Publishers, 2000.

[21] Fernand Van de Wiele Jean-Pierre Colinge, Physique Des Dispositifs Semi-Conducteurs, De Boeck Université, 1996.

[22] M. Bellodi, B. Iniguez, C.Raynaud, D. Flandre, and J.A. Martino, "Study of the deep-submicron SOI MOSFET leakage current bahavior at high temperatures", in SBMICRO 2000, XV Int. Conference on Microelectronics and Packaging, Sept 18-24 2000, pp. 304–307.

[23] T. H. Lee, The Design of CMOS Radio-Frequency Integrated Circuits, Cambridge University Press, 1998.

[24] A. Viviani L. Demeûs and D. Flandre, "High-temperature analog instrumentation system in thin-film FD SOI CMOS technology", in Hitec'98, Albuquerque, USA, June 1998.

[25] P. de Jong, Smart Sensor Systems for High-Temperature Applications, PhD thesis, Delft University, The Netherlands, Nov. 1998.

INFLUENCE OF γ- RADIATION ON SHORT CHANNEL SOI-MOSFETs WITH THIN SiO_2 FILMS.

C. CLAEYS[1,2], E. SIMOEN[1], V. G. LITOVCHENKO[3], A. EVTUKH[3], A.EFREMOV[3], A. KIZJAK[3], JU. RASSAMAKIN[3]

[1]IMEC, Kapeldreef 75, B-301, Leuven, Belgium. E-mail: cor. Claeys @ imec.be

[1]also at E.E. Dept. KU Leuven, B-3001 Leuven, Belgium

[3]Institute of Semiconductor Physics, 45 Prospekt Nauki, Kiev, 03028, Ukraine. E- mail: Evt @ book. Semicond. Kiev. ua

Abstract

The influence of 10 krad to 1 Mrad γ-irradiation on the threshold voltage and the maximum transconductance of n-channel MOS transistors, with channel lengths between 0.5 - 2.0 μm and processed in an 0.5 μm SOI technology, has been investigated. Radiation can induce both negative and positive charges in the SiO_2-Si structure. The negative charges can be annealed at T=315°C, while an additional anneal at T=415°C also removes the radiation-induced positive charges. The influence of the boron channel implantation dose on the radiation hardness of the MOS transistors is also studied. An improvement of the MOS transistor parameters after a low dose (10^5 rad) irradiation has been observed.

1. Introduction

It is well known that radiation, in particular γ-radiation, adversely affects the semiconductor device performance and its stability. There are many published data pointing out the strong influence of both the interface and the dielectric growth technique on the radiation hardness of MOS devices [1]. On the other hand, γ-irradiation and a subsequent annealing under specific conditions reveal an ordering effect [2-4]. The improvement of the material properties, i.e., a reduction of the defect concentration and recombination losses leads to a substantial improvement of the device and IC parameters. It is also important to study the influence of γ-radiation on short-

F. Balestra et al. (eds.), Progress in SOI Structures and Devices Operating at Extreme Conditions, 211–220.

channel MOSFETs with a thin gate oxide on wafers with a buried oxide layer (SOI).

This work presents investigations of the influence of a ^{60}Co radiation and subsequent low temperature annealing on the threshold voltage (V_{th}) and the maximum transconductance ($G_m = dI_d/dV_g$) of n-MOS transistors with a 0.5 ÷2.0 μm channel length, processed in an 0.5 μm SOI technology.

2. Experiment

2.1. SAMPLE PREPARATION

The LDD n-MOS transistors with LOCOS isolation have been fabricated in an 0.5 μm SOI technology on SIMOX substrates. A schematic transistor cross-section is shown in Fig. 1. The channel length L was changed in the range 0.5 ÷2.0 μm. For all transistors the channel width was W=20 μm. The 15 nm thick gate oxide was grown at T=900°C in dry oxygen. The LDD n-MOS transistors were formed using a boron channel ion implantation, through a 13 nm thick SiO_2 layer, using two different double implant conditions: (i) E=35 keV, $D_b=2x10^{12}$ cm^{-2} and E=15 keV, $D_b=2.4x10^{12}$ cm^{-2}; (ii) E=35 keV, $D_b=2x10^{12}$ cm^{-2} and E=15 keV, $D_b=3.2x10^{12}$ cm^{-2}; (where D_b is the boron channel dose). A 0.35 μm thick polycrystalline silicon layer was used as gate electrode. The drain/source regions were formed by a 100 keV arsenic ion implantation with a dose $D_{As}=2x10^{13}$ cm^{-2} (defined as lightly doped region) and $D_{As}=5x10^{15}$ cm^{-2} (defined as heavily-doped region), respectively. The MOS transistors were irradiated with γ-rays from a ^{60}Co source with a dose in the 10^4 ÷10^6 rad range. The subsequent thermal annealing of the transistors has been performed in argon at T=315°C and T=415°C.

2.2. MEASUREMENTS

The threshold voltage can be extracted from measured $I_d=f(V_g)$ curves for $V_d=0.1$ V and a gate bias corresponding to a channel current of 1 μA, where V_d is the drain voltage, I_d is the channel current, and V_g is the gate bias. The substrate and the source were grounded for all measurements. The transistor transconductance $G_m=dI_d/dV_g$ can be obtained by differentiating the $I_d=f(V_g)$ curve at $V_d=0.1V$, corresponding to the maximum G_m value. The dependencies $V_{th}=f(L)$, $G_m=f(L)$, $\Delta V_{th}=f(D)$ and $\Delta G_m= f(D)$, where $\Delta V_{th}=V_{th}-V_{tho}$ and $\Delta G_m=(G_m-G_{mo})/G_{mo}$, were determined. V_{tho} and G_{mo} are the pre-irradiation threshold voltage and transconductance, respectively, while V_{th} and G_m are the values after irradiation and annealing; D is the γ-radiation dose: 10^4, 10^5, 10^6 rad; L is the channel length. A Hewlett Packard Analyzer model HP 4145 was used for measuring the current - voltage characteristics.

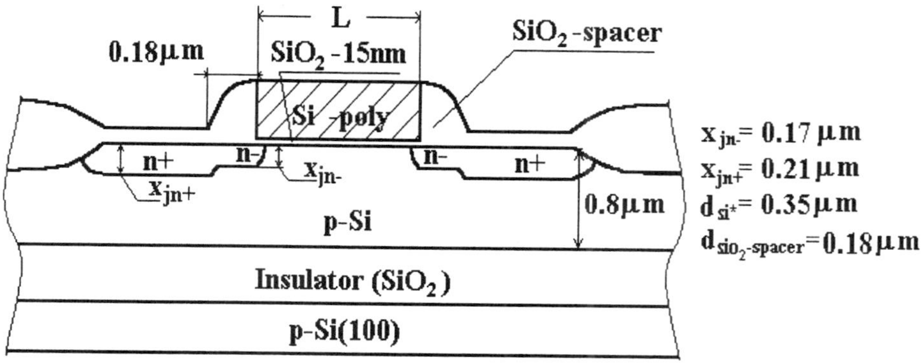

Figure 1. A cross-section of the investigated MOS transistors.

3. Results and Discussion

3.1. CHARACTERISTICS OF SHORT-CHANNEL MOSFETs

The principal MOSFET parameters being influenced by γ-radiation, are the threshold voltage V_{th} and the transconductance of the transistor $G_m=(dI_d/dV_g)|_{Vd=const}$ which determines the amplification factor of the transistor. The corresponding formula for the threshold voltage is [5]

$$V_{th}=V_{fb}+ 2\Psi_b+(2\varepsilon_s qN_A(2\Psi_b))^{1/2}/C_I \tag{1}$$

where V_{fb} is the flat-band voltage, Ψ_b is the Fermi level in relation to the Fermi level in an intrinsic semiconductor, N_A is the impurity concentration in the silicon film, ε_s is the permittivity of silicon, q is the electron charge, and C_i is the specific oxide capacity. The influence of irradiation on the V_{th} is mainly determined by the flat-band voltage V_{fb}

$$V_{fb}=\Psi_{ms}-Q_o/C_i, \tag{2}$$

where Ψ_{ms} is the difference of the work function gate – substrate, $Q_o= Q_{ox}+Q_{it}$ is the sum of the effective charge in the oxide Q_{ox} and in interface traps Q_{it}. To analyze the influence of the γ-radiation on the G_m, the next formula was used:

$$G_m = (W/L)\ \mu_n C_i\ V_d \sim \mu_n(Q_{it}), \tag{3}$$

where μ_n is the electron mobility in the channel. For short channel MOSFETs with ultrathin gate oxides, the influence of the interface trapped charge is increased. In the last formula the change of the carrier mobility is caused due to the influence of the electric field, which, in turn, is determined by the interface trap charge and its radiation-induced changes. So, the main factor which determines the parameters of short–channel MOSFETs with an ultrathin gate dielectric is the interface traps rather than the charge built-in in the dielectric layer.

3.2. INFLUENCE OF A LOW-DOSE γ - IRRADIATION

The threshold voltages of the MOS transistors are increased after a 10 or 100 krad γ-irradiation. The values of the threshold voltage versus channel length before and after a 100 krad irradiation are presented in Fig. 2a for the case of a higher boron doping in the channel (D_b=3.2x10^{12} cm^{-2}). The V_{th} increase points towards a predominant building-up of negative charge in the structure. As can be seen in Fig. 2b, the increase of the maximum transconductance after irradiation increases with decreasing channel length (8-10 % for L=0.9 μm and 15-21% for L=0.6 μm). For a 1 Mrad irradiation dose a decrease of the threshold voltage is observed, corresponding with a building up of positive charge in the structure. The dependence of the threshold voltage shift on the radiation dose is shown in Fig. 3a.

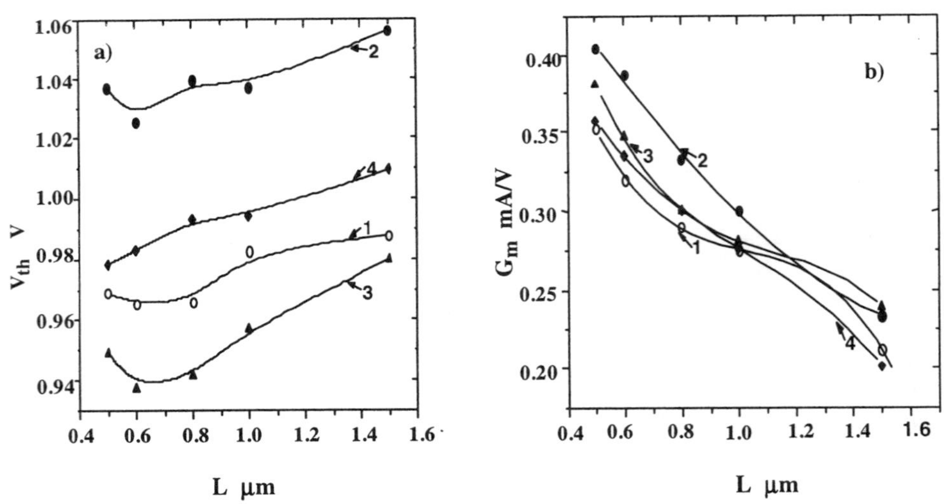

Figure 2 Threshold voltage (a) and transconductance (b) versus channel length for MOSTs with D_b=3.2x10^{12} cm^{-2}: 1(0) - initial; 2(•) – after 100 krad radiation; 3(•) - after 100 krad radiation and annealing at T=315°C; 4(♦) - after 100 krad radiation and annealing at T=315°C + T=415°C.

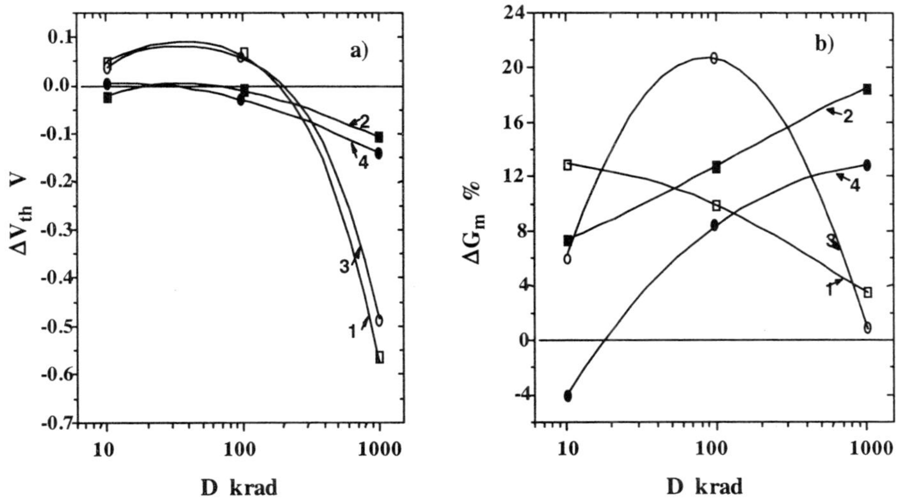

Figure 3 Shift of the threshold voltage (a) and transconductance (b) versus radiation dose for MOSTs with $D_b=3.2\times10^{12}$ cm^{-2}: L=1.5 µm: 1(•)- after radiation; 2(•) - after radiation and annealing at T=315°C; L=0.6 µm : 3(0) - after radiation; 4(•) - after radiation and annealing at T=315°C.

For the case of a lower boron doping in the channel ($D_b=2.4\times10^{12}$ cm^{-2}) the threshold voltages are decreased after irradiation, independent of the dose and the channel length. This points towards the building up of the positive charge in the MOS structure. The transconductance after irradiation is also decreased.

As can be seen from the experimental results, the level of the boron dose in the channel strongly influences the behavior of the threshold voltage and the transconductance after γ-irradiation. The traps in a MOS structure can be divided into oxide traps, border traps and interface traps [6,7]. It is known [1] that the irradiation-induced positive charge built-up in the oxide decreases with a decrease of the oxide thickness. During irradiation not only positive charge is built-up but border and interface traps are created as well. With a decrease of the oxide thickness the influence of border and interface traps on the total charge in the gate system is increased. During the measurements, there is a charge exchange between the traps and the silicon [8-11]. The electrons which compensate the positive hole charge in the oxide are localized in border and interface traps. Therefore the observation of negative charge after irradiation with a dose of 10^4-10^5 rad, for the case of a boron channel dose of 3.2×10^{12} cm^{-2}, is most likely caused by the creation of border and interface traps. These capture electrons and have therefore a higher negative charge in comparison with the positive charge of the holes which are built-in into the oxide, i.e., the total charge is negative. However, only decreasing the oxide thickness is insufficient to lead to a dominance of negative charge. The experimental results show that also the boron channel dose is important.

The measured values of positive and negative charges in function of the radiation dose is given in Table 1 for transistors with a channel length L=1.5 μm.

TABLE 1. Charge in MOS structures after radiation

Boron dose in the channel, D_b [cm^{-2}]	Charge in the MOS structure (mV)	γ- radiation dose (rad)		
		10^4	10^5	10^6
3.2×10^{12}	$+Q/C_I$	-	~0	~570
3.2×10^{12}	$-Q/C_I$	~75	~80	-
2.4×10^{12}	$+Q/C_I$	~45	~100	~1060
2.4×10^{12}	$-Q/C_I$	-	-	-

As can be seen from Table 1 the boron channel dose strongly influences the positive charge creation after irradiation. Therefore the threshold voltage and transconductance are more strongly shifted for transistors with a lower boron channel doping. The transconductance of MOS transistors with a higher boron channel doping dose is increased till a dose of 10^5 rad and for $D=10^6$ rad its further increase is small. On the contrary, for transistors with a lower boron channel doping dose the transconductance continuously decreases with increasing radiation dose. The observation of a transconductance increase after 100 krad for the case of a 3.2×10^{12} cm^{-2} boron channel dose points towards an ordering effect, allowing to improve the transistor parameters.

Simulations of the boron distribution in the silicon and the silicon oxide, by using SUPREM III, allowed to obtain the maximum boron concentrations N_b (see Fig. 4): (i) for a boron channel dose 2.4×10^{12} cm^{-2} - N_b=~2.8×10^{17} cm^{-3} in the Si and N_b=~5.5×10^{17} cm^{-3} in the SiO_2; (ii) for the case of D_b=3.2×10^{12} cm^{-2}- N_b=~3.5×10^{17} cm^{-3} in the Si and N_b=~6.8×10^{17} cm^{-3} in the SiO_2.

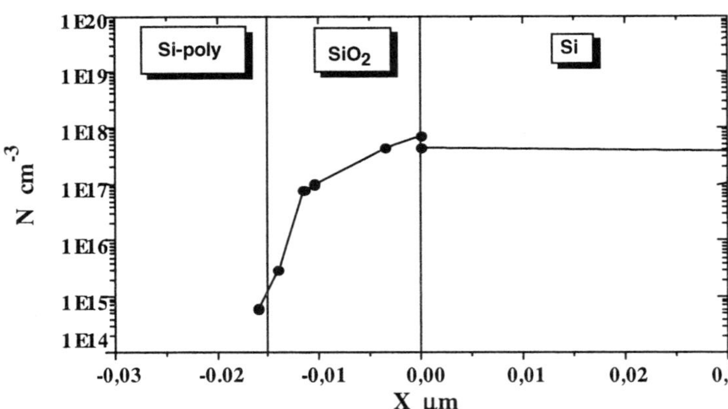

Figure 4. Calculated boron profile in Si_{poly}/SiO_2/p-Si structure processed in a 0.5 μm SOI technology for a boron implantation into the channel E=35 keV, D_b=2×10^{12} cm^{-2} and E=15 keV, D_b= 3.2×10^{12} cm^{-2}.

An increase of the boron concentration in the channel causes also an increase of the boron concentration in the oxide. About 75% of the oxide volume contains boron atoms (see Fig. 4). This forms the basis for the observed influence of the boron channel dose on the radiation hardness. The decrease of the transconductance with radiation dose for the case of D_b=2.4x10^{12} cm^{-2} and its increase for the case of D_b=3.2x10^{12} cm^{-2} (see Fig. 3) points out that the influence of the boron concentration on the creation of border and interface traps is non-monotonous.

3.3. ANNEALING

The negative charge built up after a low dose γ-irradiation (10 and 100 krad) for the case of D_b=3.2x10^{12} cm^{-2} is completely annealed after a treatment at 315°C in argon (see Fig. 2a). The curve after annealing at 315°C lies below the pre-irradiation one, pointing towards an elimination of the radiation-induced negative charge and a predominant influence of the radiation-induced positive charge. The transconductance, as a rule, is lower than after radiation but higher than the initial value and depends on the channel length (Fig. 2b). The 315°C annealing of transistors irradiated with a high dose (1 Mrad) essentially decreases the built-in positive charge resulting in a partial recovery of negatively shifted threshold voltage and increases the transconductance (Fig. 3). The annealing of transistors with a D_b=2.4x10^{12} cm^{-2} doping leads to an increase of both the threshold voltage and the transconductance.

An additional treatment at 415°C anneals the positive charge in case of a low dose irradiation of MOS transistors with D_b=3.2x10^{12} cm^{-2}. The threshold voltage and transconductance shift back to their pre-irradiation values and even a rebound effect is seen. For transistors with a boron channel dose of 2.4x10^{12} cm^{-2} the positive charge is annealed out, and the threshold voltage and the transconductance also shift towards their pre-irradiation values.

There are many experimental results pointing out that radiation induced oxide charge and interface traps can be completely annealed at 415°C or lower [1]. In ref. [12], for example, it is shown that the positive oxide charge in MOS transistors with a polysilicon gate after a ^{60}Co radiation is annealed after 30 min at 300°C. The interface states can also be annealed at the same temperature but require a 45 min anneal time.

In our case the induced negative charge is annealed at 315°C. At this temperature the induced positive charge is annealed completely in the spacer oxide and only partially in the gate SiO_2. A full annealing is observed only at 415°C. However, in case of a lower boron dose D_b=2.4x10^{12} cm^{-2} the positive charge is not fully annealed even not at 415°C. As can be seen from the experimental results, the boron channel doping promotes the annealing of the positive charge at lower temperatures, i.e., increasing the boron channel doping promotes a “softer” degradation of MOS transistors.

3.4. MODEL

The dependence of the electron mobility on the channel length is shown in Fig. 5. The values are calculated according to relation (3) by using the measured values of the transconductance. As can be seen, after 10^5 rad irradiation (D_b=3.2x10^{12} cm^{-2}) the μ_n is increased for channel lengths in the range 0.5-1.5 μm. The mobility increase is correlated with an increase of built in negative charge. In case of a 10^6 rad radiation the electron mobility practically doesn't change.

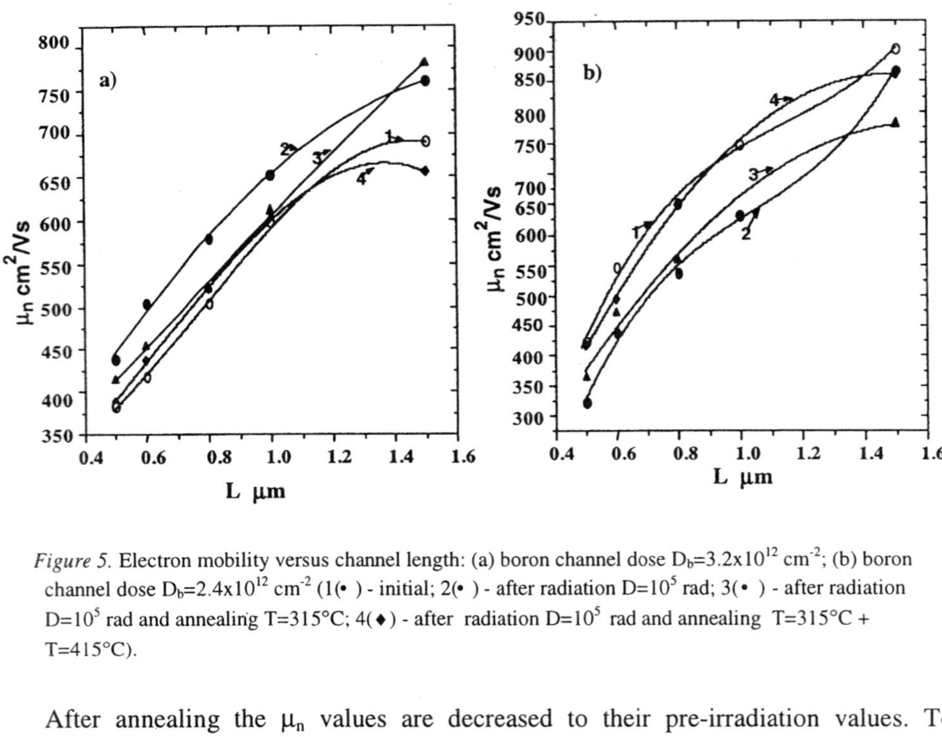

Figure 5. Electron mobility versus channel length: (a) boron channel dose D_b=3.2x10^{12} cm^{-2}; (b) boron channel dose D_b=2.4x10^{12} cm^{-2} (1(•) - initial; 2(•) - after radiation D=10^5 rad; 3(•) - after radiation D=10^5 rad and annealing T=315°C; 4(♦) - after radiation D=10^5 rad and annealing T=315°C + T=415°C).

After annealing the μ_n values are decreased to their pre-irradiation values. To explain the observed correlation between the built in negative charge and the μ_n (see Figs 2 and 5) the following model is proposed.

As shown in Fig. 4 boron impurities are built into the oxide during the gate oxidation. The boron is embedded in the SiO_2 matrix, whereby the B-O bonds are weaker than the Si-O bonds. During irradiation the B-O bonds are broken preliminary at the Si-SiO_2 interface due to the disturbing influence of the interface. After a charge exchange with the channel there are dangling bonds associated with the neutral B atom and the negatively charged oxygen O^- bonds.

The different reactions taking place in the oxide during the irradiation and the subsequent annealing are summarized in Table 2.

TABLE 2. Reactions occurring during irradiation and annealing

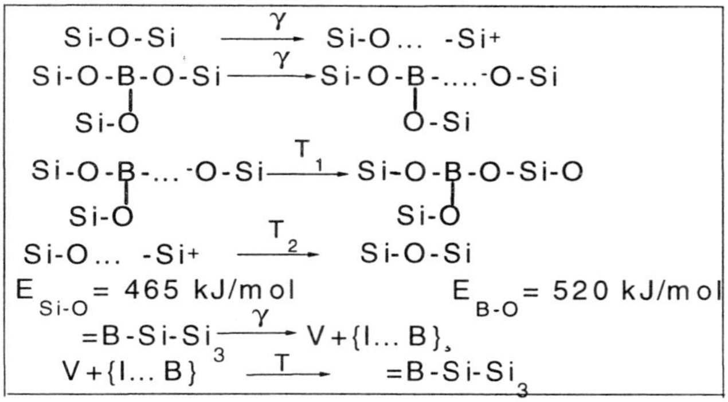

(4)
(5)
(6)
(7)
(8)
(9)

The creation of traps according to reactions (5) and (6) dominates in case of MOSFETs with a boron channel dose of D_b=3.2x10^{12} cm^{-2}. This transformation creates the negatively charged interface and border traps, which are experimentally observed. The negatively charged traps influence the electron mobility in the channel.

During annealing at T_1=315°C traps vanish partially in accordance with (6). In this case complexes with short-range-order corresponding to B_2O_3 oxide are formed. During a subsequent treatment at T_2=415°C hole traps begin to anneal intensively (7). During irradiation and annealing reactions (8) and (9) take simultaneously place in the silicon, in particular in the channel region. The formation of temporary pairs {B...I} in the samples with a higher boron concentration, promotes the annealing of radiation-induced defects during the heat treatments.

4. Conclusion

Decreasing the gate dielectric thickness and increasing the channel boron concentration strongly influence the γ-radiation hardness of MOS transistors. The boron concentration has an impact on the positive charge built up into the oxide. The positive charge is decreased in case of a higher boron concentration. After a low dose radiation (10^4, 10^5 rad) the negative charge in the oxide dominates. The part of the charge that can be annealed at a lower temperature increases with increasing boron doping dose. An improvement of the MOS transistor parameters (G_m, μ_n) after a low dose radiation and subsequent annealing has been observed.

References.

1. Ma, T.P., Dressendorfer, P.V. (1989) *Ionizing radiation effects in MOS devices and circuits,* A Wiley - Interscience Publication, John Wiley & Sons, New York.
2. Litovchenko, V.G., Kiblik, V.Ya., Litvinov, R.O. (1982) The influence of radiation - thermal treatments on semiconductor structures characteristics. *Optoelectronics and Semiconductor Technique* **1,** 69-73 (in Russian).
3. Borkovskaya, O.Yu., Dmitruk, N.L., Kiblik, V.Ya., Litovchenko, V.G. and Litvinov, R.O. (1986) Radiation stimulated annealing of structural defects in layer structures based on Si and III-V compounds. *Acta Physica Hungarica* **59 (3-4),** 327-332.
4. Lisovskii, I.P., Litvinov, R.O., Litovchenko, V.G. (1986) Effect of radiation ordering in Si-SiO_2 structures at ion and plasma-thermal treatments of surface. *Poverchnost. Phys. Chem. Mechan.* **6**, 69-75 (in Russian).
5. Sze, S.M. (1981) *Physics of Semiconductor Devices,* A Wiley - Interscience Publication, John Wiley & Sons, New York.
6. Fleetwood, D.M. (1992) Border traps in MOS devices. *IEEE Trans. Nucl. Sci.* **39**, 269.
7. Fleetwood, D.M., Winokur, P.S., Reber, R.A., Meisenheimer, T.L., Schwank, J.R., Shaneyfelt, M.R. and Riewe, L.C. (1993) Effects of oxide traps, interface traps, and border traps on MOS devices. *J.Appl.Phys.* **73,** 5058.
8. Fleetwood, D.M., Warren, W.L., Shaneyfelt, M., Devine, R. and Scofield, J.N. (1995) Enhanced MOS 1/f noise due to near-interfacial oxygen deficiency. *J. Non-Cryst. Solids* **187**, 199.
9. Fleetwood, D.M., Reber, R.A. and Winokur, P.S. (1991) Effect of bias on thermally stimulated current (TSC) in irradiated MOS devices. *IEEE Trans. Nucl. Sci.* **38,** 1066.
10. Fleetwood, D.M., Miller, S.L., Reber, R., McWhorter, P., Winokur, P., Shaneyfelt, M. and Schwank, J. (1992) New insights into radiation-induced oxide-trap charge through TSC measurements and analysis. *IEEE Trans. Nucl. Sci.* **39,** 2192-2203.
11. Reber, R.A., Fleetwood, D.M. (1992) TSC measurements of SiO_2 defect density and energy in irradiated MOS capacitors. *Rev. Sci. Instrum.* **63,** 5714.
12. Sabnis, A.G. (1983) Characterization of annealing of ^{60}Co gamma-ray damage at Si-SiO_2 interface. *IEEE Trans. Nucl. Sci.* **NS-30**, 4094-4099.

RADIATION EFFECTS IN SOI MAGNETIC SENSITIVE ELEMENTS UNDER DIFFERENT RADIATION CONDITIONS

A.D. MOKRUSHIN, N.M. OMELJANOVSKAJA, A.V. LEONOV,
V.N. MORDKOVICH, D.M. PAZHIN
Institute of Microelectronics Technology, Russian Academy of Sciences
142432 Chernogolovka, Moscow District, Russia

Current- and Hall-gate voltage characteristics, as well as C-V characteristics of Field Effect Hall Sensors (FEHS) have been measured before and after gamma-irradiation under different operating conditions. The FEHS operation during irradiation has a profound effect on these characteristics. In some operation the change of silicon layer carrier concentration was observed after light dose while the magnetic sensitivity of FEHS can be the same under doses up to 10 Mrad.

1. Introduction

The stationary or pulse irradiation is one of the most important extreme actions which should be considered at the development of microelectronics devices on SOI structures. Coincidentally with creation of such structures it has appeared a problem of their stability and hardness against ionizing irradiation. The initial stage of radiation investigations in SOI structures and devices has been adequately addressed in Nazarov's review [1]. Here different technological methods for improvement of radiation stability were considered. However the irradiation behaviour of SOI devices must be governed not only by technological factors. In particular there is good reason to believe that the radiation stability of the concrete SOI device depends essentially on operating conditions during irradiation. From this point of view the aim of our study is in general outlines to understand the manner in which the operation mode of some SOI device has influence on radiation effects and to search the ways of radiation hardness improvement in that direction.

2. Research object

In the majority of previous SOI radiation studies the Field Effect Transistors (FET) or MOS condensers had been investigated. In our case the research object was the Field Effect Hall Sensor (FEHS), combining features of FET and conventional Hall sensors [2]. A special feature of this device is that the thickness of Si layer is comparable with the width of space charge region, that appears when voltage is applied to gates. By changing the sign and magnitude of the gate potential one can control the current flowing along Si layer and the magnitude of the Hall voltage.

F. Balestra et al. (eds.), Progress in SOI Structures and Devices Operating at Extreme Conditions, 221–227.

The Fig.1 illustrates schematically the construction of our FEHS. It is manufactured on the basis of a Silicon-On-Insulator structure in which a 400 nm buried insulator layer (SiO_2) is formed by implanting oxygen ions with further annealing. The SiO_2 layer separates a 200 nm Si layer from the top of n-type silicon substrate. In Si layer the Hall cross is formed by two strips measuring 80 μm long and 30 μm wide. To form ohmic both Hall and current contacts 10 μm long sections adjoining the strip ends were doped by phosphorus with a concentration up to ~ 10^{20} cm^{-3}. All the silicon in the vicinity of the cross was thermally transformed into SiO_2. A 100 nm thick SiO_2 film was applied to the cross surface by pirolysis, above which was deposited Al film (first or top gate). For a second (bottom) gate we used a Si substrate with an Al film applied to its surface.

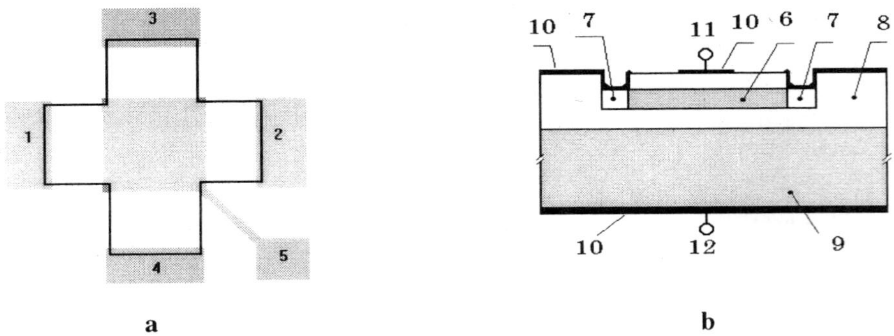

Figure1. The design of the Field Effect Hall Sensor: (a) top view: (1,2) current contacts; (3,4) Hall contacts; (5) electrode of the top gate; (b) cross section: (6) condacting layer (n-Si); (7) the ohmic contacts (n^+-Si); (8) SiO_2; (9) substrate (n-Si); (10) Al film; (11) electrode of the top gate; (12) electrode of the bottom gate.

3. Experiment

We have measured the canal current and Hall potential values of FEHS as a function of gate voltage and C-V characteristics for both FEHS MOS structures before and after gamma^{60}Co irradiation. Fig.2 shows a measurement scheme and typical current and Hall characteristics of initial FEHS for different supply voltage. The main peculiarity of these dependences is in fact that Hall voltage tends to saturation and even slightly decreases at gate voltage above supply voltage while the operating current continues to increase.

Gamma irradiations were carried out under different operating conditions of FEHS:

1. Autonomic operation when voltages are absent on all electrodes.
2. Passive operation when only gate voltages are present.
3. Active operation when gate and supply voltages are applied. In this case by gate voltage variations one can realize an active-open state of FEHS (operating current is sufficiently strong) or an active-closed state (operating current is practically absent).

One ought to note that for both measurement and irradiation processes the electrodes of top and bottom gate were connected together.

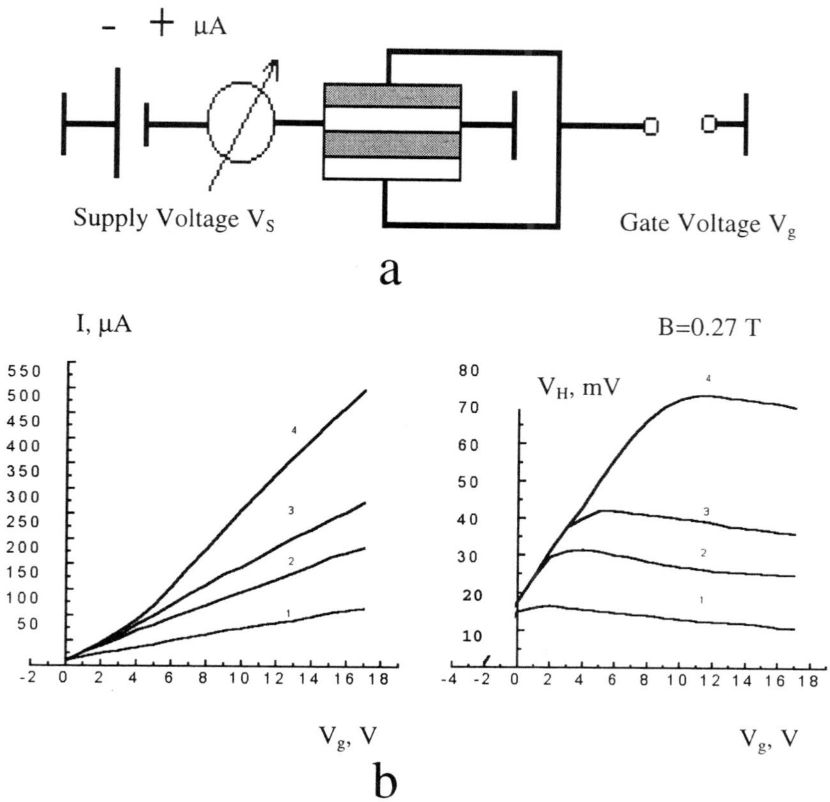

Figure2. Measurement scheme for current- and Hall- gate voltage characteristics (a) and typical dependencies for different supply voltage V_S: 1- 1.5 V, 2- 3 V, 3- 4.5 V, 4- 9 V (b).

4. Results and discussion

In order to demonstrate the essential influence of operating conditions we have selected four FEHS with identical initial characteristics and exposed them to the same dose 50 krad. During irradiation each sensor was maintained under one of the above mentioned operations. The Fig.3 presents Hall voltage and MOS capacity as the functions of the gate voltage for sensors before (curves indicated by zero) and after irradiation (curves from 1 to 4). One can see that the operation mode influences variously on behaviour of the characteristics.

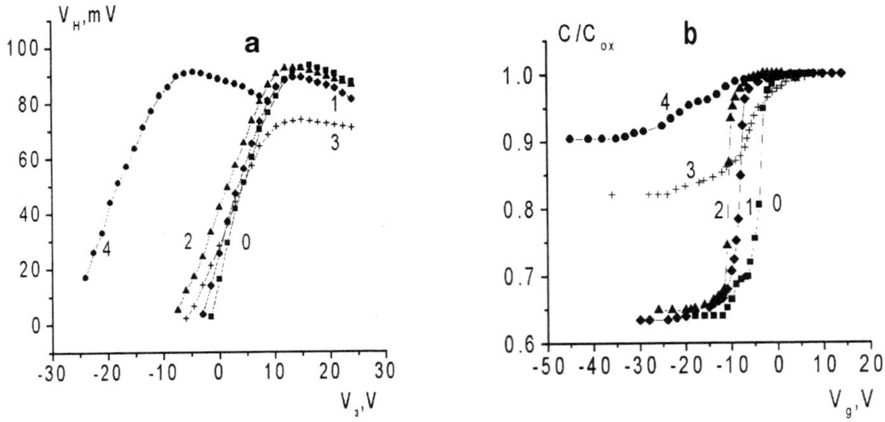

Figure3. Hall (a) and C-V (b) characteristics before and after irradiation under different operating conditions: 0- initial FEHS; 1- active-closed state: $V_g^{ir}=0$; $V_s^{ir}=9V$; 2- autonomic operation: $V_g^{ir}=V_s^{ir}=0$; 3- active-open state: $V_g^{ir}=V_s^{ir}=9V$; 4- passive operation: $V_g^{ir}=9V$; $V_s^{ir}=0$.

The most variations are observed for passive operation with positive gate voltage: curves indicated by 4 are offset to the left by 20 V along gate voltage axis. This shift is conditioned by accumulation of positive charge in silicon oxide of MOS structure as a main result of ionizing irradiation. The autonomic and active-closed operations give rather little shifts of curves (numbers 1 and 2).

The lower level of post-irradiation C-V_g curves 3 and 4 is much higher than initial one (curve 0). From this it follows that free electron concentration of silicon layer increases in result of irradiation in passive and active-open operations. Indeed, as it is demonstrated by data of direct calculations on basis of Volt-Ampere and Hall characteristics (see Fig.4), this concentration enhances in several times. From this figure we notice that qualitative alterations of concentration are similar for both passive and active operations. However the behaviour of Hall and C-V curves is different in these two cases.

Turn attention to Fig.3 once again. We can see that curves 3 as contrasted to curves 4 are not offset to the left. The Hall voltage curve has only a slope and peak value diminished in comparison with initial ones. This effect is illustrated on Fig.5, where Hall voltage characteristics are presented over a wide range of irradiation doses. As the irradiation dose is increased, the most magnitude of Hall voltage shows a decrease. This fact may be attributed to the radiation induced variation in electron mobility. A comparison between curves 1 and 4 on Fig. 5 testifies that electron Hall mobility is halved after irradiation by 10 Mrad. According to our estimates this mobility is something like 700 $cm^2/V*s$ for initial FEHS. An additional point to emphasize is that one can see the intersection for all curves on Fig. 5: at $V_g=0$ Hall voltage is independed on doses. This behaviour of Hall voltages allows to practically important conclusion: one can pick out such an operating point of FEHS that irradiation in active operation has no effect on main device characteristics. For sensor under consideration this point corresponds to gate voltage equal zero. The other sensors may have the intersection points which are different from Vg = 0 but not much. Hence under certain conditions the magnetic sensitivity of FEHS can be unchanged over a wide range of ionizing radiation doses.

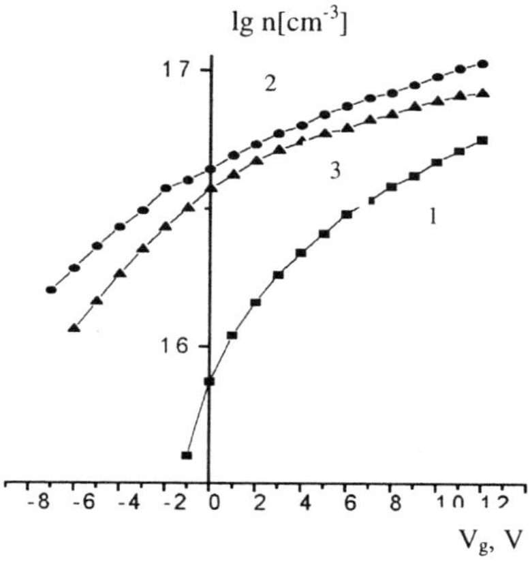

Figure 4. The results of direct calculations of free electron concentration in active silicon region: 1- initial FEHS; after irradiation in passive (2) and active-open (3) operations.

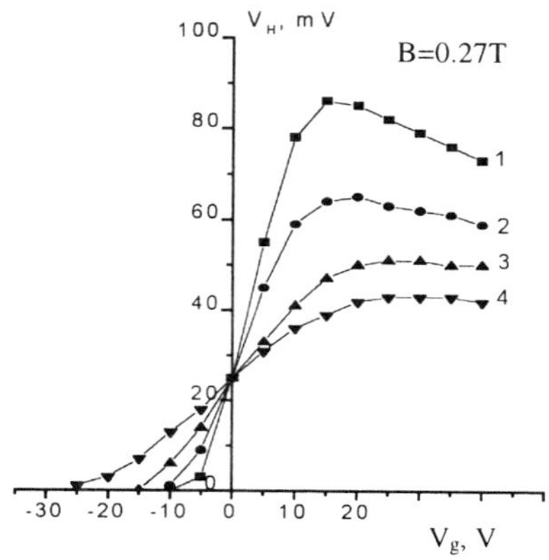

Figure 5. Hall voltage characteristics after different dose irradiation under active-open operation: 1- initial FEHS; 2- 100 krad; 3– 1 Mrad; 4– 10 Mrad.

We have also carried out measurements of the canal current and Hall voltage in relation to dose for FEHS irradiated under different operating conditions. As example Fig. 6 shows such dose dependencies in operating point determined by $V_s = 12$ V and $V_g = 0$ for four FEHS irradiated at once under operation of their own. For autonomic operation it is observed linear increase of Hall voltage with dose rise. For active-open and passive with $V_g^{ir}<0$ operations the Hall voltages change a little at all doses studied (curves 3 and 4 on Fig. 6a). For active-open operation the changes of Hall voltage and current are substantially different (curves indicated by 4).

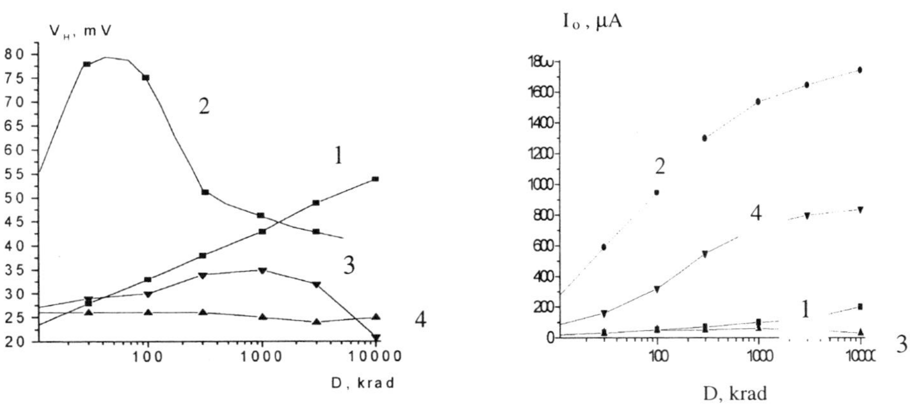

Figure 6. Dose dependencies of Hall voltage (•) and current (b) for FEHS irradiated under different conditions: 1- autonomic operation; 2- passive operation with $V_g^{ir}=9V$; 3- passive operation with $V_g^{ir}=-6V$; 4- active-open operation with $V_g^{ir}=V_s^{ir}=9V$.

The curve indicated by 2 on Fig.6a has well-marked extreme dependence moreover with increase of positive gate voltage the peak position shift in low dose direction. The extreme dose dependencies were previously observed for threshold voltage of SOI Field Effect Transistors[3-4].They were explained by combination of capture processes of radiation-induced holes and electrons in oxides.

Finally let us take up the data of repeated irradiation experiments.Turn attention to Fig.7, where C-V and Hall voltage characteristics are shown. The curves indicated by 1 correspond to the initial sensor. The first irradiation was carried out in passive operation with positive gate bias. The curves 2 are due to results of this irradiation.Then it was carried out the repeated irradiation by the same dose but with opposite in sign gate bias. We can see that the final curves 3 return nearly to their initial positions. These results can be so interpreted by the combined capture of holes and electrons in oxides. Consequently the repeated irradiation with opposite gate bias can be considered as a method of irradiation hardness enhancement of Field Effect Elements on SOI structure.

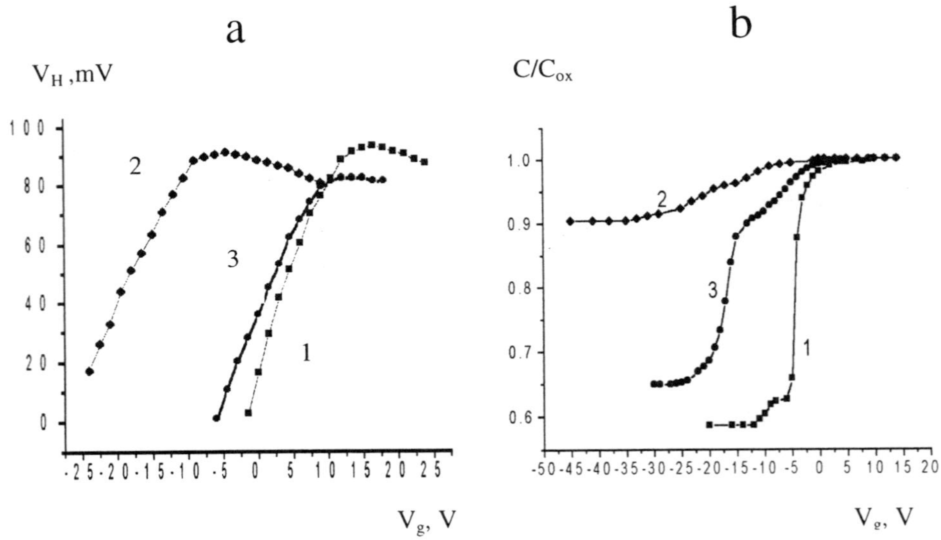

Fig. 7. Hall voltage (a) and C-V (b) characteristics for initial sensor (curves 1), after the first irradiation in passive operation with V_g^{ir}=9V (curves 2) and after repeated irradiation with V_g^{ir} = - 9V.

5. Conclusions

1. The operation modes of Field Effect Hall Sensor during irradiation have a dramatic impact on the radiation effects.
2. One can pick out such on operating point of FEHS that the irradiation in active operation has no effect on Hall voltage.
3. The variation of FEHS characteristics accompanied by irradiation in passive operation with positive gate voltage are compensated substantially by repeated irradiation with negative gate voltage. It can be considered as a method of radiation hardness improvement.

References

1. Nazarov, A.N. (1995) Problems of radiation hardness of SOI structures and devices, in J.P. Colinge et al. (eds.), Physical and Technical Problems of SOI Structures and Devices, 217-239.
2. Akimov, A.G., Barabanenkov, M.Yu., Baranochnikov, M.L., Leonov, A.V., Mokrushin, A.D., Mordkovich, V.N., Omelyanovskaja, N.M. (1998) A Controllable Resistor with features of a Field-Effect Transistor and Field Hall-Effect sensor, Instruments and Experimental Techniques 41, 706-709.
3. Flament, O., Herve, D., Musseau, D., Bonnel, P., Rafaelli, M., Leray, J.L., Morgail, S., Giffard, B., Auberton-Herve, A.J. (1992) Field dependent charge trapping effects in SIMMMOX buried oxides at very high dose, IEEE Trans. Nucl. Sci. 39, 2132-2138.
4. Paillet, P., Autran, J.L., Flument, O., Leray, J.L., Aspor, B., Auberton-Herve, A.J. (1996) X-radiation response of SIMOX buried oxides: influence of the fabrication process, IEEE Trans. Nucl. Sci. 43, 821-825.

SIMILARITY RELATION FOR *I-V* CHARACTERISTICS OF FETs WITH DIFFERENT CHANNEL SHAPE

V.N. DOBROVOLSKY[1], M. BALUCANI[2], and A.FERRARI[2]
[1]*Department of Radiophysics, Kiev University, 2 Glushkov Av., build. 5, 03127 Kiev, Ukraine*; [2]*INFM Unit E6 - Department of Electronic Engineering, Rome University "La Sapienza", Via Eudossiana 18, 00184Rome ", Via Eudossiana 18, 00184 Rome*

1. Introduction

Depending on the FET application, its channel shape may be different [1-5]. However, for all FET types the theory has been developed only in the simplest case of a rectangular transistor channel. Therefore there is a problem of calculating the characteristics of FETs with another channel shape. The similarity relation, obtained in this paper, solves the above problem.

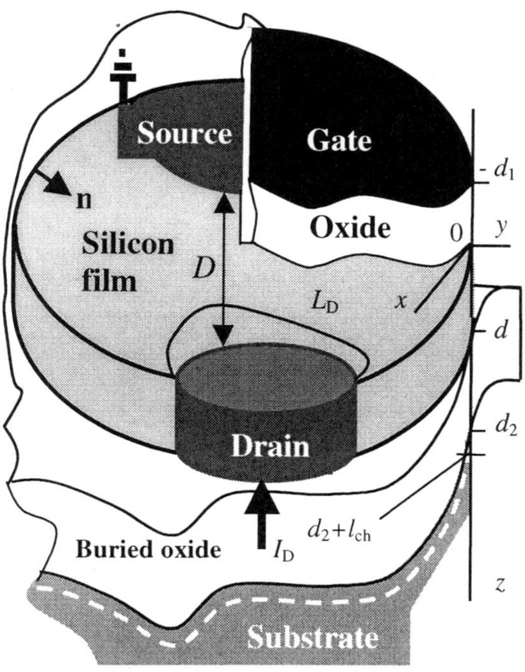

Figure 1. The SOI MOSFET channel of an arbitrary shape.

2. Basic equations

We will consider a *p*-silicon film under a SOI MOSFET gate. The film is bounded by an arbitrary shape contour (Fig. 1). The *n*-channel may be near the surface $z = 0$, or $z = d$, or the two above channels may join. A drain current I_D flows between the n^+ source and n^+ drain. Its density is

$$\mathbf{j} = -e\mu n \,\mathrm{grad}\varphi + kT\mu \,\mathrm{grad} n,$$

where e is the electron charge, k is the Boltzmann constant, T is the temperature, μ and n are the electron mobility and concentration, respectively,

F. Balestra et al. (eds.), *Progress in SOI Structures and Devices Operating at Extreme Conditions,* 229–232.

φ is the potential. The sizes D, d, d_1, (d_2-d) and l_{ch} are the minimum distance between the source and drain, the thicknesses of the silicon film, both oxide films and the space charge region in the substrate, respectively.

We mark the value corresponding to current absence by the subscript "o", and those in the source (drain) by the subscript "S" ("D"). Let us assume that $\varphi_S = \varphi_{oS} = 0$ and introduce the normalized electron Fermi quasi-level $\xi(x,y,z)$ through the expression $n = n_{oS}\exp[(e\varphi/kT)-\xi]$ [6]. Then from the above expressions and $n_{S,D} = n_{oS,oD}$ we obtain $\mathbf{j} = -kT\mu n\,\mathrm{grad}\xi$, $\xi_S = \xi_o = 0$ and $\xi_D = eV_D/kT$, where V_D is the drain voltage. If $D >> d, d_1, (d_2-d), l_{ch}$, then one can write $j_z = 0$ and $\xi = \xi(x,y)$. In this case the Poisson equation in the SOI MOSFET and its boundary conditions can be written as $\partial^2\varphi/\partial z^2 = -4\pi\rho/\varepsilon(z)$ and $\varphi|_{z=-d_1} = \varphi_{gate}$, $\varphi|_{z\to\infty} = \varphi_{subs}$, respectively. Here ρ is the space charge density, ε is the permittivity, φ_{gate} and φ_{subs} are the potential of the gate and substrate.

In the oxide films $\rho = 0$, or ρ depends either on z (if there is a built-in charge) or on $(e\varphi/kT - \xi)$ (if the surface electronic states at $z = 0, d$ serve as electron traps). In the silicon film and substrate $\rho = e(p - n - N)$, where $p = p_{oS}\exp(-e\varphi/kT)$ and $N(z)$ are the concentration of the holes and ionized impurities. From the above formulae follows that the ρ density in the SOI MOSFET may be written as $\rho = \rho(z,\varphi,\xi)$. From this functional dependence, the above Poisson equation and its boundary conditions the dependence $\varphi = \varphi(z,\xi)$ follows.

3. Interrelationship of channel and uniform wafer electric field

The surface conductivity is $Ó_{surf} = e\mu n_{oS}\int_0^d \exp[e\varphi(z,\xi)/kT - \xi]dz = \Sigma_{surf}(\xi)$. After introducing the current potential $Ø[\xi(x,y)] = (kT/e)\int_0^\xi \Sigma_{surf}(\xi)d\xi$ [7], we write the surface current density $\mathbf{J} = \int_0^d \mathbf{j}\cdot dz$ as $\mathbf{J} = -\mathrm{grad}Ø$. Substituting $\mathbf{J}$ into $\mathrm{div}\mathbf{J} = 0$ gives

$$\Delta\Psi = 0\,. \qquad (1)$$

At the silicon film boundary sections without contacts $J_n = 0$ and the equality

$$(\mathrm{grad}\Psi\cdot\mathbf{n}) = 0 \qquad (2)$$

holds, where $\mathbf{n}$ is the unit vector normal to the boundary. For the source and drain the following expressions hold

$$Ø_S = 0\,, \qquad (3)$$

$$\varnothing_{\mathrm{D}} = \varnothing(eV_{\mathrm{D}}/kT). \tag{4}$$

The drain current is given by

$$I_{\mathrm{D}} = -\int_{L_{\mathrm{D}}} (\mathrm{grad}\Psi \cdot \mathbf{n}) dL_{\mathrm{D}}, \tag{5}$$

where L_{D} is any contour connecting the drain ends.

The $\varnothing(x, y)$ value in the silicon film, through which the current I_{D} flows, can be found as a solution of equation (1) using the conditions (2), (3) and (5). Using this $\varnothing(x, y)$ and equation (4), one can determine the drain voltage V_{D} that corresponds to the current I_{D}. The $\Psi(\xi)$ relation of transistors with a differently shaped film are identical, if their physical properties and potentials φ_{gate} and φ_{subs} are the same. Below only this case will be considered.

Now we will consider a uniform wafer through which a current I_{D}^* flows. The shape and size of its contacts, as well as the contour bounding the wafer, be the same as corresponding characteristics of the transistor silicon film (Fig. 1). The wafer conductivity and thickness are σ^* and d^*, its resistance is r^*. We mark all quantities in the wafer by the superscript "*".

From $\mathbf{j}^* = -\sigma^* \mathrm{grad}\varphi^*$ and $\mathrm{div}\mathbf{j}^* = 0$ one can obtain that the equation for $\sigma^* d^* \varphi^*$ and its boundary conditions are identical to (1) - (3) and (5). Therefore via $I_{\mathrm{D}}^* = I_{\mathrm{D}}$ the equality $\Psi(x, y) = \sigma^* d^* \varphi^*(x, y)$ is valid. Using the last equality and expression (4), we write $\varnothing(eV_{\mathrm{D}}/kT) = \sigma^* d^* I_{\mathrm{D}}^* r^* = I_{\mathrm{D}}(V_{\mathrm{D}}) R^*$, where $I_{\mathrm{D}}^* r^*$ is the voltage across the wafer contacts and $R^* = \sigma^* d^* r^*$ is the wafer non-dimensional resistance. For a rectangular transistor channel its value is equal to the ratio of the channel length to channel width L/W. Using this, we write

$$I_{\mathrm{D}}(V_{\mathrm{D}}) = \frac{L}{R^* W} I_{\mathrm{D}}^{\mathrm{rect}}(V_{\mathrm{D}}), \tag{6}$$

where $I_{\mathrm{D}}^{\mathrm{rect}}$ is the current through rectangular channel. If the R^* value is known, the similarity relation (6) allows to find the *I-V* characteristics of a transistor of arbitrary channel shape, using the result for the transistor with a rectangular channel [1-6].

The resistance R^* can be found from a solution of the Laplace equation $\Delta\varphi^*(x, y) = 0$. Its solution methods are well known [8]. In some cases, the formula for R^* can be simple. For example, for the Corbino disk $R^* = [\ln(r_2/r_1)]/2\pi$, where $r_{1,2}$ are its radiuses. In addition, there is a simple analogue method of R^* definition.. Its value can be found by a measurement of the resistance of any wafer, which is geometrically similar to the transistor channel. All these wafers have non-dimensional resistance R^*. The wafer can have any convenient size and the $\sigma^* d^*$ value. Different materials, for example conductive paper, can be used for such wafer fabrication.

4. Conclusion

We obtained the similarity relation (6) for the SOI MOSFET. However when $d \to \infty$ (Fig. 1), a SOI MOSFET transforms to a bulk MOSFET. In the case of $d_1 = 0$, $(d_2 - d) \to \infty$ and when the silicon film conduction type is changed to *n*, it transforms to a JFET or MESFET. Therefore the similarity relation (6) is valid for these types of transistors too.

References

1. S.M. Sze: *Physics of Semiconductor Devices*, 2nd Edn, p. 355. Wiley-Interscience, New York (1981).
2. M. Shur, *GaAs Devices and Circuits*, p. 603. Plenum Press, New York and London (1987).
3. J.P. Colinge, *Silicon-On-Insulator Technology: Materials to VLSI*, p. 262. Kluwer Academic Publishers (1991).
4. S. Cristoloveanu and S.S. Li, *Electrical Characterization of Silicon-On-Insulator Materials and Devices*, p. 378 . Kluwer Academic Publishers (1995).
5. K.Ng. Kwok, *Complete Guide to Semiconductor Devices*, p. 677, McGraw-Hill, New York (1995).
6. V. N. Dobrovolsky and V. G. Litovchenko, *Surface Electronic Transport Phenomena in Semiconductors*, p. 220. Clarendon Press, Oxford (1991).
7. V.N. Dobrovolsky and A.N. Krolevets, *J. Appl. Phys.*, **85**, 1956 (1999).
8. P.M. Morse and H. Feshbach, *Methods of theoretical physics*, part II, p. 881, McGraw-Hill, New York (1953).

LASER-RECRYSTALLIZED SOI LAYERS FOR SENSOR APPLICATIONS AT CRYOGENIC TEMPERATURES

A. DRUZHININ, E. LAVITSKA,
I. MARYAMOVA, Y. KHOVERKO
Lviv Polytechnical University
Kotlarevsky str. 1, Lviv 79013, Ukraine

1. Introduction

Silicon-on-insulator (SOI) layers showed their perspectives of application in cryogenic electronics [1]. An aim of the studies presented was to investigate perspectives of their use in the development of mechanical and temperature sensors for operation at cryogenic temperatures. The laser recrystallization represents a technique providing adjustment of the electrical and piezoresistive parameters of polysilicon layers. That is why such a method has been used in the studies presented.

2. Carrier transport in poly-Si layers at cryogenic temperatures

Polycrystalline silicon represents a combination of small monocrystalline grains joined together by the grain boundaries (GB) being regions of disordered atoms. In terms of the carrier trapping model in the doped material mobile carriers are trapped by the energy states at the grain boundaries. As a result of such a trapping near of the grain boundaries the spatial charged (carrier depleted) regions arise as well as the potential barrier at the grain boundary.

In the approach of partial grain depletion starting from the model [2] for boron-doped p-type polycrystalline silicon the software has been developed to calculate numerically electrical and piezoresistive properties of polysilicon layers with the average grain size as a parameter. The main transport mechanism in polycrystalline poly-Si was suggested to be thermionic emission over the barriers at the grain boundaries combined with diffusion through the grain boundaries and the carrier drift through the crystallites [3]. The numerical calculations were carried out based on suggestion about the partial depletion of the grain [4].

It was shown [5] that for the variety of polycrystalline materials the electrical conductivity is very similar to that in disordered semiconductors. Depending on the average grain size, doping level and other factors different carrier transport mechanisms become dominating (from the barrier mechanism to electron percolation being usual for

F. Balestra et al. (eds.), Progress in SOI Structures and Devices Operating at Extreme Conditions, 233–237.

disordered systems). In general case an effective electrical conductivity could be written in form:

$$\sigma_{ef} = \sigma_g + \sigma_b (h / r_0), \tag{1}$$

with σ_g and σ_b as the grain and barrier electrical conductivity, correspondingly, h as thickness of the grain boundary, r_0 as effective dimension of the grain.

If the grain volume being totally or almost totally depleted, the first component in formula (1) could be neglected, $\sigma_g \to 0$, and the electrical conductivity is performed by the carrier transport through localized states at the grain boundaries. Such a condition can be realized at low doping densities and for small grain size independently on the observation temperature.

At cryogenic temperatures when a significant freezing of the carriers is expected number of free carriers within the grain volume becomes very small, excepting the case of very high doping densities (metallic-type electrical conductance). Therefore, one may consider the quantum transport by the states at GB as the main transport mechanism. Difference in the barrier heights at GBs results in a random potential relief due to the energy bands bend near the GBs. That is why this system may be considered as a heavily doped and strongly compensated semiconductor where the boundary states play a role of compensating impurities. The lower temperatures the more contribution of such quantum transport in the electrical properties; it could be described in terms of percolation theory [6].

3. Experimental

In order to study possibility to use poly-Si resistors at cryogenic temperatures with ZMR as a method to improve their parameters experimental measurements of resistance of poly-Si samples have been carried out in the range 4.2-300 K. Figs. 1 and 2 show R (T) dependencies in boron-doped laser-recrystallized SOI samples.

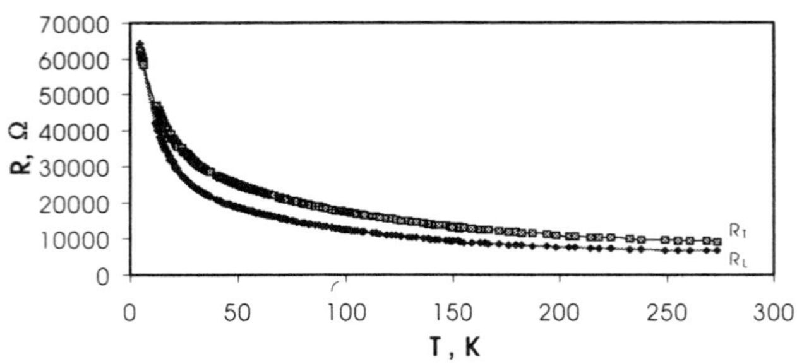

Fig.1. Temperature dependence of the resistance for laser-recrystallized poly-Si resistors with $N = 4.8 \times 10^{18}$ cm^{-3} (R_T – transversal and R_L – longitudinal resistor as related to the laser scan direction).

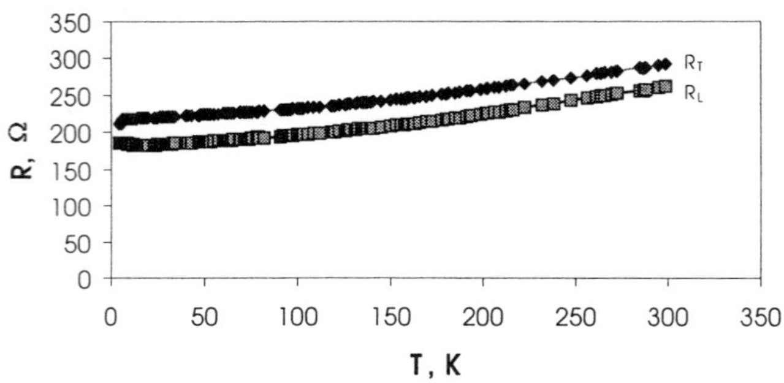

Fig.2. Low-temperature resistance of recrystallized poly-Si resistors with $N = 1.7 \times 10^{20}$ cm^{-3}. R_T and R_L represent transversal and longitudinal resistance as related to the laser scan direction.

Figure 3 shows the resistivity of the same experimental sample as in Fig.1 but before the laser recrystallization. The temperature dependence here is shown *vs* $T^{-1/4}$ axis.

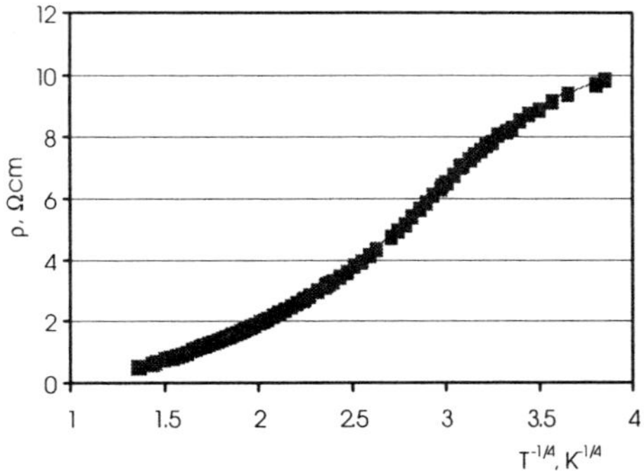

Fig.3. Resistivity of non-recrystallized poly-Si sample with $N=2.4\times10^{18}$ cm^{-3}.

It is clear from Fig.3 that in the range 8-25 K the sample follows the Mott's law $\rho\sim T^{-1/4}$. This gives us evidence of dominating hopping conductance at low temperatures. Such a moderately doped non-recrystallized poly-Si has a strong temperature coefficient of resistance and could be recommended as a thermoresistor in the wide temperature range.

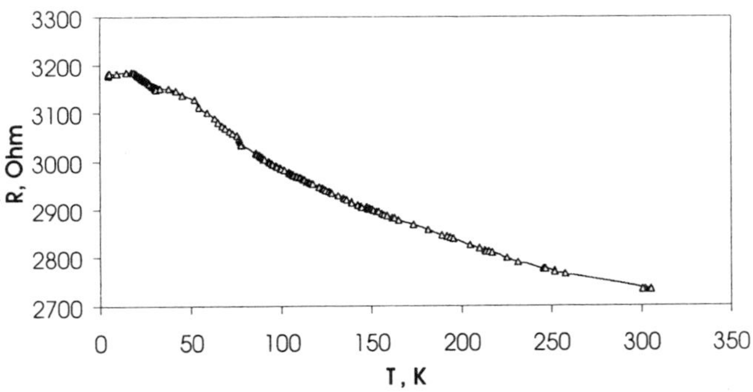

Fig.4. Resistance of non-recrystallized poly-Si resistor with $N=3.9\times10^{19}$ cm^{-3}.

Fig. 4 shows the resistance as a function of the temperature for poly-Si resistor the same as in Fig.2 but before the laser recrystallization. The electrical conductance here manifests the behavior being intermediate between the hopping mechanism and that for metallic-type conductance.

Fig.5 represents a calibrating curve for high-sensitive piezoresistor in the range 4.2-300 K. Other samples (Figs. 1-2) show a reasonable change of their resistance in the range 4.2-300 K. This result combined with theoretical predictions of their piezoresistive properties [7] let us expect possibility of their operation as piezoresistive elements at cryogenic temperatures.

The heavily doped poly-Si samples show a good linearity of their V-I curves (Fig.5). At the same time there was a certain non-linearity in moderately doped samples (Fig.6) at 4.2 K that remained significant even after the laser recrystallization.

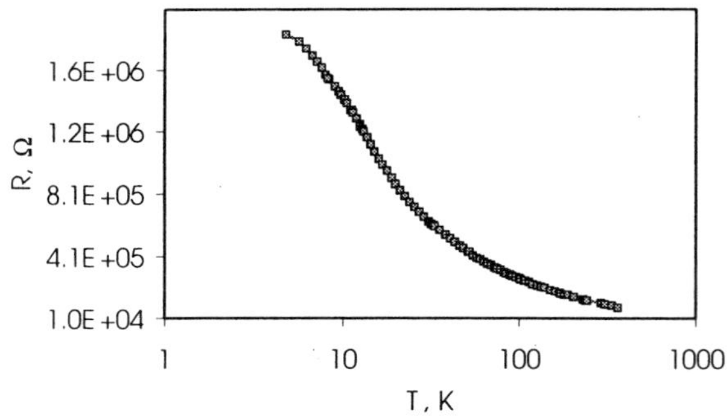

Fig.5. Calibrating curve of polysilicon thermoresistor with $N = 2.4\times10^{18}$ cm^{-3}.

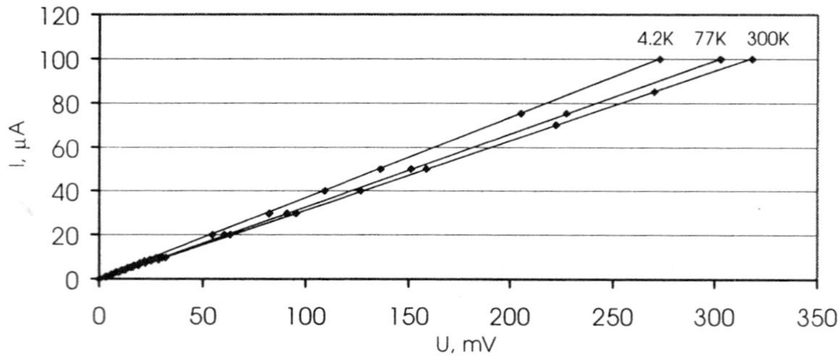

Fig.6. Experimental I-V characteristics of non-recrystallized poly-Si resistors with apparent carrier concentration at 20°C $N=3.9\times10^{19}$ cm^{-3}.

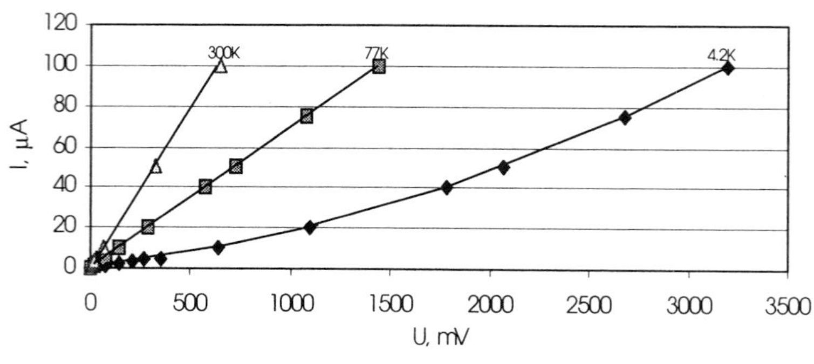

Fig.7. Experimental I-V characteristics of laser-recrystallized poly-Si resistors with apparent carrier concentration at 20°C $N=4.8\times10^{18}$ cm^{-3}.

4. Conclusions

From our theoretical and experimental studies, it could be concluded that:

- non-recrystallized moderately doped poly-Si shows high temperature coefficient of resistance and may be recommended as a temperature-sensitive element;

- laser-recrystallized moderately doped poly-Si resistors relatively weakly change their resistance in the range 77-300 K; they could be recommended for piezoresistive applications in this range;
- laser-recrystallized heavily doped poly-Si resistors could be recommended for piezoresistive applications in the wide temperature range 4.2-300 K.

References

1. Claeys, C. and Simon, E. (2000) Perspectives of silicon-on-insulator technologies for cryogenic electronics, in P.L.F.Hemment et al. (eds.), *Perspectives, Science and Technologies for Novel Silicon on Insulator Devices*, Kluwer Academic Publishers, Dordrecht, pp.233-247.
2. Druzhinin, A., Maryamova, I., Lavitska, E., Pankov, Y. and Kogut, I. (2000) Laser recrystallized polysilicon layers for sensor application: electrical and piezoresistive characterization, in P.L.F.Hemment et al. (eds.), *Perspectives, Science and Technologies for Novel Silicon on Insulator Devices*, Kluwer Academic Publishers, Dordrecht, pp.127-135.
3. Seto, J. (1975) The electrical properties of polycrystalline silicon films. *J.Appl.Phys.*, **46**, pp.5247-5254.
4. Lu, N.C.-C., Gerzberg, L. and Lu, C.-Y. (1981) Modeling and optimization of monolithic polycrystalline silicon resistors, *IEEE Trans. Electron Devices,* **ED-28**, pp. 818-830.
5. Kobka, V.G., Komirenko, R.P., Korniushin, Y.V. et al. (1982) On the electric conductivity of polycrystalline semiconductors, *Fizika Tverdogo Tela,* **16**, pp. 2176-2178.
6. Shklovskii, B.I. and Efros, A.L. (1979) *Electron Properties of Doped Semiconductors*, Nauka, Moskow.
7. Lavitska, E.N., Maryamova, I.I., Druzhinin, A.A. and Pankov, Y.M. (1996) Influence of the structure on piezoresistance of polysilicon thin films, *Inorganic Materials*, **32**, pp.1016-1018.

CHARACTERIZATION AND MODELING OF ADVANCED SOI MATERIALS AND DEVICES

F. ALLIBERT [(1,2)], J. PRETET [(1)], T. ERNST [(1,3)], J. JOMAAH[(1)], S. CRISTOLOVEANU [(1)]
(1) Laboratoire de Physique des Composants à Semiconducteurs (UMR CNRS-INPG), ENSERG, BP 257, 38016 Grenoble Cedex 1, France
(2) SOITEC S.A., Chemin des Franques, Bernin, 38926 Crolles, France
(3) STMicroelectronics, 750 rue Jean Monnet, 38920 Crolles, France

1. Introduction

The recent take off of SOI technology for VLSI applications is explained by combined effects : huge progress in SOI substrates (cost and quality), relative easiness to transpose classical CMOS processing on partially depleted SOI, and inherent gains in performance and power dissipation [1]. However, the use of SOI is far from being limited to classical CMOS and offers a whole range of new possibilities, in the domain of alternative materials as well as for conceiving innovative SOI-specific devices. Our interest in wafer characterization is focused on the Pseudo-MOSFET [2] and its applications. On the device side, we conceive, simulate and characterize ultimate or novel SOI components.

2. Materials

SOI wafers, and especially bonded wafers, offer new possibilities in material structure and device architecture. Different options such as silicon film on highly resistive silicon substrates (for RF applications), Silicon-On-Quartz, or Silicon-On-Sapphire (SOS) are already available. Other promising structures, such as Silicon-On-Nothing, compound and wide-gap semiconductor films on oxidized silicon substrates, buried structures, various types of uncommon stacking, or extremely thin active films, have already shown capability to expand the frontiers of the conventional Si technology. It is to be noticed that buried structures can take the form of a material modification prior to bonding (for example in ground plane MOSFETs) or of bonded CMOS circuitry.

For either conventional CMOS or exotic applications, the material quality is of utmost importance, since the parameters of the active film, buried oxide, silicon/oxide and oxide/substrate interfaces or the substrate underneath the BOX will determine the performance of specific structures. For this reason, and because the quality of the material is always an issue in microelectronics, wafer characterization is a main concern. A large number of methods are available for unveiling this purpose: Hall effect, spreading resistance, Pseudo-MOSFET technique [3], deep-level transient

F. Balestra et al. (eds.), Progress in SOI Structures and Devices Operating at Extreme Conditions, 239–247.

spectroscopy, characterization through devices (transistor and capacitance measurements)... Special interest was raised by the Pseudo-MOSFET (ψ-MOS) technique because it enables us to perform On-Wafer measurements with a minimum processing, thus reducing the process-induced material modifications.

2.1. PSEUDO-MOSFET TECHNIQUE

The ψ-MOS is a kind of transistor where the SOI Buried Oxide (BOX) acts as a gate oxide. It is possible to use the ψ-MOS much as one would use a MOS transistor to evaluate parameters such as carrier lifetime, doping, interface states density, threshold and flat band voltages, subthreshold swing, carrier mobility, etc. Only results concerning the mobility and interface state density are presented here because they are the most characteristic about the film and Si/BOX quality.

The ψ-MOS technique is very easy to perform. It uses the substrate as a gate and the BOX as the gate oxide, while the source and drain are formed by two needles set with a given pressure on the active film (Fig.1). The pressure applied on the needles helps piercing the native oxide and creates ohmic contacts which serve as source and drain. The electron/hole conduction takes place along the active film/BOX interface.

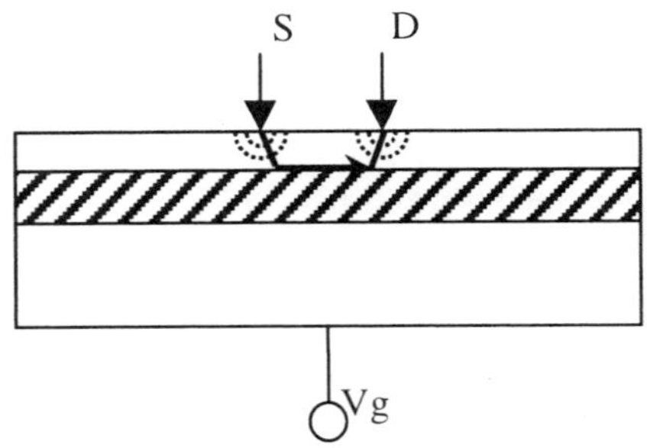

Fig.1 ψ-MOS principle.

When the substrate is biased, an inversion layer appears above the BOX (i.e. at the bottom of the silicon film) and the conduction channel can be created between source and drain, just as in a regular CMOS transistor. Figures 2 and 3 show the drain current and transconductance as a fonction of gate voltage which are perfect MOS-like characteristics. One difference between the ψ-MOS transistor and regular MOSFETs is that it can be operated either as a p-MOS or n-MOS transistor because there is no intentionally doped source or drain.

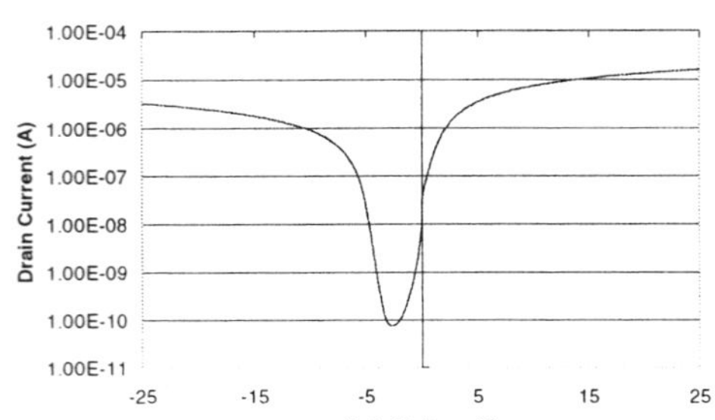

Fig.2 Drain current obtained with the ψ-MOS method in UNIBOND®.

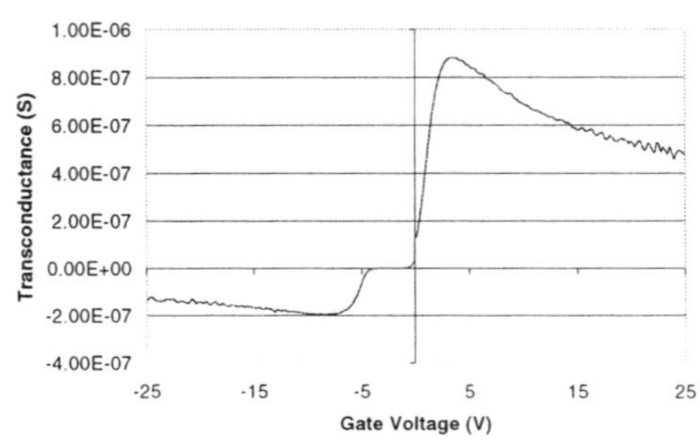

Fig.3 Transconductance of the ψ-MOS transistor in UNIBOND®.

2.2. EVOLUTION OF UNIBOND PROPERTIES

The results shown in Table 1 have been obtained on various UNIBOND® wafers arranged in chronological order: wafer A, the oldest, is several years old and wafer F, the most recent among those presented here, was processed during summer 2000.

Table 1: Pseudo-MOSFETs results in various generations of Unibond wafers.

Wafer	G_{max}-(nS)	G_{max}+(nS)	$\mu_p(cm^2V^{-1}s^{-1})$	$\mu_n(cm^2V^{-1}s^{-1})$	D_{it} $(x10^{11}cm^{-2}V^{-1})$
A[3]	-69.4	290	158	511	5.8
B[3]	-76.3	284	160	535	4.3
C[3]	-79	362	165	632	3.5
D[4]	-110	339	169	610	2.8
E	-97	446	165	745	1.9
F	-107	460	177	747	1.0

This table reflects the overall progress of the material properties. It is worth noticing the steady reduction of the interface states density (D_{it}) at the active film/BOX interface. This feature and the improved quality of the film explain the increase in carrier mobility (μ_n) and transconductance peak (G_{max}, measured for Vd=100mV) in particular for electrons.

2.3. EXTENSION TO RECENT SOS MATERIALS

The very thick sapphire insulator normally prevents the application of the ψ-MOS to characterize SOS wafers (the necessary operating gate voltages would be in the 10-100 kV range). In our case however, the sapphire substrate was thinned down to 30μm (Fig.4) by etch-back, grinding and mechanical polishing [4]. The thinned insulator substrate allowed us to operate the ψ-MOS transistor using reasonable voltages, from 0 to 1100V. "Classical" drain current curves are presented on Figure 5.
The results obtained for these specimens are: $29cm^2/Vs$ electron mobility at the silicon/sapphire interface, and interface state density of $1.5.10^{12}cm^{-2}eV^{-1}$. The difference between the threshold and flat-band voltages is about 1400V.

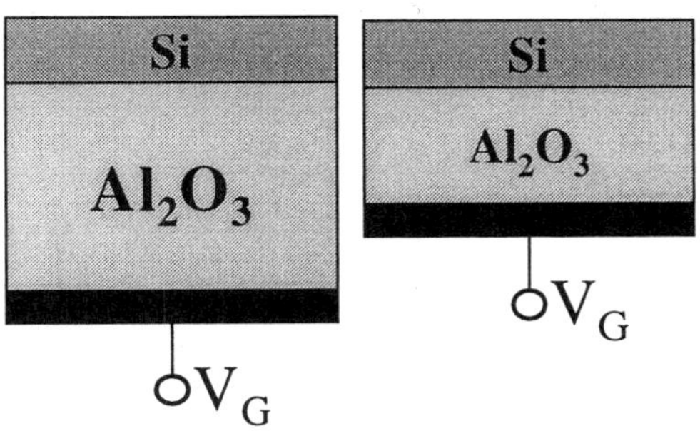

Fig.4 a) Standard and b) thinned SOS structures.

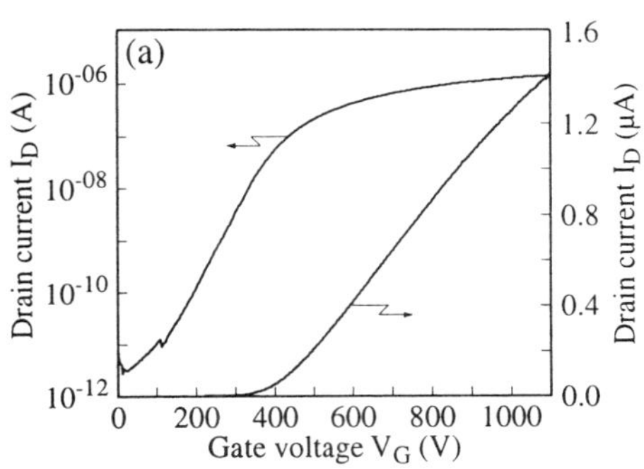

Fig.5 Drain current in a SOS wafer

These results show that the electron mobility is still affected by lateral stress and interface defects, in spite of the significant progress achieved by solid-phase re-epitaxial growth and annealing.

3. Devices

The advantages of SOI over bulk-silicon substrates have been demonstrated in terms of performance, reliability [1] and new opportunities (scaling, architecture) for the future generations of devices. With the progress of SOI technologies, we explore the limits of SOI using Fully-Depleted SOI MOSFET's with very short channel (< 0.1 μm), narrow channel (< 0.3 μm), ultra-thin Si film (< 20 nm), thin and/or low-K buried oxide, low doping and mid-gap metal gate. New architectures for ultimate SOI transistors such as Dynamic-Threshold (DT-MOS), Ground-Plane (GP) or Double-Gate (DG) SOI MOSFET are also under investigation. In the following sections, we will discuss and illustrate, with selected examples, contributions in characterization, simulation and modeling.

3.1. NARROW-CHANNEL MOSFETS

The interest for SOI technology in the field of low-power and low-voltage VLSI application has led to the need for very small devices. Small geometries mean short but also narrow channel devices. If short-channel effects have extensively been studied, there are relatively few results on narrow-channel effects. We investigate narrow-channel effects in fully depleted, LOCOS isolated n-MOSFETs devices where the peculiar lateral topography leads to a parasitic conduction path between source and drain. The main effect of this parasitic conduction, which takes place along the highly inhomogeneous edges of the Si island, on the device characteristics (Figure 7) is the reduction of the threshold voltage and field effect mobility with decreasing width [5]. The experiments also demonstrate an improvement in the floating body effects and breakdown voltage in narrow MOSFETs. The thinner film and thicker gate oxide of the lateral transistor explain these results. Neither short-channel/narrow-channel coupling nor special front-gate/back-gate coupling was observed. In ultra thin (15nm) SOI films, both short-channel and narrow-channel effects are attenuated.

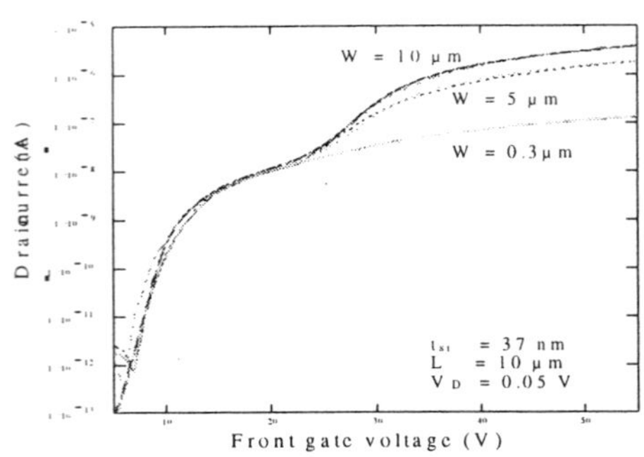

Fig.6 Subthreshold characteristics for various widths W. Experimental (dotted lines) and simulated (solid lines) curves.

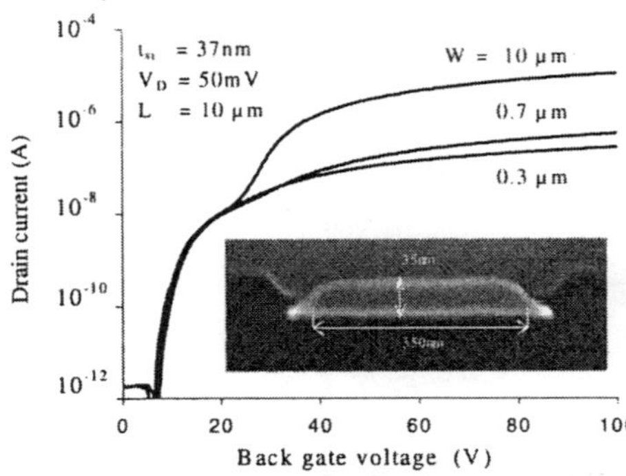

Fig.7 Back gate subthreshold characteristics influenced by the parasitic sidewall transistor (shown in the insert) [5].

A physical compact model based on the variation of film and oxide thickness along the width direction helps to simulate the experimental trends. The model reproduces the front and back gate characteristics (Figure 6). This systematic exploration of the various narrow-channel effects is of high interest for low-power/voltage CMOS applications and for further optimization of the lateral isolation techniques.

3.2. DYNAMIC-THRESHOLD MOSFET

3.2.1. Static characteristics

The SOI Dynamic-Threshold MOSFET (DT-MOSFET) is a good candidate for the future ultra-low voltage (< 0.6 V) CMOS applications [6]. This transistor, with the body directly connected to the gate (Figure 8), offers enhanced current drive, speed and V_t scalability. The principle of this device is that when a voltage is applied on the gate, it also acts on the body, thus dynamically reducing the threshold voltage which in turn increases the subthreshold slope and drain current. On the other hand, when the gate is not biased, the threshold voltage remains high, maintaining a low leakage current.
Systematic measurements of DT-MOSFET characteristics have been performed to explore and quantify its enhanced performance. It was found that the DIBL, V_t roll-off and transconductance are greatly improved (as compared with the floating body or the body-tied devices, more than 35 % gain in transconductance at a lower V_G, Figure 9). A simple compact model, allowing circuit designers to anticipate the performance of the DT-MOSFET and to simulate its main parameters has been developed [7]. This model yields an enhancement factor • (calculated from the transistor's technological parameters) that describes the DT-MOSFET's improvements as follows:

$$V_{T,DT} \approx \frac{V_{T,BG}}{1+\alpha} \; ; \; S_{DT} \approx \frac{S_{BG}}{1+\alpha} \; ; \; \mu_{DT} \approx (1+\alpha)\mu_{BG} \; ; \; \theta_{DT} \approx (1+\alpha)\theta_{BG}$$

where DT indicates the characteristics for the DT-MOSFET and BG those for a body grounded transistor.

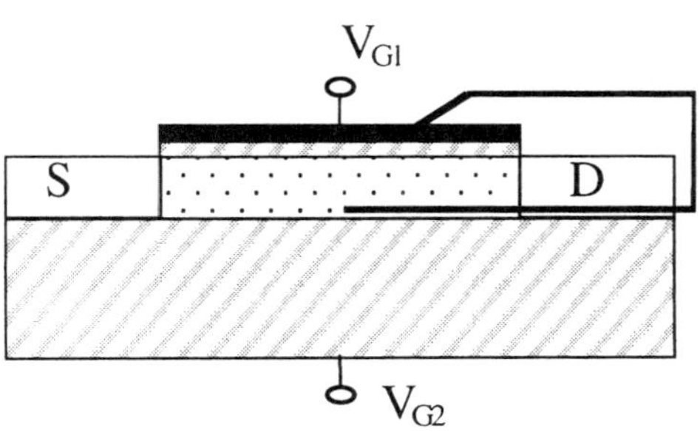

Fig.8 Dynamic-threshold SOI MOSFET.

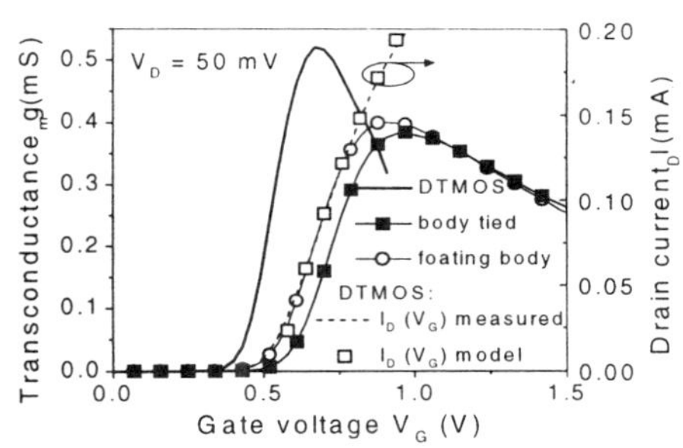

Fig.9 Transconductance and drain current versus gate voltage in DT, body-tied and floating body MOSFETs with L = 0.25 μm and W = 10 μm [7].

3.2.2. Noise measurements

In order to use DT-MOSFETs in wireless communication, it is important to analyse the behaviour of the low frequency noise [8].

N- and P-channel dynamic-threshold Unibond MOSFET's, fabricated at LETI-Grenoble, have been used in this study. The relevant technological details are: front gate oxide thickness t_{ox} of 4.5 nm, buried oxide thickness t_{box} of 400 nm and final film thickness t_{si} of 100 nm. Due to the high-leakage current when the body is strongly forward biased, a small size current limiter (clamping transistor) is added between gate and body in DTMOS structure. Note that the clamping transistor complicates the design but it prevents the body potential (V_{bs}) to exceed 0.65V and allows to operate at 1V gate bias without high gate current.

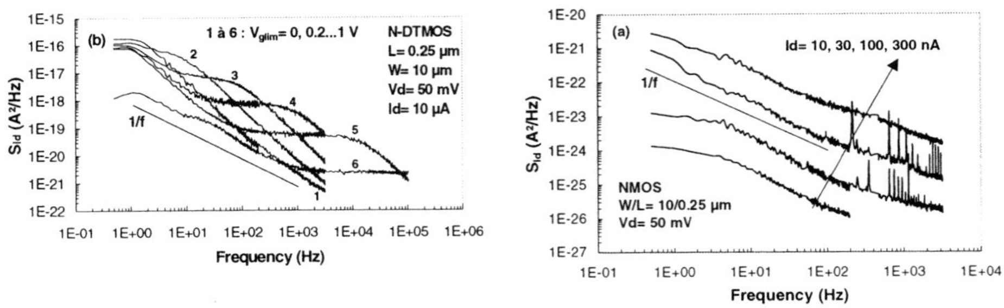

Fig.10 Drain current power spectral or 0.25 µm N-channel DTMOS without (a) density versus frequency in ohmic region and with (b) limiter current.

In Figure 10a, the drain current power spectral density (S_{Id}) versus frequency is plotted in ohmic region (V_d=50 mV), at various drain currents, for 0.25 µm N-channel DTMOS without clamping transistor. 1/f spectra are obtained, whatever the gate current is. Similar results have also been found for P-channel DT-MOSFETs. However, for DTMOS with a clamping transistor, S_{Id} (f) shows Lorentzian spectra (Fig. 10b), which are strongly dependent on the gate voltage. The plateau level of the Lorentzian spectrum decreases and the corner frequency increases when V_{glim} (gate voltage of the current limiter) increases. This behaviour is quite similar to the noise overshoot (at high V_d) induced by the kink effect in partially-depleted SOI, which is due to the impact ionisation mechanism [9]. In our case, V_d is too weak to induce an impact ionisation current, however, the clamping transistor current flows through the body inducing an excess noise presenting Lorentzian spectra. When V_{glim} is between zero and 1V, the body potential is fixed by the clamping transistor current and, consequently, this current increases the body potential inducing a direct biasing source-body junction and a kink-like excess noise, even at low drain voltage. At V_{glim}=1V, the clamping transistor is ON, therefore, the body is directly connected to the gate. In this case, a quasi 1/f behavior is recovered (Fig. 10b).

3.3. GROUND PLANE MOSFET

In SOI transistors, the fringing fields within and underneath the BOX are the cause of a supplementary drain-induced barrier lowering (DIBL) [10]. The effect of the lateral penetration of the field is equivalent to an undesirable biasing of the transistor's back interface: the back channel becomes slightly inverted which lowers the threshold voltage. Numerical simulations (Figure 12) have shown that the introduction of a

Ground-plane (that is, a conductive plane kept at a zero potential, Fig. 11), located just below a preferably thin BOX, can attenuate this phenomenon. The motivation for the Ground-plane is that the substrate underneath the BOX should not be depleted and that the substrate/body capacitance should be greater than the drain/body capacitance (which is true if $t_{BOX} < L/2$) [11].
Figure 12 shows that if the body doping is an important parameter in the reduction of DIBL for "thick" films, it becomes irrelevant when the thickness decreases under about 150Å (for L=0.1μm), though the Ground Plane is still effective. In other words, the requirement for a high doping is relaxed which is greatly beneficial in terms of mobility and transconductance.

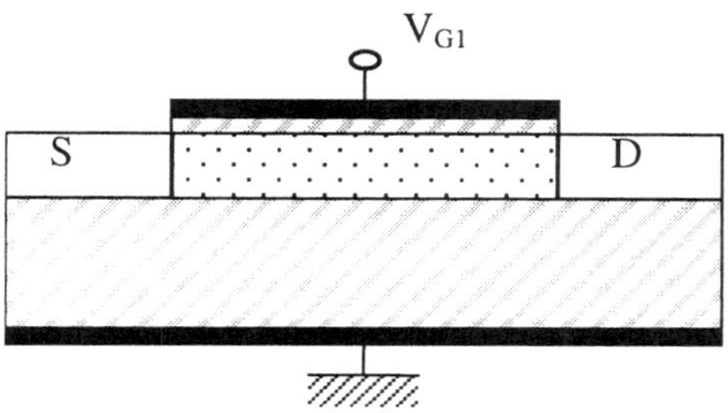

Fig.11 Ground-plane SOI MOSFET.

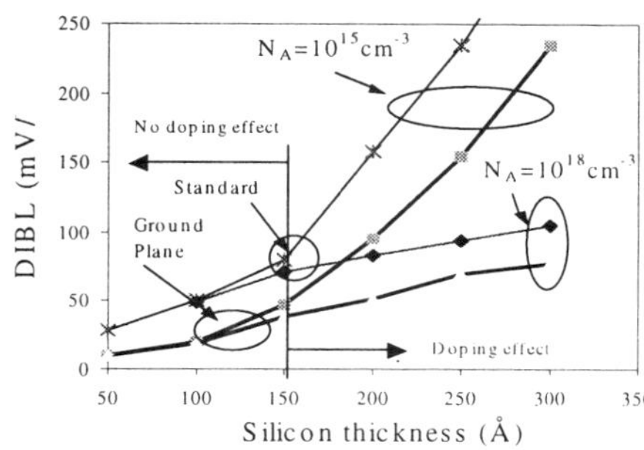

Fig.12 DIBL versus silicon thickness in standard and ground plane structures (L = 0.1 μm) for high and low doping.

3.4. EXTREMELY THIN SOI MOSFET

The recent progress in wafer uniformity and device processing allows the fabrication of extremely thin (1-5 nm) SOI MOSFETs (Fig. 13), which offer very attractive prospects in terms of performance and quantum properties.
These devices work on the same model as the regular fully-depleted MOSFETs but with much thinner Si films. The $I_D(V_G)$ curves are typical (Figure 14), although peculiar strong couplings appear between the front and the back gates and between the film and the substrate. Interesting thickness-related effects on the carrier mobility, threshold voltage, subthreshold swing and transconductance are also observed [12]. The threshold voltage decreases linearly when V_{G2} is increased, showing no saturation region, thus demonstrating that the film is too thin to be accumulated at the back interface. A consequence of this is a constant subthreshold swing. An exception to this phenomenon can be observed at low V_{G2} when the substrate underneath the BOX becomes depleted. This depletion leads to an apparent increase of the BOX thickness, resulting in a decrease of its apparent capacitance. The substrate depletion effects could be seen most clearly when the device was operated using the back gate as the main gate.
The special distribution of minority carriers in ultimately thin film devices is governed by quantum confinement and allows the development of new device concepts such as volume inversion and quantum operation at room temperature. In particular, the threshold voltage increases as the film thickness is reduced below 10nm.

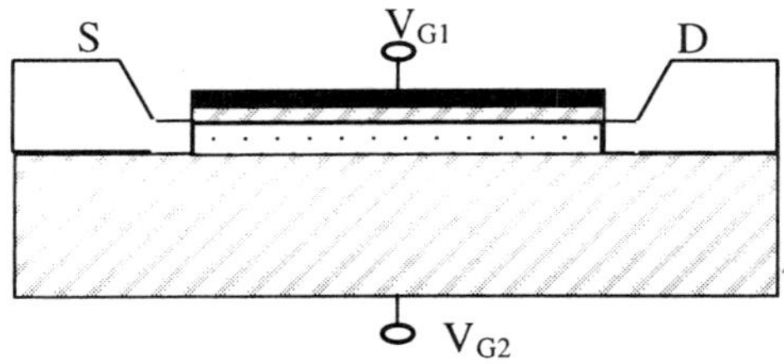

Fig.13 Ultra-thin SOI MOSFET.

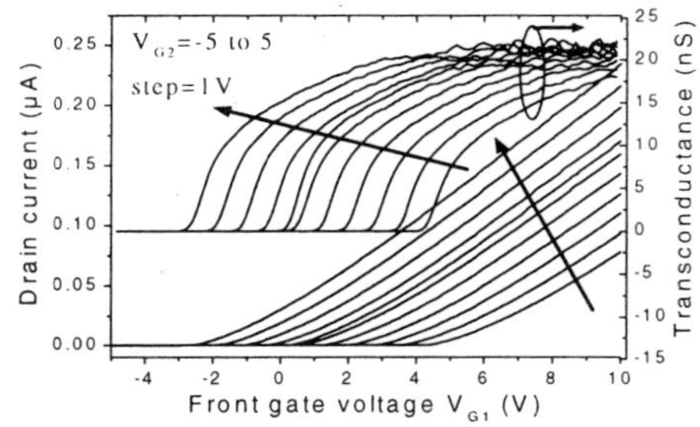

Fig.14 Drain current and transconductance versus front gate bias in a 1nm thick transistor with L = 30 μm and W = 100 μm [12].

3.5. DOUBLE-GATE MOSFET

The peculiarity of Double-Gate transistors (DG-MOSFETs) is that the top and bottom gates are biased simultaneously. $V_{G2}=\frac{t_{ox2}}{t_{ox1}}V_{G1}$: this accounts for the slight difference in front and back oxide thickness and guarantees rather identical surface potentials (Figure 15).

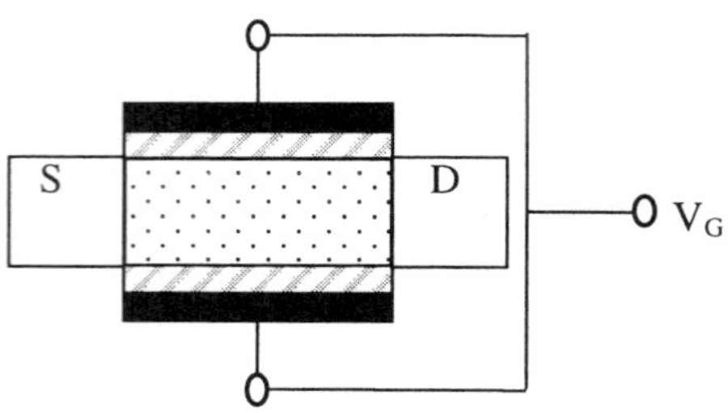

Fig.15 Double gate SOI MOSFET.

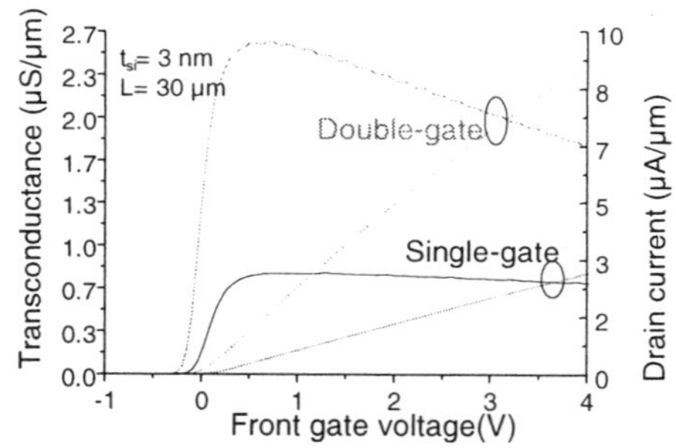

Fig.16 Drain current and transconductance versus front gate bias for a 3 nm thick transistor (W = 3 μm et L = 30 μm). In double gate regime, V_{G2} = 1.24□V_{G1} [12].

In a partially-depleted transistor with a thick enough active layer, two independent channels are formed close to the interfaces, while the central region of the film remains undepleted [1]. The resulting current is nothing but the sum of the contributions of two independent transistors. On the other hand, in fully depleted transistors with a thin enough film, controlling the channel from both sides at the same time forces most of the carriers to flow through the middle of the film (especially around the threshold voltage), according to the volume inversion concept [13]. The electrons are then submitted to a lower electric field and to less scattering on the surface roughness. Figure 16 shows the comparison between a 3nm thick Double-Gate MOSFET operated with only one or both gates. The transconductance and current in Double-Gate mode exceed twice the values in Single-Gate mode. This demonstrates that a DG-MOSFET is more than the sum of two classical transistors. Our calculations, based on the resolution of Poisson and Schrodinger equations, tend to show that this improvement is due to the migration of the flow of carriers towards the middle of the body, where the mobility is presumably improved.

4. Conclusion

SOI enables the design of transistors that may succeed beyond the frontiers of the conventional CMOS technology. SOI circuits deserve a dedicated technology and design to push the limits of microelectronics. As the quality of the material improves and the technology is better controlled, the processing of these alternative devices becomes more and more realistic.

Acknowledgements: This work has been performed at the Center for Projects in Advanced Microelectronics (CPMA), Grenoble, France. The CPMA is operated by CNRS, LETI-CEA, INP Grenoble and INSA Lyon.

References

[1] Cristoloveanu, S. and Li, S.S. (1995) *Electrical Characterization of SOI Devices*, Kluwer, Boston.
[2] Cristoloveanu, S., & al (May 2000) *IEEE TED* **47-5**, 1018-1027.
[3] Munteanu, D. (Sept. 1999) Thèse INPG.
[4] Hefyene, N. (July 1999) DEA INPG.
[5] Pretet, J. & *al IEEE Int. SOI Conf., in* press Oct. 2000.
[6] Assaderaghi, F. & al (Mar. 1997) *IEEE TED*, **44**, 414-422.
[7] Ernst, T.& al Proc. ESSDERC'99.
[8] Hsu, T.L., Tang, D.D.L. and Gong, J. (October 1999) Low-Frequency Noise Properties of Dynamic-Threshold (DT) MOSFET's, *IEEE Elect. Dev. Lett.*, **20-10**, 532-534.
[9] Haendler, S., Jomaah, J., Balestra, F., Pelloie, J.L., Raynaud C. and Boussey, J. (1999) Thorough Investigation of Kink-Realated Excess Noise in Deep-Submicron SOI N-MOSFETs on Unibond Substrate, SSDM Int. Conf., Tokyo, Japan, 364-365.
[10] Ernst, T. and Cristoloveanu, S. (Oct. 1999) IEEE Int. SOI Conf., 38-39.
[11] Ernst, T.& al (Jan. 2000) ULIS'2000 Workshop.
[12] Ernst, T.& al (Oct. 1999) IEEE Int. SOI Conf., 92-93.
[13] Chen, J. & al (1992) IEDM'92 Tech. Digest, 35.

MODELING AND MEASUREMENTS OF GENERATION AND RECOMBINATION CURRENTS IN THIN-FILM SOI GATED-DIODES

T.E. RUDENKO, and V.I. KILCHYTSKA
Institute of Semiconductor Physics, National Academy of Sciences of Ukraine
03028 Kiev 28, Prospect Nauki, 45, Kiev, Ukraine

1. Introduction

Generation-recombination (g-r) parameters are of great importance for SOI devices, because they determine junction leakage, bipolar effects, and various floating body effects. On the other hand, g-r parameters are the measure of the material and processing quality. In SOI devices, g-r parameters are generally determined from the transient drain current response of a SOI MOSFET to front- or back-gate pulses [1-3]. These measurements usually allow extracting the carrier lifetime without details as to the origin of g-r mechanisms. In this work, we discuss the gated-diode method based on d. c. measurements, which provides more detailed information about g-r processes [1,4]. The gated-diode method is widely used in a bulk-Si technology, because it enables easily to separate volume- and surface-related component [4]. In thin-film SOI devices, the behavior of g-r currents is more complicated due to interface coupling, the specific free carrier concentration distribution, and superposition of volume and surface components. Generally numerical simulation is required for the correct determination of g-r parameters in thin-film SOI devices.

In this work, we present a detailed analysis of generation and recombination current behavior in a thin-film SOI device, based on modeling of volume and surface components at various bias conditions. An approach, which allows for the determination of g-r parameters in thin-film SOI devices without numerical analysis, is developed. The proposed approach is applied for characterizing g-r processes and extracting g-r parameters in UNIBOND SOI devices.

2. Samples

Experimental results presented in this study were obtained on p^+-p-n^+ gated-diodes fabricated at the University of Louvain-La-Neuve (Belgium) on standard UNIBOND substrates with 85-nm-thick silicon film and 380-nm-thick buried oxide. The body doping was from $N_A=10^{16}$ to 10^{17} cm^{-3}, the gate oxide thickness was d_{of}=30 nm. The device length and width were, respectively L=3 μm, W=1600 μm.

F. Balestra et al. (eds.), Progress in SOI Structures and Devices Operating at Extreme Conditions, 249–261.

3. Simulations

Simulations were performed on the assumptions of the single-level trap with mid-gap energy $E_t = E_i$, uniform doping and trap distributions throughout the device body, and zero flat-band voltages at both Si-SiO$_2$ interfaces. For forward bias conditions, a low forward bias across the p-n-junction is considered, which assumes low carrier injection and recombination components to dominate over diffusion.

Following the Shockley-Read-Hall model, the recombination (generation) rate may be expressed as:

$$R(x) = \frac{1}{\tau_r} \frac{n(x)p(x) - n_i^2}{n(x) + p(x) + 2n_i}, \tag{1}$$

where τ_r is the carrier recombination (generation) lifetime. For the simplicity it is assumed that $\tau_n \approx \tau_p \approx \tau_r$. The net generation-recombination current is obtained by integrating R(x) over the whole Si film.

For low g-r currents, the quasi-Fermi levels can be assumed approximately constant along the device channel. The separation between the quasi-Fermi levels for electrons and holes is given by the applied forward bias V_F (or reverse bias V_R), so that the product (np) is:

$$np = n_i^2 \exp\left(\frac{qV_F}{kT}\right). \tag{2}$$

4. Forward current and extraction of recombination parameters

4.1. SIMULATED RECOMBINATION CURRENT COMPONENTS

Fig. 1 shows the calculated recombination current components in a thin-film SOI MOS device as a function of the front-gate voltage V_{gf} for various back-gate voltages V_{gb}. All current components in Fig.1 are normalized to their maximum values, so that presented curves do not involve the carrier lifetime or surface recombination velocity values and represent only the relative variation of each of the components with gate voltages. It can be seen from Fig.1, that all of components dramatically change with both front- and back-gate biases. The reason is that in a thin-film SOI, the potential and carrier concentration distributions and thus the recombination rate distribution strongly change with both front and back surface potentials. Below we consider each of the components in more detail.

For front-gate operation, the front-surface component appears as a peak being suppressed in strong accumulation or inversion and maximum in depletion, similarly to that in a bulk device. The only difference is that due to interface coupling this peak shifts linearly along the V_{gf} axis with V_{gb} (Fig.1b).

The behavior of the volume component in a SOI device is essentially different from that in a bulk device. As can be seen from Fig.1a, at proper conditions the volume component can also appear as a peak, similar to surface recombination. As the front surface is accumulated or inverted, volume component is exhibited by the plateau which changes with V_{gb} and reaches saturation when one interface is in strong accumulation and the opposite interface is in strong inversion.

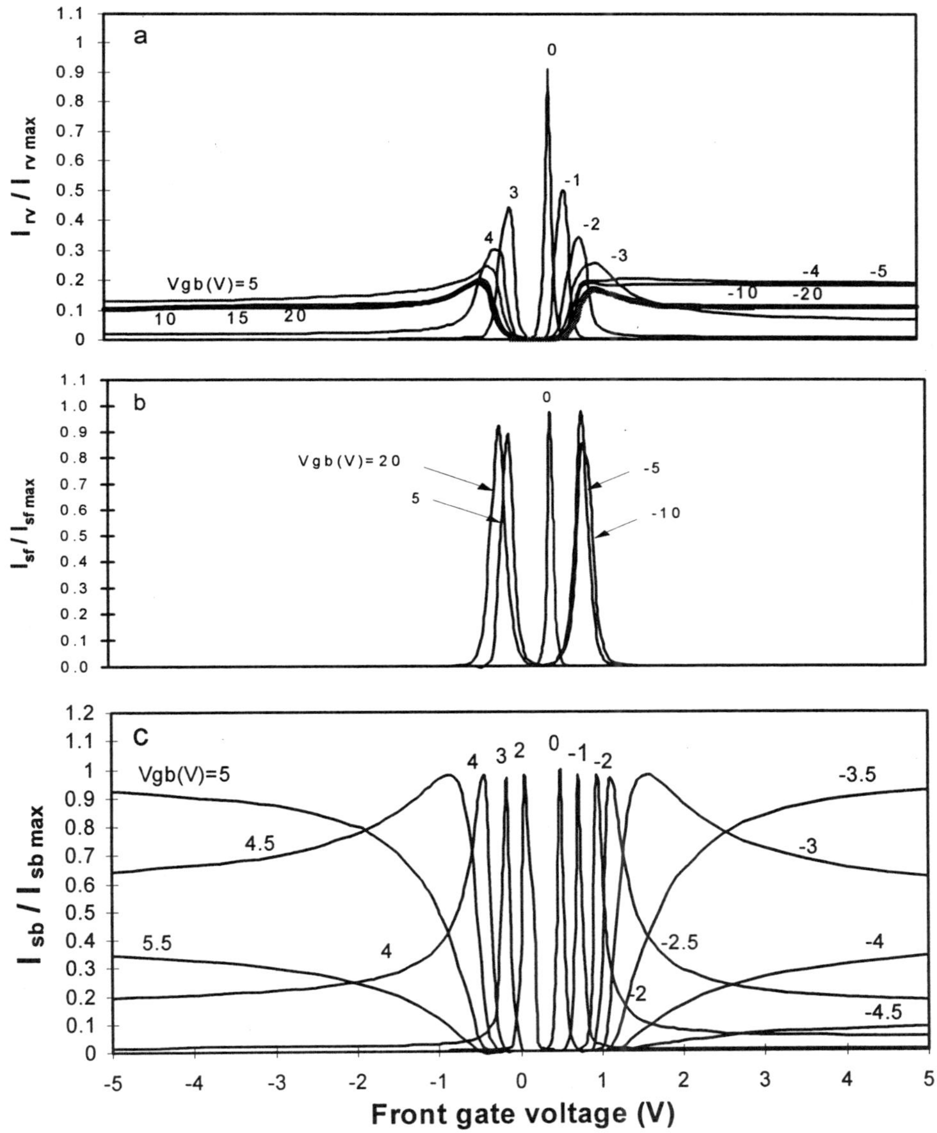

Figure 1. Calculated recombination current components versus the front gate voltage for various back gate biases : (a) volume recombination; (b) the front surface recombination; (c) the back surface recombination. All current components are normalized to their maximum values. (V_F=-0.2 V, d_{Si}=85 nm, d_{of}=30 nm, d_{ob}=400 nm, $N_A=5\times10^{16}$ cm^{-3}).

For front-gate operation, the back-surface component can be revealed by the peak, as the front-surface recombination, or by the plateau similar to the volume component. This plateau reaches the maximum value when the back interface is depleted. In contrast to volume component, the back surface-related plateau can be totally suppressed by driving the back interface into strong accumulation or inversion (Fig.1c).

In a real device, the above components are superimposed on one another, and depending on the Si film and interface quality the relation between them may be rather different. Thus an interpretation of experimental characteristics and extracting recombination parameters in a thin-film SOI device is expected to be very complicated and essentially different from the conventional bulk-Si case. However, close inspection of the potential and recombination rate distributions shows that data analysis is simplified for two type of conditions: 1) for double-gate regime, or conditions close to double-gate regime; 2) for conditions when one interface is in strong inversion and the opposite interface is in strong accumulation.

In double-gate regime or when the difference between the front and back surface potentials is kept small, in thin or low-doped SOI films the potential and recombination rate is in fact constant across the Si film thickness, as illustrated by Fig.2. For these conditions, all of the recombination components are exhibited by the peak (see the central peak in Fig.1a, b, c). The variation of volume and surface recombination current components in the double-gate regime is shown in Fig.3. For low forward biases, recombination rate along the device channel is also in fact constant. Thus recombination occurs with a constant rate throughout the device, so that the maximum recombination current can be expressed as the product of the maximum recombination rate R_{max} and the device volume [5, 6]:

$$I_{R\max} = q \cdot d_{Si} \cdot W \cdot L \cdot R_{\max}, \tag{3}$$

where q is the electron charge, d_{Si} is the Si film thickness. From (1)-(3) we can obtain the following simple expression for the maximum recombination current:

$$I_{R\max} = \frac{q \cdot d_{Si} \cdot W \cdot L}{2\tau_r} \cdot n_i \exp\left(\frac{qV_F}{2kT}\right) \tag{4}$$

Eqn. (4) allows one easily to extract the recombination lifetime from the experimental current peak observed in double-gate regime (or at the conditions close to double-gate regime). However, the extracted lifetime will have the meaning of the effective lifetime because it will involve the contributions from recombination in the Si film volume and at both interfaces. More detailed analysis of the recombination current in double-gate regime is presented in [6].

In single-gate regime, an analysis is simplified when during scanning the voltage at one gate, the opposite gate is kept in strong accumulation or inversion. This case is illustrated by Fig.4. Then only two components are available, namely, volume component shown up as a plateau, and, for front-gate operation, the front-surface component exhibited by a peak (the back-surface recombination is suppressed). The role

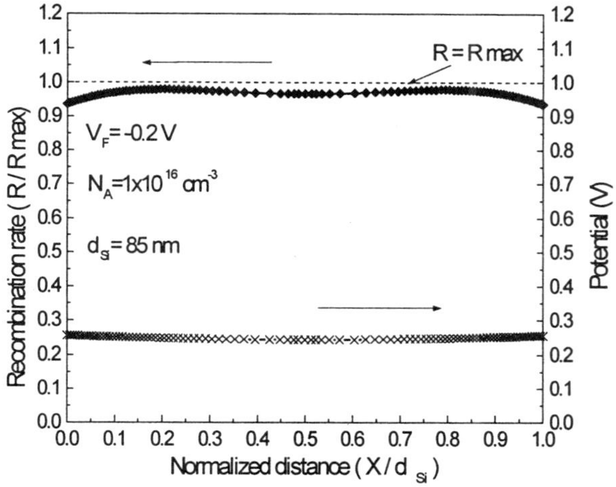

Figure 2. The potential and recombination rate distributions across the Si film thickness in double-gate regime for the conditions of maximum recombination.

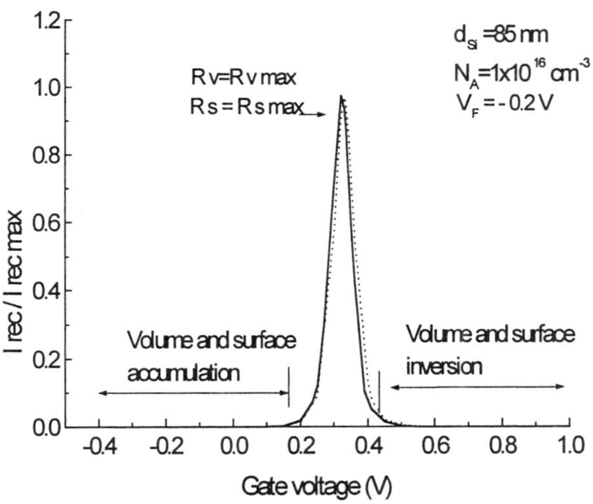

Figure 3. The gate voltage variation of volume and surface recombination components calculated for double-gate regime.

of both gates is interchangeable. In the plateau range, one of interfaces is in strong accumulation and the other is in strong inversion. The potential and recombination rate distributions in the Si film for given conditions are shown in Fig.5. For these conditions, recombination occurs mainly in the central part of the Si film thickness, where $n \approx p$.

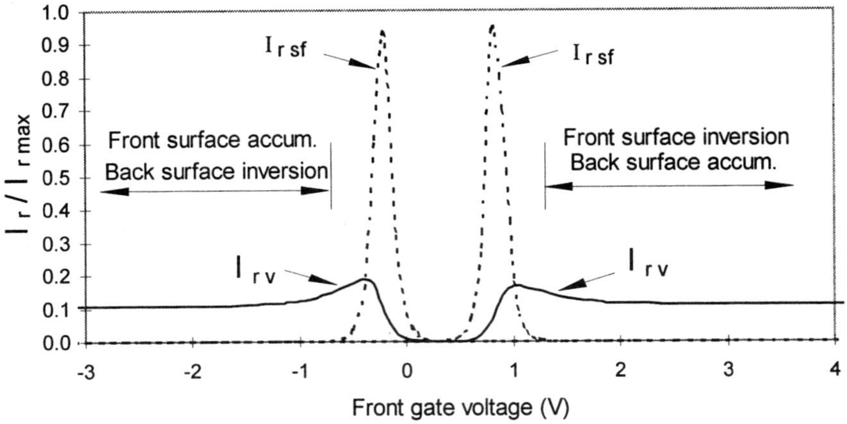

Figure 4. The front-gate voltage dependencies of volume recombination (solid line) and the front-surface recombination (dashed line) when back surface is kept in strong inversion (V_{gb}=20 V) or strong accumulation (V_{gb}= -20 V). Calculations are performed for V_F= -0.2 V, N_A=10^{16} cm^{-3}, d_{Si}=85 nm, d_{of}=30 nm, d_{ob}=380 nm.

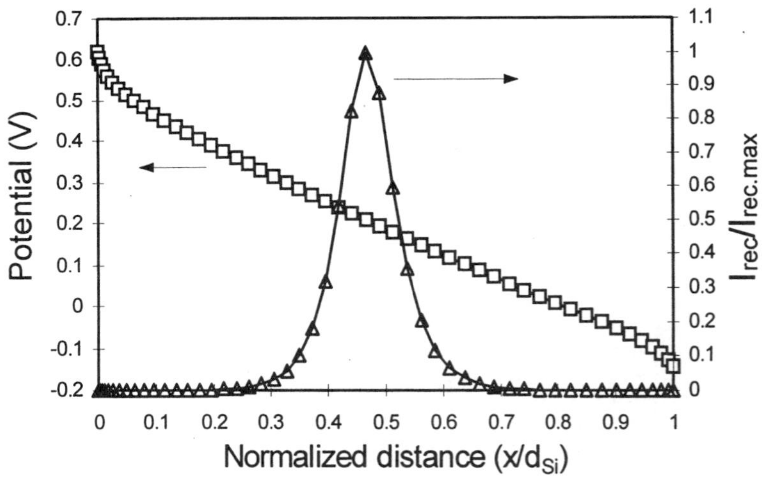

Figure 5. Potential and recombination rate distributions across the Si film thickness when the front interface is in strong inversion with the back interface in strong accumulation (V_F= -0.2 V, N_A=10^{16} cm^{-3}, d_{Si}=85 nm).

It is seen from Fig.5 that for these conditions, a linear potential variation across the film thickness can be assumed, so that one can define the effective recombination width as:

$$d_{reff} \approx d_{Si} \frac{4kT/q}{2\varphi_F - V_F}, \tag{5}$$

where $\varphi_F = (kT/q)\cdot\ln(N_A/n_i)$ is the Fermi potential. Then the total recombination current inside the Si film can be given by:

$$I_{Rv} = q\frac{n_i}{\tau_{rv}}\cdot d_{reff}\cdot W\cdot L\cdot\exp\left(\frac{qV_F}{2kT}\right). \tag{6}$$

In this case, the lifetime value extracted from the current plateau would reflect the quality of the Si film, because the contributions from both interfaces are eliminated.

4.2. RECOMBINATION CURRENT MEASUREMENTS

Fig.6 shows the experimental forward current as a function of the front-gate voltage at various back-gate biases in a UNIBOND SOI gated-diode. From the above analysis it follows that the central peak, which is observed when both interfaces are depleted, represents the sum of all recombination current components. The plateau to the left and to the right from the peak may be due to recombination in the Si film volume or due to recombination at the back interface. It can be seen from Fig.6 that biasing the back interface into strong accumulation or inversion sharply decreases the plateau current indicating that, in a given case, the plateau range is due to the back surface recombination, while the volume component is negligible. Peaks observed when the back interface is kept in accumulation or inversion, in accordance with the previous analysis are due to the front surface recombination. Comparing the value of the central peak, the front-surface-related peaks, and maximum plateau current representing the back surface component, one can state that, in this instance, recombination current is

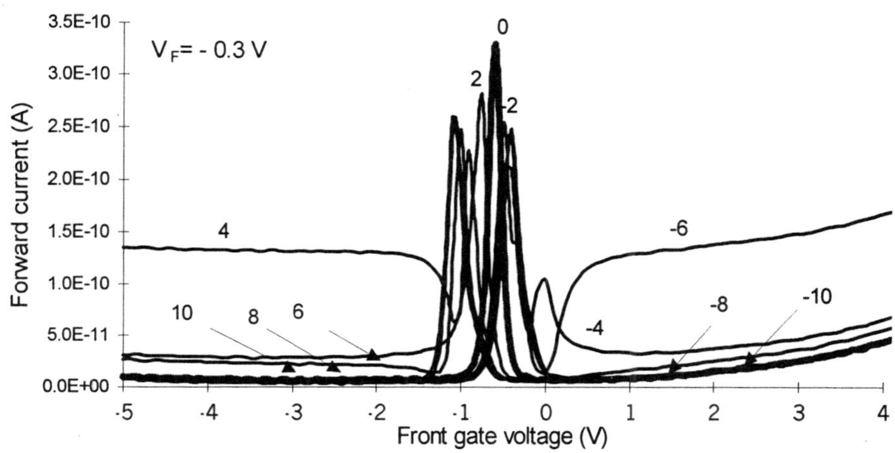

Figure 6. Forward current vs. the front gate voltage at various back gate biases in the UNIBOND n^+-p-p+ gated diode (T=25^0 C, W/L=1600/3, N_A=3x10^{16} cm^{-3}).

entirely determined by two interfaces with a negligible contribution from the Si film volume. Note that the front surface recombination is more than twice is higher that the back surface recombination. This conclusion is entirely supported by back-gate measurements shown in Fig.7. The extracted surface recombination velocities at the front and at the back interfaces are 6 cm/s and 2.7 cm/s, respectively. The effective recombination lifetime extracted from the central peak in Fig.6 is 1 μs.

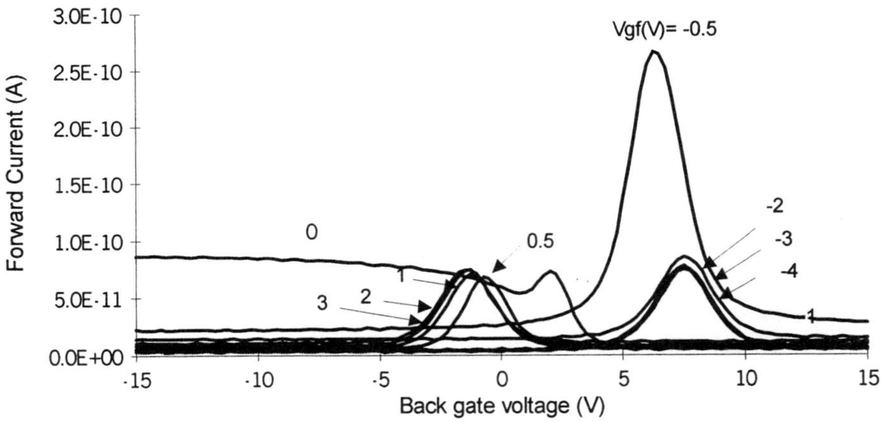

Figure 7. Forward current vs. the back gate voltage at various front gate voltages (V_F= - 0.3 V).

5. Reverse current and extraction of generation parameters

5.1. SIMULATED GENERATION CURRENT COMPONENTS

Fig.8 shows generation current components in a thin-film SOI device calculated as a function of the front-gate voltage for various back-gate voltages. For front-gate operation, the behavior of the front surface component shown in Fig.1b, is similar to that in a bulk device in the sense, that it "switches on" when $\varphi_{sf} \approx \varphi_F$, and "switches off" when $\varphi_{sf} \approx 2\varphi_F+V_R$. The only difference is that these points (rise and fall edges) shift linearly along the V_{gf} axis with the back-gate bias due to interface coupling. Special attention must be given to the SOI-specific behavior of the volume component (see Fig.8a). Note that the volume component sharply decreases with onset of inversion or accumulation, similar to surface generation, which gives rise to steps down on the curves. However, the height of these volume-related steps strongly depends on the back gate bias. The reason is that, in a thin-film SOI device, with onset of inversion or accumulation not only the surface generation switches off, but also generation in the significant part of the Si film, where the free carrier concentration exceeds the intrinsic concentration n_i. Because the carrier concentration distribution in the Si film changes with the back gate bias, the width of the "dead zone" for generation also changes with V_{gb}. If the difference between the front- and back-surface potentials is kept small ($\leq\varphi_F$),

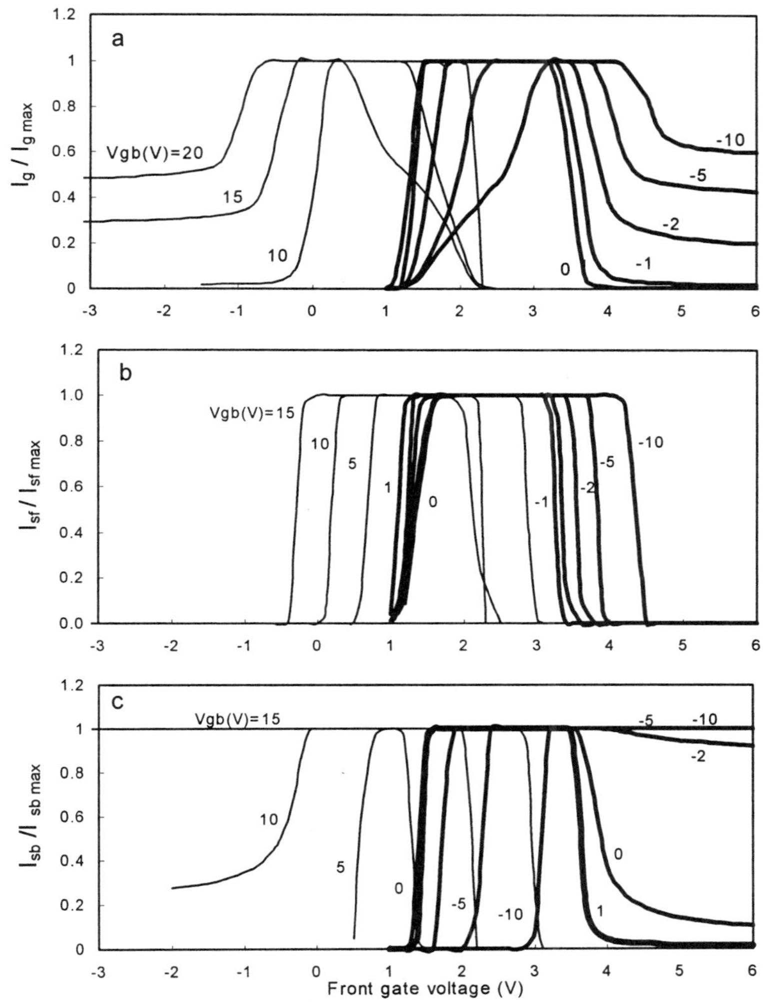

Figure 8. Generation current components calculated as a function of the front-gate voltage at various back-gate voltages for V_R=2 V: (a) volume generation; (b) the front surface generation; (c) the back surface generation (d_{Si}=85 nm, d_{of}=30 nm, d_{ob}=400 nm , $N_A=5\times10^{16}$ cm^{-3}. All current components are normalized to their maximum values).

with accumulation or inversion the whole Si film under the gate will represent the "dead zone" for generation. This situation is always realized in double-gate regime. In this case, the gate-voltage dependencies of all of the components will have the rectangle-like shape with the plateau level corresponding to their maximum contributions and negligibly small values outside of the plateau region, as shown in Fig.9. The width of the plateau is determined by the Si film doping and the reverse bias V_R. For these conditions, the plateau level of the experimental reverse current curve can be used for the evaluation of the effective generation lifetime $\tau_{g\ eff}$, which would involve all of the components with their maximum contributions, similar to the forward current peak in double-gate regime.

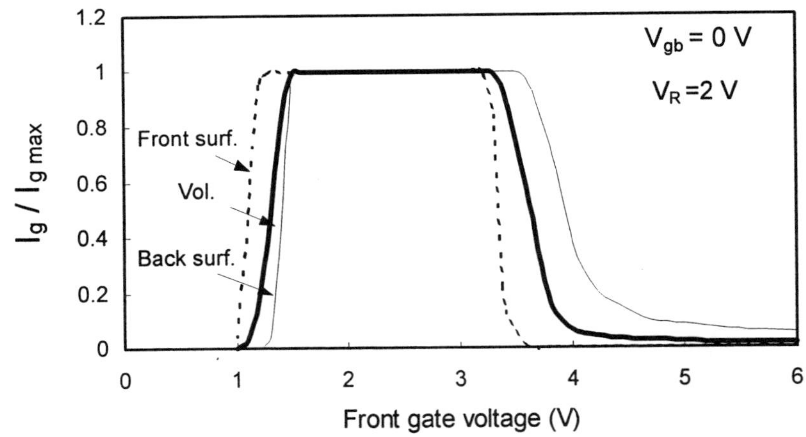

Figure 9. Generation current components vs. the front-gate voltage when the difference between front and back surface potentials is kept small (for the conditions close to double-gate regime).

The surface generation component can be extracted from the rise edge of the reverse current curve, when the opposite gate is biased in accumulation or inversion. This case is illustrated by Fig.10.

Volume generation lifetime can be determined from the reverse current, if conditions are reached such that one of interfaces is in strong inversion and the other is in strong accumulation. As this takes place, the potential varies linearly across the Si film thickness with the potential difference between interfaces $\Delta\varphi_{sf} = 2\varphi_F + |V_R|$. In this case, the generation layer width is easily obtainable from the following expression:

$$d_{geff} = d_{Si} \frac{V_R}{2\varphi_F + |V_R|}. \tag{7}$$

The overall generation current in the Si film can be given as:

$$I_{gen} = q \cdot \frac{n_i}{\tau_{gv}} \cdot d_{geff} \cdot W \cdot L \tag{8}$$

The lifetime value extracted from Eqn. (8) will reflect only the Si film quality, because for the conditions being considered generation at both interfaces is eliminated. It should be noted that the higher reverse bias, the higher gate voltages are needed to provide these conditions.

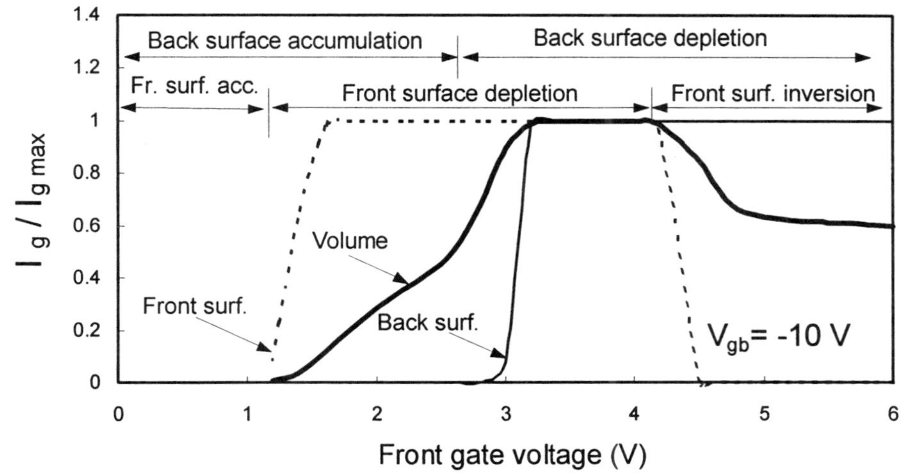

Figure 10. Generation current components in a reverse-biased n^+-p-p^+ SOI gated-diode at V_{gb}= -10 V (V_R=2 V). When the front surface generation is turned on, the back interface is accumulated.

5.2. GENERATION CURRENT MEASUREMENTS

Fig.11 shows the experimental reverse current versus the front gate voltage measured in a n^+-p-p^+ UNIBOND SOI device at various back gate biases. According to the previous analysis, a step that observed at the large negative V_{gb} (the back interface is accumulated) or large positive V_{gb} (the back interface is inverted) represents the front surface generation component. The second step, observed when the back interface becomes depleted, represents the back surface generation component. The central curve

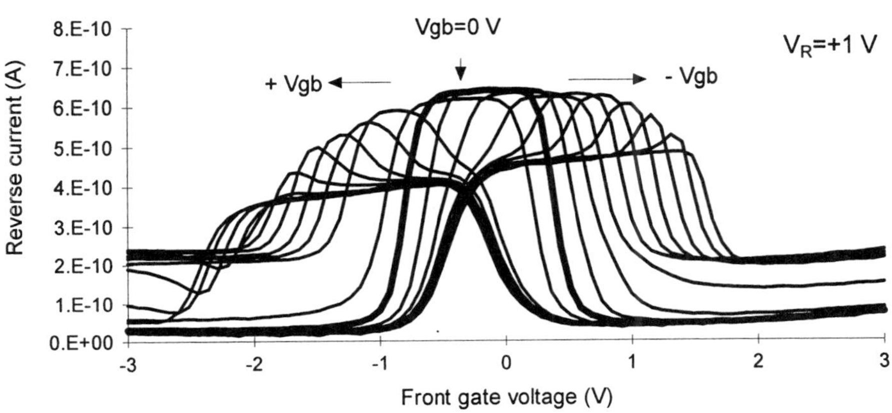

Figure 11. Experimental reverse current in a UNIBOND SOI gated-diode versus the front gate voltage for various back gate biases from -20 V to 20 V with a back gate bias step equal to 2 V (V_R=1 V, T=150^{0}C, W/L=3/1600 μm).

having the rectangle-like shape is measured under conditions close to double-gate regime. Its maximum value is equal to the sum of all components. In a given case, the maximum generation current presents the sum of two surface components with the contribution from the front interface being more than twice higher than that from the back interface. This agrees well with back-gate measurements shown by Fig.12.

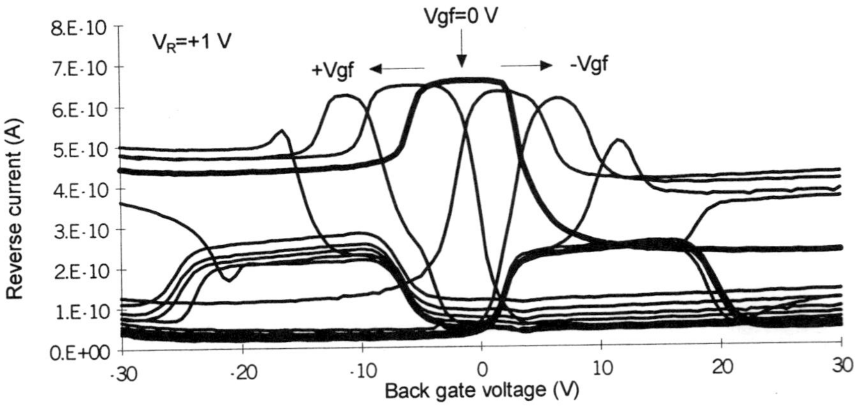

Figure 12. Experimental reverse current in a UNIBOND SOI gated-diode versus the back gate voltage for various front gate voltages from V_{gf}= -4.5 V to 4.5 V with a front gate voltage step equal to 0.5 V (V_R=1 V, T=150°C, W/L=3/1600 μm).

The effective generation lifetime extracted from the maximum generation current is equal to 1.3 μs (T=150°C). However, it does not reflect the Si film quality, because, in a given device, it is determined by both surfaces. The surface generation velocities, extracted from the experimental data in Fig.11 and 12 (T=150°C), are 4.4 cm/s and 2 cm/s at the front and at the back interfaces, respectively.

6. Conclusions

A systematic analysis of generation and recombination current behavior in a thin-film SOI gated-diode, based on an accurate modeling of volume and surface g-r components, is presented. It is shown that, for the case of the thin-film SOI devices, the gated-diode measurements should be interpreted with a great care. There are two types of conditions when data interpretation is simplified and does not require any numerical simulation. One of them is realized when the difference between front and back surface potentials is kept small, that is, at the conditions close to double-gate regime. Then volume and both surface g-r components exhibit similar gate-voltage dependencies being suppressed in accumulation and inversion and maximum in depletion. In this case, the maximum experimental forward (or reverse) current can be used for the extraction of the effective recombination (or generation) lifetime, which will involve the maximum contributions from the Si film volume and both surfaces.

The data analysis is also simplified for the conditions when one of interfaces is in strong inversion, while the other one is in strong accumulation. As this takes place, the contributions from both interfaces are eliminated, and the overall g-r current in the Si film can be expressed assuming a linear potential variation across the film thickness. In this case, the lifetime value extracted from the experimental reverse or forward current should reflect the quality of the Si film.

Using the gated-diode method, it has been shown that in thin-film UNIBOND SOI devices generation and recombination currents are entirely determined by interfaces with a negligible contribution from the Si film volume, which confirms the high quality of UNIBOND wafers.

Acknowledgment

This work was partially supported by NATO Linkage Grant #SST.CLG.975925. The authors also thanks to the staff of the Microelectronics Laboratory, UCL, for providing the samples and research facilities as well as to Professor Vincent Bayot for his altruistic co-ordination of the project.

References

1. Cristoloveanu, S. and Li, S. (1995) *Electrical Characterization of Silicon-On-Insulator Materials and Devices,* Kluwer Academic Publishers, Norwell.
2. Yasuda, N., Taniguchi, K., Hamaguchi, C., Yamaguchi, Y., and Nishimura, T. (1992) New carrier lifetime measurement method for fully depleted SOI MOSFET's, *IEEE Trans. Electron Devices* **39**, 1197-1201.
3. Sinha, S. P., Zalevski, A., and Ioannou, D. E. (1994) Investigation of carrier generation in fully depleted enhancement and accumulation mode SOI MOSFET's, *IEEE Trans. Electron Devices* **41**, 2413-2418.
4. Grove, A., Fitzgerald, D. (1966) Surface effect on P-N-junction characteristics of surface space charge regions under nonequlibrium conditions, *Sol. State Electron.* **9,** 783-806.
5. Rudenko, T. E., Rudenko, A. N., and Lysenko, V.S. (1995) Electrical properties of ZMR SOI structures: Characterization techniques and experimental results, in *Physical and Technical Problems of SOI Structures and Devices,* Kluwer Academic Publishers, Norwell, pp. 169-179.
6. Ernst, T., Cristoloveanu, S., Vandooren, A., Rudenko, T., and Colinge, J.-P. (1999) Recombination current modeling and carrier lifetime extraction in dual-gate fully-depleted SOI devices, *IEEE Trans. Electron Devices* **46**, 1503-1509.

DEFECT CREATION MECHANISMS DUE TO HOT-CARRIERS IN 0.15 μm SIMOX MOSFETs

P. DIMITRAKIS, J. JOMAAH*, F. BALESTRA*, G.J.PAPAIOANNOU
University of Athens, Solid State Physics Section, Panpistimiopolis Zografos, GR-15784 Athens, Greece
**LPCS-ENSERG, BP 257, 38016 Grenoble cedex 1, France*

1. Introduction

Silicon-On-Insulator MOSFETs have unique advantages over bulk Si MOSFETs making them to be an excellent alternative for low voltage and low-power applications. The current IC scaling trend and the performance obtained on SOI-CMOS devices predict a one or two generations SOI advantage over bulk-CMOS in terms of speed-power [1].

A lot of research has been done on device degradation due to hot-carrier (HC) stressing, since it is one of the major challenges for sub-quarter micron gate length transistors. The degradation mechanisms in SOI devices are more complex than bulk Si and their lifetime are strongly dependent on bias conditions and SOI substrates [2-4]. The off-state operation (stand-by mode) is of special interest for low voltage and low power ICs. In this case, (i.e. $V_g \leq V_{th}$) the off-state current must be minimized as low as possible. Under these conditions, the injection of hot carriers into the oxide is strongly influenced by the low vertical electric field.

The aim of this contribution is to present a comprehensive study of the bulk and interface defects that are created after HC stress. The generation mechanisms of these defects will be identified.

2. Experimental Procedure

Low dose SIMOX MOSFETs with 4.5nm gate oxide, 80nm buried oxide and 100nm silicon film tested (fig.1.). The devices were partially depleted (PD) with edges. During stress, the source electrode was grounded, the drain bias was V_{ds}=+3V and the front gate bias V_{gf} near the threshold voltage of the front channel V_{thf} and back gate voltage V_{gb} (A) 0V, (B) +20V and (C) –20V. The total stress time was 50000s. MOSFETs with two different channel doping concentrations were tested (1) $8 \times 10^{17} cm^{-3}$ and (2) $3 \times 10^{17} cm^{-3}$.

Static characteristics were used for aging evaluation. In order to observe the masking of defects near the drain region, measurements in normal and reverse mode (source and drain interchanged) were performed. In reverse mode, the defects near the drain region are always inside the channel, while in normal mode they are not active if the channel is pinched-off. I-V measurements performed under the following conditions: front channel characteristics obtained with back interface in accumulation, while back channel characteristics obtained with V_{gf}=-3V, i.e. accumulated front interface.

F. Balestra et al. (eds.), Progress in SOI Structures and Devices Operating at Extreme Conditions, 263–268.

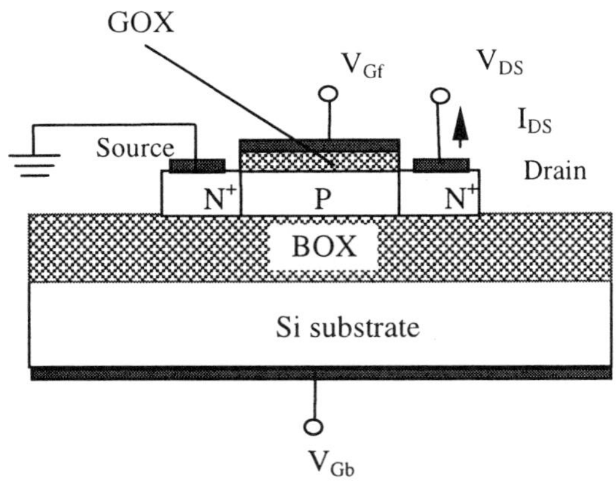

Figure 1. Typical four terminal SOI N-MOSFET.

Defect characterization performed by applying Current Deep Level Transient Spectroscopy (DLTS) [5]. Two different gate voltage pulse conditions used for defect evaluation in stressed transistors while V_d has the constant value of 50mV. Applying the first kind of voltage pulse on the gate, the transistor is kept in "OFF" state for 100ms and then return to the "ON" state (quiescent bias). During the "OFF" state, the Si film majority carrier traps are filled with holes, which are emitted when the pulse is finished. In order to suppress drain current overshoot due to floating body [6], large pulse voltages are used [7]. Applying the second kind of voltage pulse, the transistor is kept in "ON" state for 100ms and then return to the "OFF" state (quiescent bias). During this pulse the bulk and interface traps are filled with electrons and the recorded transient after the end of the pulse is mainly due to the emission of captured electrons. Both kinds of pulses are used for defect investigation in front and back channel. The above DLTS experiments applied on both gates: front and back. When the one gate is pulsed the other was floating.

Calculating the slope of output conductance $g_d = dI_d/dV_d$ in the weak inversion region monitored the evolution of interface states in the front channel. The expression for g_d is [8]

$$g_d = g_{d0} \exp(-\frac{q}{kT}\frac{C_{ox}+C_d}{C_{ox}+C_d+C_{it}}V_d) \tag{1}$$

where C_d , $C_{it}=qD_{it}$ stand for depletion and interface states capacitance respectively, g_{d0} is the ohmic region conductance, all the other symbols have their usual meaning. The slope of the curve $\ln(g_d)=f(V_d)$ is a straight line with slope proportional to the inverse of interface states capacitance ($A=C_{ox}/(C_{ox}+C_d+C_{it})$). The expression for g_{d0} is

$$g_{d0} = \frac{kT}{q}\frac{W}{L}\mu_0 C_d \exp\left(\frac{q}{kT}A(V_{gf}-V_{thf})\right) \tag{2}$$

3. Results and Discussion

3.1 LIFETIME CALCULATIONS

The accelerated device lifetimes were calculated in all cases from ΔV_{thf}, $\Delta g_m/g_{m0}$ and $\Delta I_{dss}/I_{dss0}$ versus time. The lifetime criteria were 50mV shift of ΔV_{thf} and 10% change of $\Delta g_m/g_{m0}$ and $\Delta I_{dss}/I_{dss0}$. A logarithmic dependence with stressing time was observed for all the above quantities and stress conditions (fig.2), except for B2 case i.e. lower doping and back interface in strong inversion [9]. For this case, the aging of device characteristics was a power law with time ($\Delta X/X=A\cdot t^n$). The exponent n was 0.17. In table 1, the device lifetimes calculated by different methods are summarized. Data in this table show clearly that the lower the doping, the longer the lifetime.

TABLE 1. Calculated device lifetimes using different methods.

Model	Lifetime (s)					
	A1	A2	B1	B2	C1	C2
ΔV_{thf}	1164	2596	9185	39563	894	271158
$\Delta g_m/g_{m0}$	449	616	2150	2311	256	109
$\Delta I_{dss}/I_{dss0}$ (V_{gf} =+1V)	36	233	198	1882	35	2451
$\Delta I_{dss}/I_{dss0}$ (V_{gf} =+1.5V)	285	1488	4052	8367	490	671009

3.2 INTERFACE AND BULK DEFECTS

3.2.1 *Analysis of static characteristic curves*

The shift polarity of the monitoring parameters indicates the type of the injected carriers for each stress conditions. In all cases, hot electrons degrade the front and back channels. The only exceptions are C1 and C2 where the back channel degraded by hot holes. This is expected (V_{gb}=-20V) and is a direct indication that each channel is degraded separately. On the other hand, the data presented in table 1 show that there is a strong influence of front channel aging characteristics on back gate bias V_{gb}.

According to relation (1), the $\Delta C_{it}/C_{it0}$ was calculated as a function of stress time. The results are presented in figure 3. The influence of the back interface degradation on V_{gb} is obvious and its maximum effect takes place when the back interface is accumulated (case C). For this case, the interface states creation mechanism is dependent on stress time by a power law relation and is the only case also where lower doped devices are degraded more rapidly than those with higher doping.

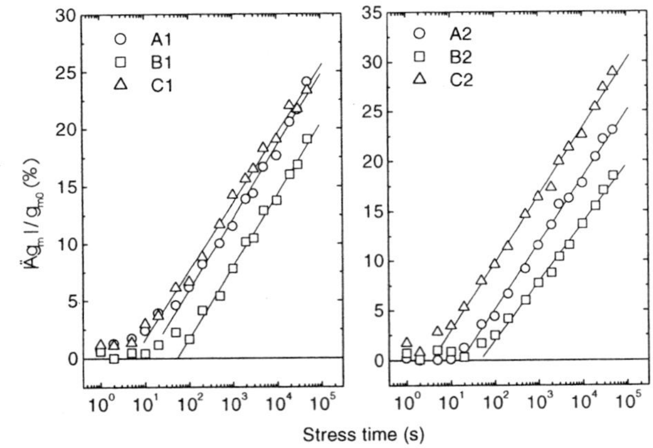

Figure 2. The degradation of front channel maximum transcoductance g_m.

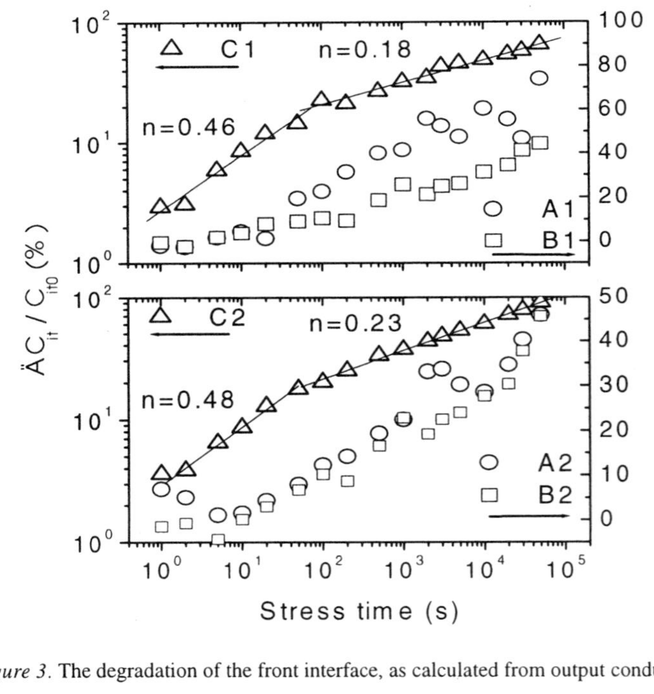

Figure 3. The degradation of the front interface, as calculated from output conductance.

3.2.2 *DLTS studies*

In tables 2 and 3, the results for two different gate pulses, used for deep level characterization as has been previously mentioned, are presented.

TABLE 2. Deep Levels detected in the "front channel" for two different gate pulses.

A. Pulse Mode: ON-OFF-ON

Electron Traps				Hole Traps			
Normal		Reverse		Normal		Reverse	
$E_A=E_C-E_T$ (meV)	σ_n (cm^2)	$E_A=E_C-E_T$ (meV)	σ_n (cm^2)	$E_A=E_T-E_V$ (meV)	σ_p (cm^2)	$E_A=E_T-E_V$ (meV)	σ_p (cm^2)
577 ± 48	$4.1\cdot10^{-17}$	628 ± 64	$4.2\cdot10^{-16}$	577 ± 7	$6.4\cdot10^{-17}$		
600 ± 33	$2.7\cdot10^{-17}$	617 ± 23	$3.5\cdot10^{-16}$	632 ± 17	$1.4\cdot10^{-16}$		
662 ± 26	$5.6\cdot10^{-15}$			732 ± 15	$3.7\cdot10^{-15}$		
667 ± 35	$1.4\cdot10^{-15}$			735 ± 15	$3.5\cdot10^{-15}$		

B. Pulse Mode: OFF-ON-OFF

Electron Traps				Hole Traps			
Normal		Reverse		Normal		Reverse	
$E_A=E_C-E_T$ (meV)	σ_n (cm^2)	$E_A=E_C-E_T$ (meV)	σ_n (cm^2)	$E_A=E_T-E_V$ (meV)	σ_p (cm^2)	$E_A=E_T-E_V$ (meV)	σ_p (cm^2)
640 ± 20	$4.5\cdot10^{-16}$	617 ± 9	$1.6\cdot10^{-16}$	541 ± 20			
		635 ± 28	$3.1\cdot10^{-16}$				

TABLE 3. Deep Levels detected in the "back channel" for two different gate pulses.

A. Pulse Mode: ON-OFF-ON

Electron Traps				Hole Traps			
Normal		Reverse		Normal		Reverse	
$E_A=E_C-E_T$ (meV)	σ_n (cm^2)	$E_A=E_C-E_T$ (meV)	σ_n (cm^2)	$E_A=E_T-E_V$ (meV)	σ_p (cm^2)	$E_A=E_T-E_V$ (meV)	σ_p (cm^2)
454 ± 29	$3.6\cdot10^{-14}$			255 ± 16	$1.4\cdot10^{-19}$		
561 ± 28	$3.3\cdot10^{-16}$						

B. Pulse Mode: OFF-ON-OFF

Electron Traps				Hole Traps			
Normal		Reverse		Normal		Reverse	
$E_A=E_C-E_T$ (meV)	σ_n (cm^2)	$E_A=E_C-E_T$ (meV)	σ_n (cm^2)	$E_A=E_T-E_V$ (meV)	σ_p (cm^2)	$E_A=E_T-E_V$ (meV)	σ_p (cm^2)
				426 ± 89	$2.4\cdot10^{-16}$	304 ± 20	$1.1\cdot10^{-18}$
				640 ± 12	$4.7\cdot10^{-16}$	629 ± 8	$2.9\cdot10^{-16}$
				754 ± 16	$7.1\cdot10^{-16}$		

The majority of the induced defects were detected after B and C cases. There is no correlation between front and back channel deep levels. Electron and hole traps are introduced simultaneously, regardless of the stress conditions. Front channel hole traps seems to be created by an ion implantation-like mechanism. This is proved by the identification that E_V+0.735 eV is probably the II683 [10] trap (fig. 4). On the other hand, back channel electron defects are induced probably by a mechanism similar to that of electron irradiation. One of these defects, the E_C-0.454 eV, has almost the same Arrhenius signature with the trap EL5 [11] (fig. 4). It is worthwhile to notice that in reverse mode less defects than in normal mode were detected and, additionally, their cross section is slightly modified. This is an indication that the cross section is dependent on the electric field.

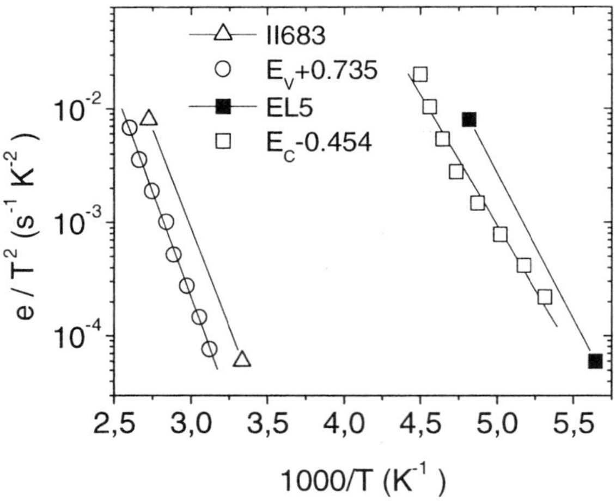

Figure 4. Back channel electron trap with Arrhenius plot similar to EL5 and front channel hole trap with Arrhenius plot similar to II683.

4. Acknowledgments

This work has been financially supported by Greek Secretariat of Research and Technology, in the frame of Greek-French bilateral exchange program PLATO.

5. References

1. Adan A.O., Naka T., Ueda T., Seguchi Y., Kagisawa A. (1997) SOI devices status for low-power IC applications, S.Cristoloveanu (ed.). *VIII Internat. Symp. Silicon-On-Insulator Technology & Devices,* ECS, **PV 97-23**, *340-351.*
2. Renn S.H., Pelloie J.L., Balestra F. (1996) Hot-carrier effects in deep submicron thin film SOI MOSFETs. *IEDM Technical Digest, 877-880.*
3. Renn S.H., Pelloie J.L., Balestra F. (1997) Hot-carrier-indused degradation and lifetime prediction in off-state operation of deep submicron SOI N-MOSFETs, in S.Cristoloveanu (ed.). *VIII International Symposium on Silicon-On-Insulator Technology and Devices,* ECS, **PV 97-23**, *283-288.*
4. Renn S.H., Raynaud C., Pelloie J.L., Balestra F. (1998) Comparison of Hot-carrier effects in deep submicron n- and p-channel partially- and fully-depleted UNIBOND and SIMOX MOSFETs. *IRPS98,* IEEE, *203.*
5. Cristoloveanu S., Li S.S. (1995) *Electrical Characterization of Silicon-On-Insulator Material and Devices.* Kluwer Academic Publishers, Dordrecht.
6. Kato K., Taniguchi K. (1986) Numerical analysis of switching characteristics in SOI MOSFTEs. *IEEE Trans.Electr.Dev.* **ED-33**, *133-139.*
7. McLarty P.K., Ioannou D.E., Hughes H.L. (1988) Deep states in silicon-on-insulator prepared by oxygen implantation using current deep level transient spectroscopy. *Appl.Phys.Let.* **53**, 87*1-873.*
8. Ghibaudo G. (1995) Static measurements and parameter extraction, in H.Haddara (ed.) *Characterization methods for submicron MOSFETs,* Kluwer Academic Publishers, Dordrecht.
9. Dimitrakis P., Jomaah J., Balestra F., Papaioannou G.J. (2000) Hot-carrier effects in 0.15• m Low Dose SIMOX N-MOSFETs. *III International Conference Advanced Semiconductor Devices & Microsystems* (to be presented).
10. Sah C.T. (1977), *Semiconductor Silicon 77,* **2**, 868.
11. Properties of Si (1988), INSPEC, The Institution of Electrical Engineers, London and New York.

DEFECTS AND THEIR ELECTRONIC PROPERTIES IN HIGH-PRESSURE-ANNEALED SOI STRUCTURES SLICED BY HYDROGEN

V.P. Popov*[1], A.K. Gutakovskii[1], L.N. Safronov[1], I.E. Tyschenko[1], S.K. Zhuravlev[1], A.B. Talochkin[1], A.I. Antonova[1], O.V. Naumova[1], V.I Obodikov[1], A. Misiuk[2], J. Bak-Misiuk[3], J. Domagala[3], A. Romano-Rodrigues[4], A. Bachrouri[4]

[1] Institute of Semiconductor Physics, RAS, 630090 Novosibirsk, Russia

[2]Institute of Electron Technology, [3] Institute of Physics, Polish Academy of Sciences, Al. Lotnikow 46, 02-668 Warsaw, Poland
[4] University of Barcelona, Dept. of Electronics, Marti i Franques, 1, E-08028 Barcelona, Spain

1. Introduction

In the fundamental phenomenon of blistering observed during implantation of gas ions into silicon, there remains a long-standing controversy concerning the mechanism of decay of the over-saturated solid solution of gas atoms and intrinsic point defects. The interest to this phenomenon has been drastically renewed in recent years due to its practical significance since blistering adds to the toolbox of SOI technology a useful method known as "Smart Cut" [1].

Several mechanisms of silicon splitting with implanted hydrogen were proposed at atomic scale. In one of the first reported study [2], the ascending diffusion of atomic oxygen was considered as an effect underlying the phenomenon of interest. Based on IR-absorption data, Weldon at al. [3] suggested that aggregation of mobile multi-vacancy clusters formed during implantation and their subsequent transformation into an ordered structure of tetrahydrated vacancies $(VH_4)_n$ is the main reason for blistering. Pantellides and co-workers [4] examined this model theoretically and arrived at the conclusion that internal surfaces are formed due to ($3VH_4$*) species, where the supersript * denotes that three tetrahydrated vacancies occupy the sites in the silicon lattice separated by two interatomic distances. In this model, the growth of extended {111} and {100} defects was attributed to reconstruction and migration of hydrated divacancies (($2VH_4$*) – V_2H_6) occurring with ejection and subsequent re-incorporation of H_2 molecules. The authors of [4] invoke the notion about mobile H_2 molecules contradicting however to reported theoretical and experimental results [5].

The data of [6] count in favor of participation of intrinsic defects in nucleation and subsequent growth of microcavities filled with the gas of implanted substances (hydrogen or helium). In [6], an assumption was put forward about participation of hydrated defects (in particular, vacancy- (VH_2) or interstitial-related complexes (IH_2)) in hydrogen diffusion to extended defects (platelets and microvoids) and their mutual annihilation at platelet walls accompanied by ejection of molecular hydrogen. This process is capable in principle of providing a constant excess pressure inside a growing

F. Balestra et al. (eds.), Progress in SOI Structures and Devices Operating at Extreme Conditions, 269–288.

microcavity necessary for its enlargement. If this process is limited by diffusion of intrinsic defects, it is probable that variation of the concentration and mobility of these defects during changing their position in crystal lattice by, say, applied external hydrostatic pressure would be useful for gaining information about the type of defects responsible for blistering. In particular, this approach was successfully used by Turnbull with co-workers [7,8] for studying the solid-phase crystallization of silicon and germanium.

The observed Arrhenius dependence of the latent period of blistering on the temperature [1] makes it possible to invoke the phenomenological theory of transitional state having a certain activation volume ΔV*. In this case, the expectation time for the occurrence of blistering should depend exponentially on pressure:

$$\frac{1}{t} = \frac{1}{C_H} \exp\left(\frac{-E_a}{kT}\right) \qquad (1),$$

where $E_a = E_{a0} + P\,\Delta V^*$.

Judging from the sign and value of ΔV*, one can estimate the contribution of vacancies, interstitials and dangling bonds to the process that limits blister formation. Besides, it was shown [9] that formation of multi-vacancy defects in silicon is suppressed under conditions of a high hydrostatic pressure. The fact that no blistering, extended platelike defects and hydrated internal surfaces were observed during annealing under high pressure provides evidence for the assumed involvement of vacancies into the phenomenon under study. In the present work, single silicon wafers as well as SOI structures implanted with hydrogen and were subjected to heat treatments under elevated external argon pressure. The aim was to gain better insight into the microscopic mechanism of blistering by revealing both the type of hydrated point and extended defects limiting silicon splitting and the role of internal pressure in formation of micro-voids and SOI technology. In the case of SOI structures, the external pressure provides additional possibilities for controlling the formation of regular bonds during wafer bonding and material splitting during growth of the internal pressure in the layer over-saturated with hydrogen.

2. Experimental

In experiments we used lightly doped (100) and (111) silicon wafers. These wafers were cut from Czochralski-grown silicon ingots and ingots grown by the float-zone technique, respectively. Hydrogen was implanted into the wafers at room temperature in the form of H_2^+ ions. The ions had the energy 130 keV and the implantation dose ranged from $3\text{x}10^{15}$ to $5\text{x}10^{16}\,\text{cm}^{-2}$. The mean penetration depth of the ions into silicon was 0.65 μm. Samples cut from irradiated silicon wafers and SOI structures were given heat treatments in the temperature range 450 –1200°C under atmospheric or high (1.5 GPa) pressure in argon ambient. To raise the concentration of residual hydrogen by a factor of 2-5, a certain part of the samples was additionally implanted with protons having the energy 18-24 keV. The total flux of protons in this case was $5\text{x}10^{16}$ - $3\text{x}10^{17}\,\text{см}^{-2}$.

The structure of samples was studied using high-resolution X-ray diffractometry

(HRXRD), Rutherford backscattering under channelling conditions (RBS/C), transmission electron microscopy (TEM), Raman spectroscopy (RS), Fourier transform IR absorption (FTIR), and secondary-ion mass-spectroscopy (SIMS). Photoluminescence (PL) and spectral ellipsometry (SE) were additionally used to study the optical properties of the samples. Shear and bulk concentrations of charge carriers were determined using capacitance and Hall measurements in the Van-der-Paw geometry.

3. High pressure effects in H implanted silicon

3.1. SUPPRESSION OF BLISTERING IN SI:H UNDER HP >0.5 GPA

The main result concerning the annealing of samples implanted with hydrogen under the pressure above.1 GPa is the absence of blistering at annealing temperatures > 400°C. At the same time, in samples annealed at 450°C during 10 hours under atmospheric pressure, 0.5 – 1.0x10^6 blisters/cm^2 formed on the surface and 14 –18 % of their total area was delaminated. (Fig.1).

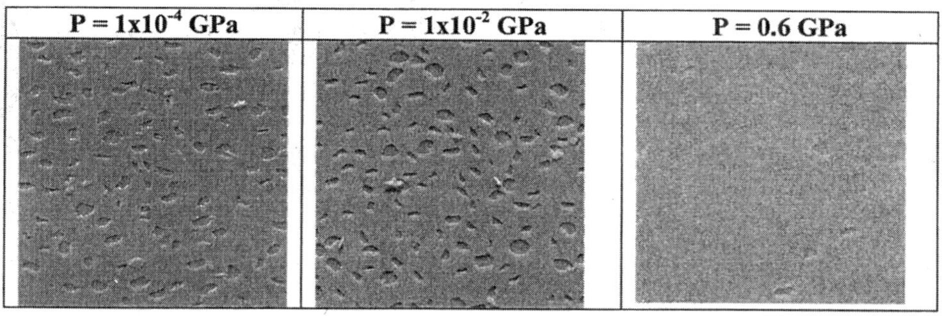

Figure 1. Suppression of blistering in Si:H during HP-HT treatments at 450°C, 10h, under pressure P.

The structure of crystals, as well as the defects and mechanical strains in annealed samples, were studied using a high-resolution diffractometer MRD produced by Philips. Figure 2 shows the distribution of the scattered intensity of X-rays in the vicinity of the 333 reflection of (111) FZ wafers after HP-HT treatments. The widened width of the peak and greater intensity of scattered X-rays around it is indicative of the occurrence of severe damages and considerable deformations in implanted samples. The 450°C-anneal not only fails to diminish these quantities but even results in an asymmetry of the halo presumably caused by predominance of tensile stresses, which attain their highest after heat treatments under P= 1.5 GPa during t=0.5 h (Fig. 2c). The X-ray-diffraction titling curve near the 444 reflection measured on the same silicon wafer is shown in Fig.3. This curve was measured in the double configuration using an analyzing crystal installed ahead of the detector. In this curve, a shoulder on the slope of the 444 peak is clearly seen arising due to tensile stresses after the annealing under high pressure. These stresses are caused both by introduced lattice defects and by an increase in the internal pressure caused by implanted hydrogen. The stresses in the damaged layer were estimated by assessing the relative change in the distance between atomic planes: $\varepsilon = \frac{\Delta a}{a} = \frac{a_L - a_S}{a_S}$, where a_L and a_S are lattice parameters in the damaged layer and in the substrate, respectively. The measured values for (111) wafers were higher than those for the (100) ones. The largest values of observed tensile stresses $\varepsilon = 0.70\%$ are known to correspond to the change of lattice parameter expected for the external pressure 0.5 Gpa. In its turn, this value corresponds to the ultimate shear strength of silicon single-crystals [10].

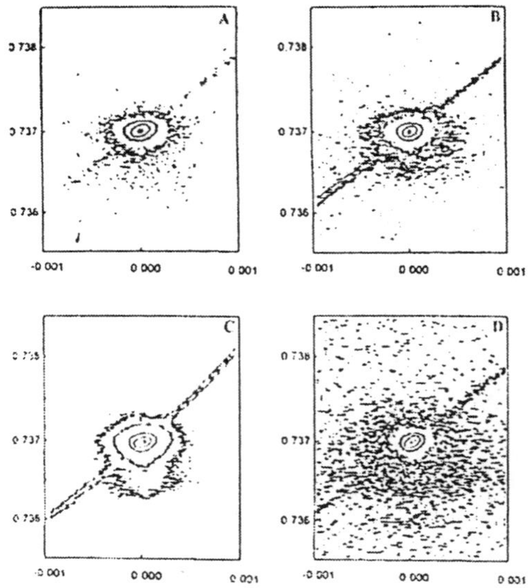

Figure 2. Reciprocal-space Bragg maps in the vicinity of the 333 reflection of FZ-Si:H: samples: a - sample implanted with hydrogen dose $4.5x10^{16}cm^{-2}$, b - sample implanted with the same dose and subsequently annealed at 450°C under AP (10^{-4} GPa) during 10 h, c - the same sample as in (b) but annealed under HP (1.2 GPa) during 0.5 h, d - the same sample as in (c) but annealed during 10 h.

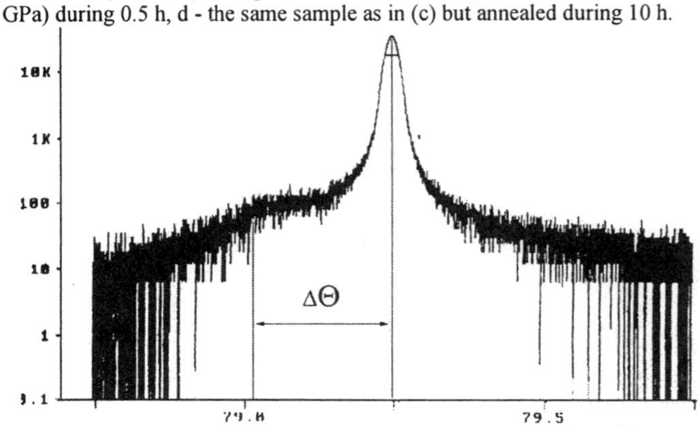

Figure 3. HRXD titling curve near the 444 reflection of the (111) Fz-Si sample implanted with hydrogen up to the shear concentration $4.5x10^{16}cm^{-2}$ and annealed at 450°C during 10 hours under the pressure 1.2 GPa in argon atmosphere. The shoulder on the slope of the main peak correspond to stresses □ in the damaged layer ranging from 0 to 0.697%.

3.2. TRANSFORMATION OF STRUCTURE OF SI:H DURING HT-HP TREATMENT

RBS/c measurements (Fig. 4) showed that, the formation of blisters gives rise to an increased yield of back-scattered ions in spectra from channel numbers higher than 100. This increase is caused by violation of channeling conditions when the silicon lattice at the slopes of blisters is titled by an angle larger than the critical one for channeling of analyzing ions. When no blisters form, the spectra of annealed and virgin crystals in this part of channels coincide with one another perfectly well (Fig. 4).

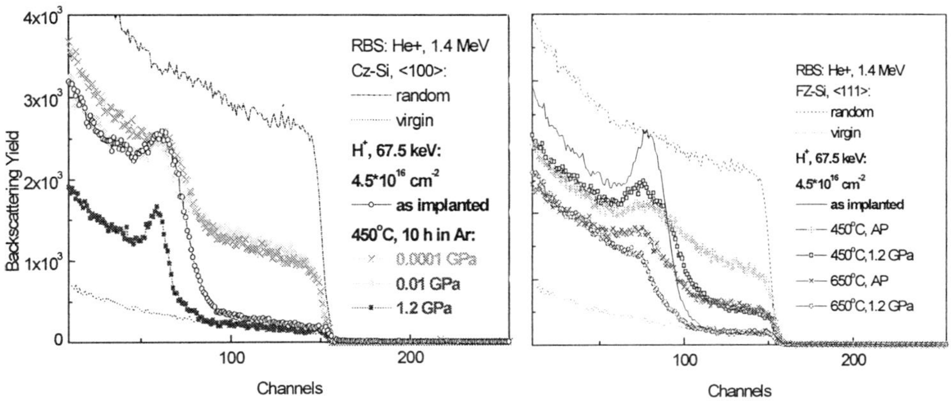

Figure 4. RBS/c spectra of Fz-Si and Cz-Si samples (left and right, respectively) implanted with hydrogen dose $4.5x10^{16}cm^{-2}$ and annealed under various pressures during 10 hours.

Figure 5. TEM cross-sectional photomicrographs of (100) Cz-Si wafers implanted with hydrogen and annealed under atmospheric (10^{-4} GPa) and high (1.2 GPa) pressure at 450°C during half an hour.

Similar data were obtained in [11] with the help of SEM and TEM. TEM data show (Fig.5) that here, in contrast to the samples annealed under pressures below 0.6 GPa, no extended microcracks were formed responsible for blistering and splitting-off of damaged layers. Only defects smaller than 50 nm were found abundant in high concentrations.

Close examination of TEM microphotographs showed that {111} and {100} platelets not transformed into blisters and dislocation loops were the main defects in samples annealed at 450 – 550°C and 600 – 1200°C, respectively. These data correlate fairly well with HRXD data (Fig. 6) [11-13]. The reciprocal-space Bragg maps in the vicinity of the 333 reflection for the samples annealed at T ≥ 650°C are indicative of the absence of substantial stresses in the samples. Only in the sample annealed under the pressure 0.6 GPa we observed changes in the X-ray intensity in the region under consideration. These changes are known to be typical for mosaic crystals. For samples treated under other conditions, a broad background in RBS spectra was found to be more pronounced, which is indicative of the formation of a great number of extended defects.

In samples annealed under P= 1.2 GPa, no evidence for the formation of crystal blocks was found, and the concentration of extended defects was lower. Thus, the pressure above 1.2 GPa prevents both the formation of blisters and exfoliation of damaged layers, and allows one to preserve the initial planar (111) silicon-wafer surface intact during annealing

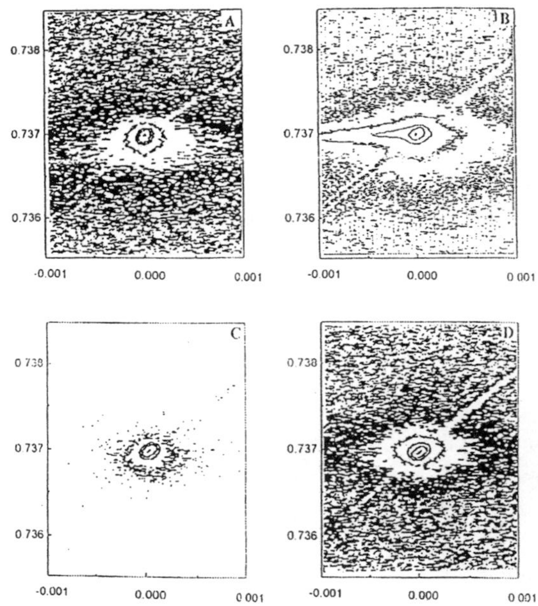

Figure 6. Reciprocal-space Bragg maps of 333 reciprocal lattice sites for FZ-Si:H: samples: a - sample implanted with the hydrogen dose 4.5x10^{16}cm^{-2} and annealed at 650°C under AP (10^{-4} GPa) during 10 h, b - the same sample as in (a) but annealed under HP (0.6 GPa), c - the same sample as in (a) but annealed under HP (1.2 GPa), d - sample implanted and annealed at 1100°C under HP (1.2 GPa) during 5 h (d).

3.3. RESIDUAL HYDROGEN, SI-H BONDS, AND ELECTRON CONCENTRATIONS

SIMS data are shown in Fig. 7. They show that the high pressure leads to a slight increase in the concentration of residual hydrogen in the implanted layer ($5x10^{15}cm^{-2}$). During the annealing under high pressure, the concentration of hydrogen in the surface layer also increases presumably due to oxidation of samples by residual oxygen. With the increase in pressure under which the annealing was carried out, the concentration of donors increases, the effect being observed not only in the damaged region (Fig. 8) but also in the bulk of samples.

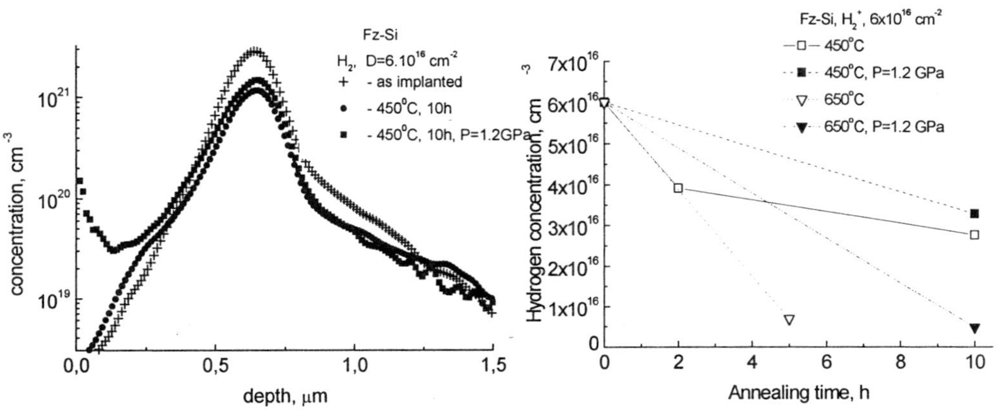

Figure 7. SIMS profiles of hydrogen concentration in as-implanted samples and samples annealed at 450°C during 10 hours under atmospheric and high pressure (left) and profiles of the shear concentration of hydrogen vs the annealing time (right) for various pressures and temperatures.

The enhanced formation of hydrogen-related donors and thermal donors provide evidence for participation of vacancy-related species in this process. Yet, it remains unclear whether the vacancies immediately enter the donor structure or they only promote nucleation of donors. Further insight into the nature of donor centers can be gained by considering data on their thermal stability. Dependences of the shear concentration of donors near the wafer surface on the annealing temperature are shown in Fig. 9 [6]. This concentration attains its highest ($5x10^{12}cm^{-2}$) at annealing temperatures around 400-450°C. The maximum concentration attained is by four orders of magnitude smaller than the concentration of implanted oxygen. As the pressure rises, the concentration of donors in the damaged layer, in contrast to the sample bulk, drops by one order of magnitude [14]; however, the n-type of conductivity was retained even after the 1200°C-anneal under the highest pressure. In Fz-Si, the concentration of donors increases only within the 1-μm-thick damaged layer, and it only weakly depends on pressure.

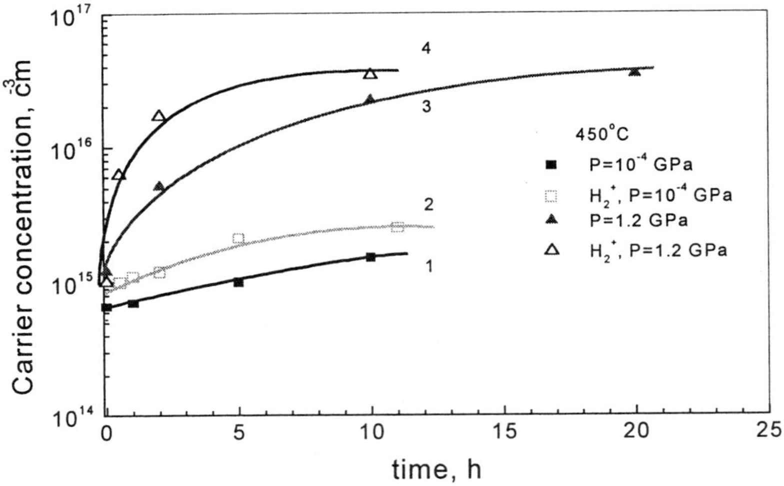

Figure 8. Time dependence of bulk carrier concentration in the as-grown Cz silicon crystal, after hydrogen implantation, and in HP-annealed samples.

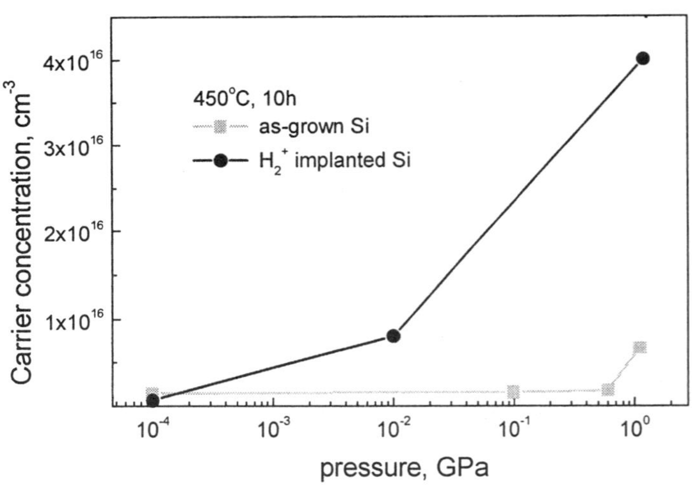

Figure 9. Pressure dependence of bulk carrier concentration in the as-grown Cz silicon crystal, after hydrogen implantation, and after subsequent HP HT treatment.

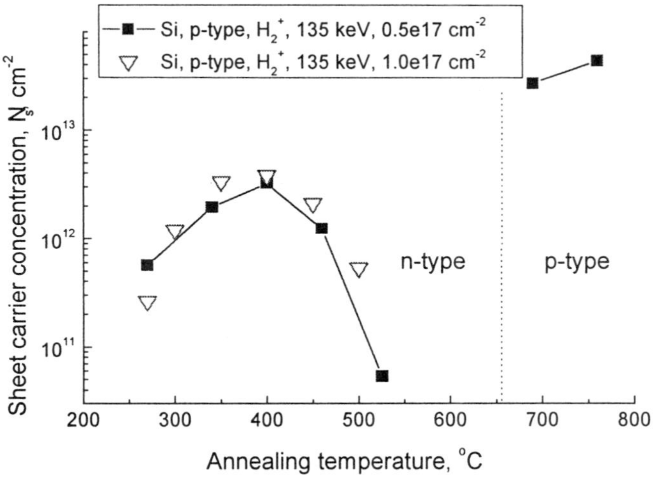

Figure 10. Temperature dependence of the free-carrier concentration in the upper silicon layer after hydrogen implantation

Since the molecular hydrogen in silicon is immobile below ~500°C [5], its diffusion cannot be the reason for blister formation at these temperatures. Hence, two reasons for retardation of blistering remain to be considered. First, the external pressure compensates for the internal one and increases the value of the critical pressure for blister formation. Second, the proportion between concentrations of vacancy and interstitial defects changes.

Let discuss these reasons in more detail. From the date of [5] and our data [15], it follows that the enhanced formation of blisters in the presence of helium is caused exclusively by the excess pressure inside microcavities. This pressure increases the concentration of hydrogen captured by ruptured Si-Si bonds. Then, the compensation of the internal pressure by the external one should diminish the concentration of Si-H bonds. Indeed, according to IR data [15], in experiments with pressure the concentration of these bonds at temperatures 400 - 500°C is about 10^{16} cm^{-2} lower than under atmospheric pressure. At the same time, the total concentration of hydrogen after heat treatments at these temperatures is higher by the same value (Fig.7). It means that the excessive hydrogen is contained under elevated pressure in optically inactive species, presumably in the form of H_2 molecules. Yet, the excessive hydrogen is obviously not contained inside microcavities since the concentration of Si-H bonds at internal surfaces is lower in samples annealed under high pressure. Indeed, since the volume of microcavities cannot be increased with the surface area being decreased, we may conclude that either the gas molecules are contained inside cavities at higher pressures, or they are not contained inside cavities at all. Suppose that the molecular hydrogen is contained inside microcavities at elevated pressure and behaves as a perfect gas. Then, from the condition of rough equality of inside areas [15] and from the known values of radii of semi-spherical domes r_1 = 10 μm under standard pressure and radii of disk-like 1 nm-thick nanocavities r_2 = 100 nm for P=1.5 Gpa we obtain that the ratio of pressures P_2/P_1 should run at least into 10^{11}. It follows from here that the absence of blistering under the high pressure P_2 = 1.2 GPa should be attributed to the capture of gaseous hydrogen outside cavities rather than to the decrease in the excess pressure by $P = P_1 - P_2$.

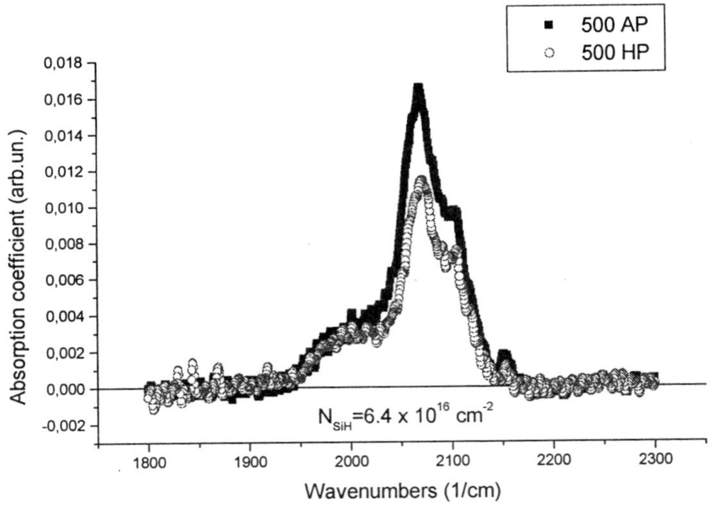

Figure 11. FTIR spectra of Si:H stretching modes for samples annealed at 500°C under AP (10^{-4} GPa) and HP (1.2 GPa) in argon ambient

This hypothesis is also favored by data which show that the concentration of donor centers in the damaged layer decreases with increasing pressure. The increased concentration of point and extended defects (Figs. 2 and 3) after high-pressure anneals results in a smaller mean distance between hydrogen molecules and these defects. Thus, variation of both concentrations and types of intrinsic defects is a more likely reason for retardation of blister formation during annealing under high pressure.

Let us consider now variation of concentration of defect centers during high-pressure annealing. As noted above, theoretical considerations show that the concentration of multi-vacancy complexes should decrease with increasing pressure [9]. Under the assumption that the work on attaching an interstitial atom to silicon surface equals zero, the activation volume for the formation of vacancy in silicon was shown to be negative [16]. It implies that the concentration of free vacancies increases with increasing pressure. It seems that the capture of hydrogen molecules by vacancies under conditions of a high hydrostatic pressure retards both the coagulation of hydrogen molecules and the formation of blisters near the free surface of silicon or disintegration of bonded wafers.

4. Internal and external pressure during fabrication of SOI structures

4.1. DEFECTS AND STRAIN IN AS-SLICED SOI STRUCTURES

High-resolution x-ray diffraction (HRXRD) is a very effective and non-destructive tool for the characterization of the SOI structures [17,18]. In the case of bonded wafers the layer and substrate diffraction peaks are separable. It is believed [19] that this simplifies the interpretation of the diffraction spectra. HRXRD is capable to accurately measuring the film thickness, tilt and twist of film planes with respect to the substrate, extract lattice constant **a** and strain in silicon film.

SOI structures were fabricated by bonding of silicon wafers using a new variant of Smart-Cut technology [20]. Ordinary as-bonded SOI structures are annealed at high temperature (1100°C) for removal of hydrogen, radiation defects and strength of bonding interface. The transformation of structure in as-bonded and annealed SOI wafers was investigated by HRXRD method before such annealing. High-resolution diffraction measurements in double and triple configuration were carried out using a four-crystal Ge (220) Bartels –type monochromator in primary $CuK\alpha_1$ beam and a channel-cut double reflection Ge (220) analyzer for the diffracted beam. ω is the angle between the incident beam and wafer surface. 2θ is the angle between incident beams and reflected beams, φ denotes a rotation of the sample around the axis perpendicular to the samples surface. The (004) spots were measured in the double and triple-configuration modes (i.e. with full diffracted beam and with an analyzer in front of the detector that accepts diffracted radiation over a range of 12 secs and stops (ω-scans) or scans two times faster the sample speed ($2\theta/\omega$-scans, Fig.12). The (113) spots were measured using ω-scans. Measurements of rocking curve of symmetrical (004) and asymmetrical (113) reflections enabled us to calculate the strain, tilt and twist of film planes with respect to the substrate planes. The strain in the SOI film was determined by Fewster method [21] and also extracted from the reciprocal space map for other reflections. The thickness was estimated for the high quality film from the Pendellosung fringes.

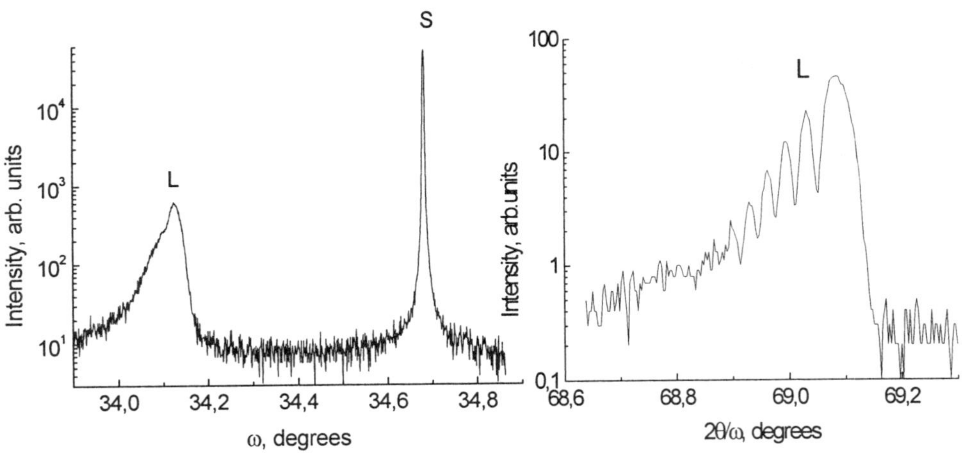

Figure 12. ω-scan of the 004 reflection in double configuration (a) and $2\theta/\omega$-scan (b), from the SOI-A structure. Peaks S and L correspond to the substrate and top silicon layer, respectively.

The SOI- structures were fabricated by bonding of Czochralski-grown silicon wafers with (100) orientation, p-type of conductivity (20 Ωcm) and 4 in. in diameter with using of hydrogen–cut technology [4]. This method indicates the bonding of hydrogen – implanted silicon wafer with insulator covered substrate. The heat treatment of connected wafers at relatively low temperature (400-600°C) causes a delamination of the thin silicon layer from implanted wafer and transfers it on fully oxidized substrate. The final SOI structure was obtained by high temperature annealing (1100°C for 1 hour). As-bonded SOI (lets mark it SOI-A) as well as final SOI structure (SOI-B) annealed under AP, SOI structure annealed under HP (SOI-C), and as implanted Si wafer (H_2^+, 100 keV, $4x10^{16}$ at.cm^{-2}) were investigated by HRXRD after annealing at different HP-HT treatment.

The rocking curve in double configuration (ω-scan) for the SOI-A, the 2θ/ω-scan of the (004) reflections for three SOI structures are shown in Figs.12 and 13. The peaks from the top silicon layer and the substrate are separated in the ω-scan (Fig.12). This separation of peaks is connected either to the strain in the SOI top layer or with the tilt between the layer and the substrate. The tilt angle between the top layer and the substrate was estimated independently from the two rocking curves, measured by the rotating sample around φ - axis with minimal and maximal distance between the substrate and layer peaks. The tilt value was calculated from the value of $(\Delta\omega_{\varphi-} - \Delta\omega_{\varphi-})/2$, where Δω is the difference between the angular position of the substrate peak and that of the top silicon layer. For the SOI-A sample the tilt was equal to 0.5±0.05° due to unintentional misalignement of the initial silicon wafers.

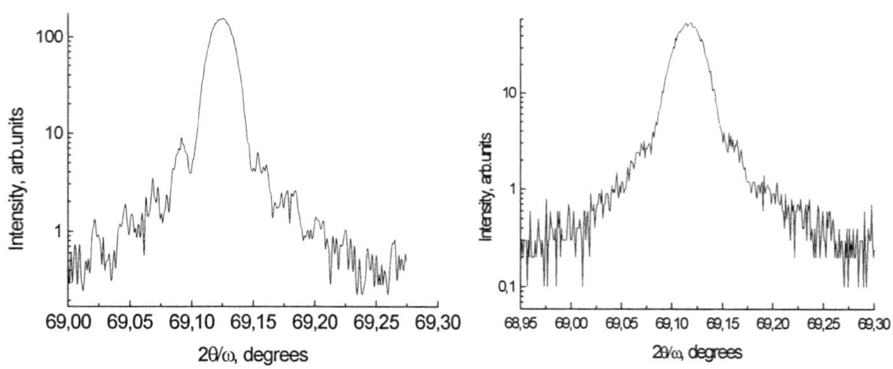

Figure 13. 2θ/ω-scan of the 004 reflection for the SOI structures annealed at 1100°C under AP (left) and HP 1.2 GPa (right).

The strains $\frac{\Delta a}{a}$ for the samples are given in the Table 1. The strain value is high in the as-bonded SOI structure. The residual strain for final SOI structures is at least one order of magnitude lower than that for SOI processed by the BESOI method [19] and below a sensitivity of the method employed.

The twist between the top Si layer and the substrate for SOI was estimated from the difference of azimuth (φ) positions with the maximum intensities of asymmetrical 113 reflections for the substrate and the top layer. The twist angle was found to be (2.5±0.03)°.

4.2. TRANSFORMATION OF SOI STRAIN AT HP-HT TREATMENT

The thickness Pendellosung fringes for SOI-A were visible for the layer peak only at the small angle side of the rocking curve (Fig. 12). This asymmetry may be connected with the fluctuation of the lattice parameter due to the depth distribution of the hydrogen–related and radiation defects or can be caused by the roughness of the Si/SiO_2 interface if the layer and the substrate were not contacting at all points of the interface resulting in incomplete adhesion. The thickness fringes are very well resolved for SOI-B (Fig.13). The Si layer thickness calculated from the fringes was equal to 0.43±0.05 μm in a good agreement with the ellipsometry data. As it is seen from Table 1, the FWHM values for the silicon layer of the SOI-C, SOI-B, and SOI-A structures are higher than that of the typical silicon single crystal. Broadening of the rocking curve in double configuration may be caused by the fluctuation of the lattice constant (via presence of defects and local strains) and by dislocations (mosaic-like structure) created in the sample. But the transmission electron microscopy (TEM) investigation shows that either dislocation loops or network are not observed in top silicon layer [22].

Table 1. The FWHM and $\frac{\Delta a}{a}$ values for HP-HT treated samples obtained in double and triple configurations.

Samples	FWHM, arcsec, double configuration		FWHM, arcsec, triple configuration		$\frac{\Delta a}{a} \times 10^{-3}$
Si:H as-implanted	27				
Si:H, 1100°C, AP	15		9		-
Si:H, 450°C, 1.2 GPa	24		9		7.0±0.02
Si:H, 1200°C, 1.5 GPa	23		9		-
	Si film	Substrate	Si film	Substrate	
SOI, 450°C, AP	135	15	80	9	0.5±0.02
SOI, 1100°C, AP	60	15	48	9	0.02±0.02
SOI, 1100°C, 1.2 GPa	125	15	108	9	0.15±0.02

Larger FWHM values of Si films in triple configuration in comparison with the substrate value are caused by increase in FWHM due to final thickness effect (but the computed value of FWHM for this thickness is only 25 arcsec) and mosaic-like structure. The reason for such mosaic structure in the film lattice may be connected with incomplete adhesion at the bonding and different thermal expansion during thermal treatment needed for removing of the defects. Lesser FWHM values of Si film in double configuration for SOI-C than for SOI-A does not agree with a larger values in triple one. This fact reflects mosaicity or microcrystalline nature of high pressure annealed SOI structure.

4.3. MICROSTRUCTURE OF HP-HT TREATED SOI FILMS

The internal structure of SOI layers was analyzed using TEM measurements. These measurements were carried out on a JEOL 4000EX electron microscope. Dark-field microphotographs of HP-HT-treated SOI structures obtained using low-intensity beams are shown in Fig. 14. The full absence of damages in the top Si layer after the 1100°C-anneal is also confirmed by the electron microdiffraction pattern shown in the insert to the left photograph. After the annealing under P=1.2 GPa (see the right part of Fig. 14), numerous microcrystallites are clearly seen differing in orientation from the surrounding matrix. Additional reflections appear in HREM patterns including the forbidden 110 ones.

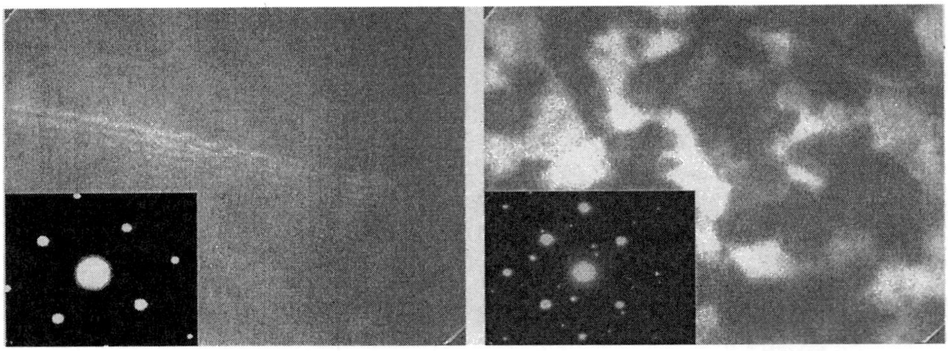

Figure 14. TEM microphotographs and ED patterns of the top Si layer of SOI structures annealed at 1200°C, under AP (left) and HP (right) in argon ambient. The total shown area here is 200x270 nm.

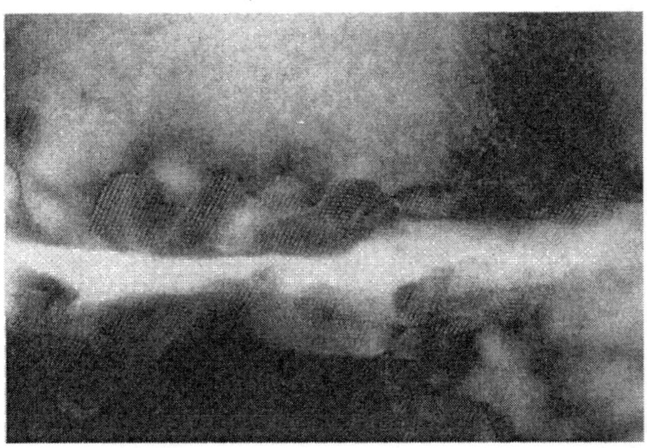

Figure 15. Microtwins at the walls of microcrack in H implanted Si after HP-HT treatment (650°C, 1.2 GPa).

The reason for their appearance may be the double diffraction on microcrystallites or possible formation of Si microcrystallites with an internal structure different from the cubic hexagonal diamond lattice, e. g., with the wurtzite structure. In turn, the reason for the formation of this lattice may be clusterization of hydrated point defects on extended imperfections such as platelets or microcracks near the maximum in the concentration profile of implanted hydrogen in the top Si layer. This assumption is corroborated by the cross-sectional TEM microphotographs of Si implanted with hydrogen and annealed at 650°C under argon pressure 1.2 GPa (Fig.15).

Twin microcrystallites with a network of moire strips growing-through from crack walls into matrix in the form of truncated octahedrons indicate that the presence of hydrogen on microcrack walls and inside them is a condition necessary for their formation. More exact data about the structure and grain-boundary angles are obviously required to devise an adequate model for the formation of microcrystallites at microscale.

5. Defect annealing and changes in electrical and optical properties of Si:H and SOI:H films

5.1.CHANGES IN THE RESISTIVITY AFTER HP-HT TREATMENTS

The top silicon layers in all bonded SOI structures had the n-type conductivity due to the formation of shallow donors in the presence of hydrogen. The annealing of the H-implanted silicon at 450°C under high pressure increases the concentration of donor centers in the implanted layer. Yet, the resistivity increases because of the decreasing electron mobility. Under conditions of a high temperature (1100°C) and pressure the mobility increases, with the n-type conductivity type kept unchanged although the free-carrier concentration drops by more than two orders of magnitude. Effects similar to those in the bulk material (Table 2, [22]) are also observed in SOI structures [23]. HP-HT treatment increases in the donor concentration in comparison with AP annealing.

Table.2. Resistivity and concentration of charg e carriers in the bulk Si and in SOI structures

Samples	Resistivity, $\rho_v x10^{-2}$ Ohm*cm *n-type,* p-type		$p_s x10^{13}, cm^{-2}$ $n_s x10^{13}, cm^{-2}$		$p_v x10^{16}, cm^{-3}$ $n_v x10^{16}, cm^{-3}$
Si:H, 450C, AP	*15*		*0.3*		*8*
Si:H, 450C, 1.2GPa	*790*		*3*		*76*
Si:H, 1100C, AP	90		0.9		0.02
Si:H, 1100C, 1.2GPa	*23*		*9.9*		***0.2***
	Si film	Substrate	Si film	Substrate	Si film
SOI, 450C, AP	*1*	*15*	*1*	*3*	*25±5*
SOI, 450C, 1.2GPa	**15**	*5*		*9*	-
SOI, 1100C, AP	*60*	10	*0.02*	9	***0.4±0.1***
SOI, 1100C, 1.2GPa	**125**	90	-	1	-
SOI:H, 1000C, 1.2GPa	*4.7*	90	***108***	0.9	***8.5±1.0***

However, the type of conductivity for SOI film after HP-HT treatment cannot be revealed by Hall-effect measurements because of its local changes over the sample area. Last row in the table means an additional hydrogen implantation to avoid uncertainty in the type of carriers in SOI. If we assume, that vacancy participates in shallow donor formation, than it should have negative activation volume according to equation (1). The last conclusion coincides with the result obtained by Antonelli et al [16].

5.2 HYDROGEN IMPLANTATION INTO SOI FILMS.

To increase the concentration of hydrogen-related donors, we additionally implanted Si and SOI samples before heat treatments with 20-keV protons. The implantation doses were $6x10^{16}$ to $2x10^{17}cm^{-2}$. These samples demonstrate similar behavior of the electron concentration as a function of pressure. These dependencies confirm results obtained on annealed SOI structures. Unlike the case of silicon, the high-temperature anneal under high pressure diminishes the concentration of donors by one order of magnitude in comparison with heat treatments under atmospheric pressure (Table 2).

5.3.HYDROGEN AND DEFECT TRANSFORMATIONS.

Apparently, the natural reason for the decrease in the concentration of donors in top Si layers of SOI structures may be the decrease in hydrogen concentration [23]. In view of this, SIMS measurements were carried out to immediately monitor hydrogen concentration (Fig.16). As follows from the SIMS data, the residual hydrogen content in SOI layers is higher than in the bulk silicon. It means that the reason for the decrease in donor concentration in SOI structures during annealing under pressure is elastic stresses rather than changes in the hydrogen concentration. This conclusion correlates well with the data on IR absorption on Si-H vibrational modes measured in the bulk silicon. IR absorption data are compared with SIMS data in Fig. 17.

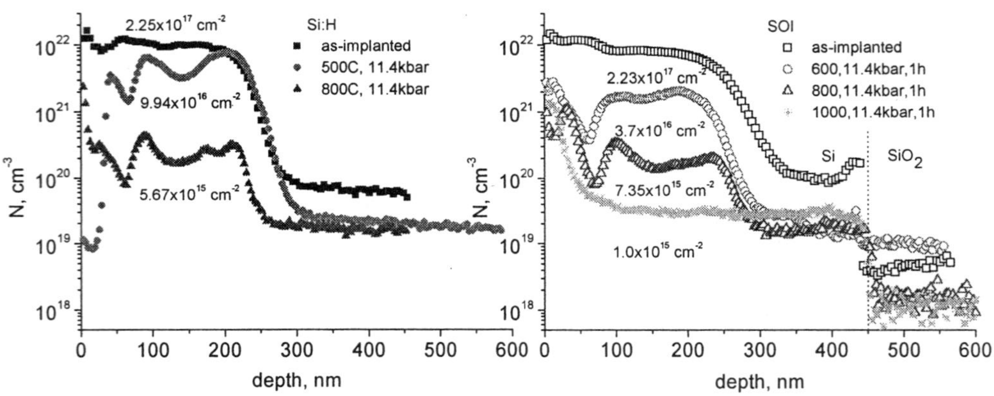

Figure 16. SIMS profiles of hydrogen concentration in HP_HT-treated Si:H and SOI:H samples.

The enhanced decay of the solid solution of interstitial oxygen in top layers of SOI structures at temperatures around 1100°C may be an additional phenomenon underlying the decrease in the concentration of donors.

5.4. VISIBLE LUMINISCENCE FROM H OVERSATURATED SILICON LAYERS.

To study photoluminescence (PL) properties of Si:H layers of Si and SOI wafers, we used samples implanted with 24-keV hydrogen ions up to doses $(0.6\text{-}3.0)x10^{17}$ cm^{-2}.

Hydrogen implantation was carried out using an ion-plasma source. After implantation, samples were annealed at temperatures ranging from 200 to 1000°C during 1 hour in N_2 or Ar ambient at pressures 1 or 11.4 kbar. PL emission spectra we excited at room temperature with 337-nm radiation of an N_2-laser. As-implanted samples showed a weak broad peak of PL intensity between 400 and 800 nm (Fig. 18).

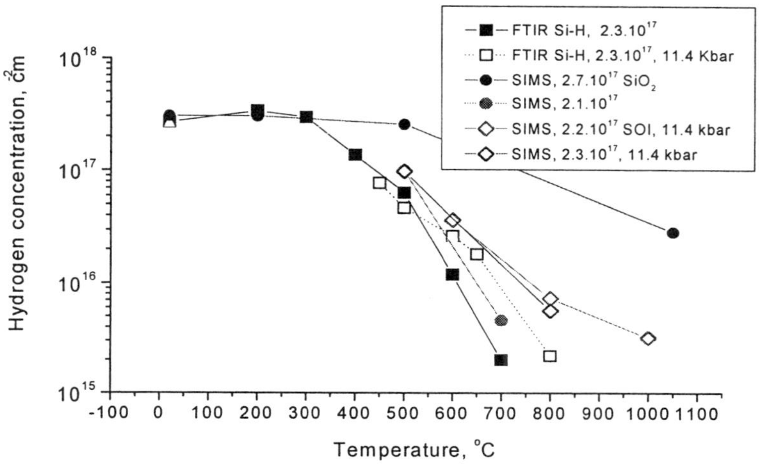

Figure 17. Residual concentration of implanted hydrogen after heat treatments under atmospheric and high (11.4 kbar) pressure.

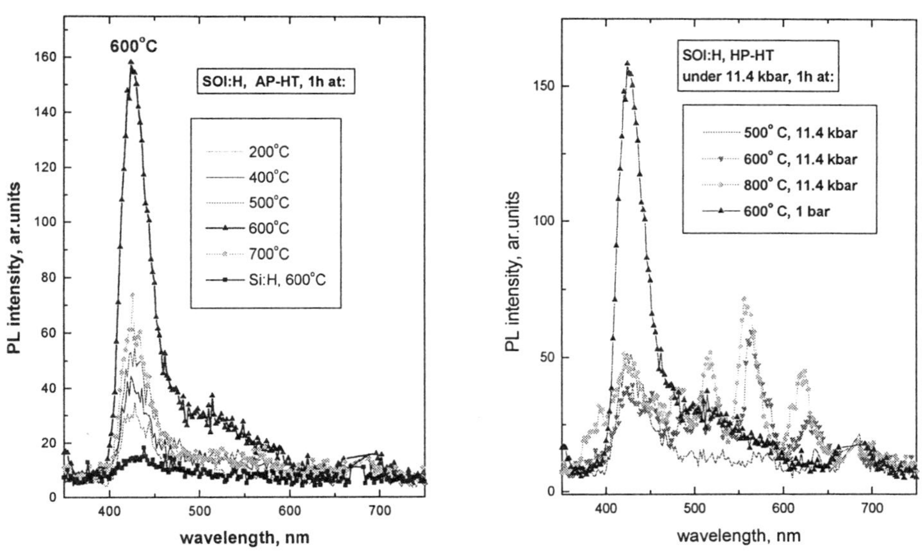

Figure 18. Increase in luminosity during AP annealing and interference in HP-HT treated SOI:H samples.

The PL intensity was found to increase with increasing annealing temperature. The highest PL intensity was exhibited by samples annealed at 600°C. As the hydrostatic pressure during annealing was raised to 11.4 kbar, PL peaks at 420 and 570 nm started dominating in PL spectra of heat-treated SOI and Si wafers, respectively. Unusual phenomenon of strong interference at the interval 450 – 750 nm was found in the SOI wafers after high pressure treatment. The nature of this interference is still unclear, but HP-HT treatment can be used to stimulate the luminosity and to obtain sharp lines instead of broad PL spectra in hydrogen implanted silicon wafers.

6. Summary

Silicon splitting with implanted hydrogen is a pure mechanical phenomenon. The internal pressure due to hydrogen aggregation provides a positive strain up to 0.7% and leads to development of microcracks responsible for blister formation. External hydrostatic pressure higher than 0.5 GPa is sufficient for avoiding blistering during heat treatment. The strain in as-sliced and HP-HT-treated SOI structures and the concentration of residual hydrogen in them were found to be higher than in volume Si wafers, implanted to the same fluence. The local strains caused by residual hydrogen seem to be the reason for the greater thermal stability of introduced donor centers and twinned mictocrystals. In our opinion, vacancy aggregation under high external pressure is suppressed and hydrated vacancy and interstitial form nuclei for a new phase or microtwins at the walls of platelet defects directly observed by cross-sectional TEM. HP-HT treatments offer a promising means for obtaining H-sliced SOI structures with modified physical properties.

7. Acknowledgements

The authors would like to thank Dr. F. Garner for fruitful discussion. These efforts were partly supported by the ISTC Grant no. 563.

8. References

1. M. Bruel. (1996) Application of hydrogen ion beams to Silicon On Insulator material technology. *Nucl. Inst. Meths. B,* **108**, 313-319.
2. C.M. Varma. (1997) Hydrogen-implant induced exfoliation of silicon and other crystals. *Appl. Phys. Lett.* **71**, 3519.
3. M. K. Weldon, V.E. Marsico, Y.J. Chabal, A. Agarval, D.J. Eaglesham, J. Sapjeta, W.L. Brown, D.C. Jacobson, Y. Caudano, S.B. Christman, E.E. Chaban.. (1997) On the mechanism of the hydrogen-induced exfoliation of silicon. *J. Vac. Sci. Technol. B* **15**, 1065-1073.
4. F. A. Reboredo, M. Ferconi, S. T. Pantelides. (1999) Theory of the nucleation, growth, and structure of hydrogen-induced extended defects in silicon. *Phys. Rev. Let.,* **82**, 4870-4873.
5. H. Takaba, A. Endou, A. Yamada, M. Kubo, K. Teraishi, K.G. Nakamura, K. Ishioka, M. Kitajama, A. Miyamoto. (2000) Tight-binding molecular dynamics

study of hydrogen molecule inside silicon crystal. *Jpn. Journ. Appl. Phys.*, **39,** 2744-2747.

6. V.P. Popov, V.F. Stas. (1999) Processes leading to delamination of thin layers in hydrogen irradiated silicon. In J. Weber and A Mesli (eds) *Defects in silicon: hydrogen.* E-MRS Symp. Proceed., Elsevier, p.187.
7. G.Q. Lu, E. Nygren, M.J. Aziz, D. Turnbull, C.W. White. (1989) Pressure enhanced solid phase epitaxy of silicon. *Appl. Phys. Lett.*, **54,** 2583-2586.
8. G.Q. Lu, E. Nygren, M.J. Aziz, D. Turnbull, C.W. White. (1990) Pressure enhanced solid phase epitaxy of germanium. *Appl. Phys. Lett.*, **56,** 137-139.
9. N. Cuendet, T. Halioioglu, W.A. Tiller (1996) The energies of microvoid formation in Si as a function of applied hydrostatic stress. *Appl. Phys. Lett.*, **69**, 4071-4072.
10. E. P. EerNisse and S. T. Picraux. (1977) Role of integrated lateral stress in surface deformation of He-implanted surfaces. *J. Appl. Phys.* **48**, 9-17.
11. A. Misiuk, H.B. Surma, I.V. Antonova, V.P. Popov, J. Bak- Misiuk, M. Lopez, A. Romano-Rodriguez, A. Barcz, J. Jun. (1999) Effect of external stress applied during annealing on hydrogen- and oxygen-implanted silicon. *Sol. State Phen.* **69-70,** 345-348.
12. A.Misiuk, J.Bak-Misiuk, A.Barcz, A.Romano-Radriguez, I. Antonova, V. Popov, C.A.Londos, J.Jun. (2000) Effect of annealing at argon pressure up to 1.2 GPa on hydrogen-implanted and hydrogen-plasma-etched single crystalline silicon, *Int. Journ. of Hydrogen Energy.* In press.
13. J. Bak- Misiuk, I.V. Antonova, A. Misiuk, J. Domagala, V.P. Popov, V.I. Obodnikov, J. Hartwig, A. Romano-Rodriguez, A. Bachrouri. (2000) Strain in hydrogen and oxygen implanted silicon and SOI structures annealed at high pressure. In Proceed. of Int. Conf. Synchr. Rad., Jaszowez, Poland.
14. I. V. Antonova, V. P. Popov, D. V. Kilanov, E. P. Neustroev, and A. Misuk. (1999) Pressure-induced formation of thermal donor centers in silicon after oxygen ion bombardment. *Semiconductors*. **33**, 1049-1053.
15. V. P. Popov, D. V. Kilanov, I.V. Antonova. Blistering and electronic properties of silicon coimplanted with helium and hydrogen. To be published.
16. A. Antonelli, E. Kaxiras, D.J. Chadi. (1998) Vacancy in silicon revisited: Structure and Pressure Effects. *Phys. Rev. Lett.*, **81,** 2088-2091.
17. G.M. Cohen, P.M. Mooney, E.C. Jones, K.K. Chan, P.M. Solomon, H-S.P. Wong. (1999) Characterization of the silicon on insulator film in bonded wafers by high resolution x-ray diffraction. *Appl. Phys. Lett.* **75,** 787-789.
18. S. Kimura, A. Ogura. (1998) Precise Measurement of Strain Induced by Local Oxidation in Thin Silicon Layers of Silicon-on-Insulator Structures . *Jpn. J. Appl. Phys.* **37**, 1282-1284.
19. T. Iida, T. Itoh, D. Noguchi, and Y. Takano. (2000) Residual lattice strain in thin silicon-on-insulator bonded wafers: Thermal behavior and formation mechanisms. *J. Appl. Phys.*, **87,** 675-681.
20. V.P. Popov, I.V. Antonova, V.F. Stas, L.V. Mironova, E.P. Neustroev, A.K. Gutakovskii, A.A. Franzusov, G.N. Feofanov. (2000) in P.L.F. Hemment (eds.) *Perspectives, Science and Technologies for Novel SOI Devices,* Kluwer, pp.47-54.
21. P.F.Fewster and N.L.Andrew. (1995) Absolute Lattice-Parameter Measurement. *J. Appl. Crystallogr.*, **28,** 451-458.

22. I.V. Antonova, V.P. Popov, V.F. Stas, A.K. Gutakovskii, A.E. Plotnikov, V.I.Obodnikov. (1999) Splitting and electrical properties of the SOI structure formed from the heavily boron doped Silicon with Using of the smart-Cut technology. *Microelectronic Engineering* **48**, 383-386.
23. I. V. Antonova, V. F. Stas', V. P. Popov, V. I. Obodnikov, and A. K. Gutakovski. (2000) Electrical Conductivity of Silicon-on-Insulator Structures Prepared by Bonding Silicon Wafers to a Substrate Using Hydrogen Implantation. *Semiconductors,* **34,** 1054-1057.

DC AND AC MODELS OF PARTIALLY-DEPLETED SOI MOSFETS IN WEAK INVERSION

DANIEL TOMASZEWSKI[1)], KRZYSZTOF DOMAŃSKI[1,2)], LIDIA ŁUKASIAK[2)], AGNIESZKA ZARĘBA[2)], JAN GIBKI[2)], ANDRZEJ JAKUBOWSKI[2)]

1) Institute of Electron Technology,
Al.Lotników 32/46, 02-668 Warszawa, Poland

2) Institute of Microlectronics and Optoelectronics,
Warsaw University of Technology,
ul.Koszykowa 75, 00-662 Warszawa, Poland

1. INTRODUCTION

Subthreshold range is critical for proper operation of analog SOI ICs in harsh environment. Partially depleted (PD) SOI MOSFETs are particularly sensitive to these conditions, because of the parasitic phenomena like kink-effect. This regards to operation in steady-ŝtate .and small-signal excitations conditions. Since integrated circuits based on PD SOI MOSFETs are widely fabricated the problem of reliable small-signal modelling is relevant.

There are works, which state, that PD SOI MOSFETs require non-qasi-static (NQS) models [1]. Moreover it seems that models of PD SOI MOSFETs operating in saturation range require improvement to obtain better consistency with non-saturation range models.

An attempt to prepare DC and AC companion models of the PD SOI MOSFETs, which fulfil the requirements mentioned above is presented in this paper.

2. MODEL

The model is similar to the one valid for strong inversion conditions [2]. It accounts for the following physical phenomena:

- Diffusion of the minority carriers at the Si-SiO_2 interface
- Diffusion (accompanied by thermal recombination) of the minority carriers in the floating body
- Thermal generation/recombination in the space-charge areas of the drain-body and source-body junctions
- Avalanche ionization of the carriers in the areas of high electric field, i.e. in the depletion region of the drain-body junction and in the "pinch-off" region.
- Displacement currents (in case of AC analysis)

F. Balestra et al. (eds.), Progress in SOI Structures and Devices Operating at Extreme Conditions, 289–298.

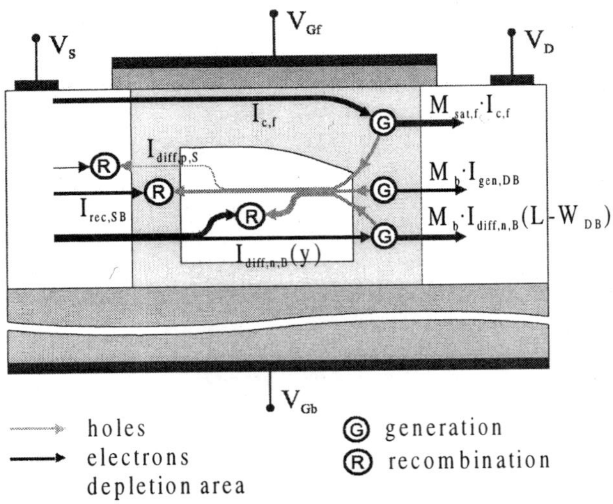

Fig.1 Approximation of spatial distribution of currents components in PD SOI MOSFETs

Therefore in the thin Si film there are several streams of the carriers which are explicitely related to the phenomena mentioned above. These streams are schematically shown in Fig.1. Only one of them, i.e. diffusion current at the SiO_2-Si interface is accounted for description of operation of bulk MOSFETs in normal conditions. Other ones are parasitic current components. However in case of the SOI MOSFETs in spite of their small magnitude as compared with the interface current they strongly influence the electrical behaviour of the body. They are responsible mainly for the so-called "kink-effect" [3]. In the bulk MOSFETs they should be diminished, mainly due to the fact, that they increase total leakage current and power consumption of CMOS integrated circuits.

2.1. BASIC EQUATIONS

Similarly to the case of PD SOI MOSFETs operation in strong inversion body-source voltage $v_{BS}(t)$ is chosen as a key variable of this model. It is obtained as a solution of the equation (1), which expresses a general charge neutrality or current continuity condition.

$$i_S(t)+i_{Gf}(t)+i_D(t)+i_{Gb}(t)=0 \tag{1}$$

Under the assumption that time-variation of transistor bias may be considered as small-amplitude sinusoidal wave-form a sinusoidal steady-state analysis (**S^{3}A**) method [4] may be used to obtain the solution of the equation (1). According to this approach all the variables in the model, dependent as well as independent on the space coordinate may be expressed as superposition of DC and AC components. The AC components are products of complex pre-exponential factors, so-called phasors and of exponential factors periodically dependent on time. This is expressed with formula (2).

$$v(y,t) = V(y) + v^*(y) \cdot e^{j\omega t} \tag{2}$$

This method allows to derive from (1) two continuity equations (3) and (4) which are solved to obtain DC and AC components of the $v_{BS}(t)$ voltage.

$$I_S + I_{Gf} + I_D + I_{Gb} = 0 \tag{3}$$

$$i_s^* + i_{gf}^* + i_d^* + i_{gb}^* = 0 \tag{4}$$

The main advantage of this approach is that DC as well as AC models of the PD SOI MOSFETs are fully consistent, i.e. they account identically for the same physical phenomena.

Since the presented paper is a continuation of the previous work related to small-signal modelling of the PD SOI MOSFETs in the strong inversion range [2], now only the analytical model of the diffusion current at the Si-SiO$_2$ interface will be presented in detail. Other current components which must be taken into account for the purpose of physically correct description of the PD SOI MOSFETs operation were presented in that work.

2.2. DIFFUSION AT THE SI-SIO$_2$ INTERFACE

This transport of the minority carriers along the weakly inverted channel is described with the following set of current continuity and transport equations [5].

$$\begin{cases} \dfrac{\partial i_{c,f}(y,t)}{\partial y} = W \cdot \dfrac{\partial q_{c,f}(y,t)}{\partial t} \\ i_{c,f}(y,t) = W \cdot \mu_{c,f} \cdot V_t \cdot \dfrac{\partial q_{c,f}(y,t)}{\partial y} \end{cases} \tag{5}$$

Here $i_{c,f}(y,t)$ and $q_{c,f}(y,t)$ denote spatial distributions of the current and mobile charge density in the channel at the front interface, $\mu_{c,f}$ denotes low-field mobility of the carriers, V_t denotes thermal voltage. If a parasitic conduction path at the bottom Si-SiO$_2$ interface exists it may be modelled with the analogous set of equations.

Integration of equations (5) with respect to y-coordinate within this part of the channel, where the so-called *Gradual Channel Approximation* (GCA) conditions are fulfilled leads to the sets of equations (6a) and (6b) which allow to get solutions of (5) at the source and drain boundaries of the channel, i.e. the expressions for currents at both ends. In the steady-state domain the solutions of (6a) and (6b) are obviously identical.

$$\begin{cases} i_{c,f}(L_{eff},t) - i_{c,f}(y,t) = W \cdot \displaystyle\int_y^{L_{eff}} \dfrac{\partial q_{c,f}(y',t)}{\partial t} dy' \\ \displaystyle\int_y^{L_{eff}} i_{c,f}(y',t)\, dy' = W \cdot \mu_{c,f} \cdot V_t \cdot \left(q_{c,f}(L_{eff},t) - q_{c,f}(y,t)\right) \end{cases} \tag{6a}$$

$$\begin{cases} i_{c,f}(y,t) - i_{c,f}(0,t) = W \cdot \int_0^y \frac{\partial q_{c,f}(y',t)}{\partial t} dy' \\ \int_0^y i_{c,f}(y',t) dy' = W \cdot \mu_{c,f} \cdot V_t \cdot \left(q_{c,f}(y,t) - q_{c,f}(0,t)\right) \end{cases} \tag{6b}$$

Approximate solution of the equations sets (6a) and (6b) may be obtained using the **S³A** approach mentioned above. After the separation of steady-state and small-signal components of these equations the DC model of channel current and AC models of drain and source currents may be obtained.

2.3. A MODEL OF THE DC COMPONENT OF THE CHANNEL CURRENT

The mutually equivalent DC components of the equations (6a) and (6b) may be expressed in with formula (7).

$$\begin{cases} I_{c,f}(L_{eff}) - I_{c,f}(y) = 0 \\ \int_y^{L_{eff}} I_{c,f}(y') dy' = W \cdot \mu_{c,f} \cdot V_t \cdot \left(Q_{c,f}(L_{eff}) - Q_{c,f}(y)\right) \end{cases} \tag{7}$$

The first conclusion (obvious from elsewhere) arising from the above formulae is that the DC component of the channel current is constant along the channel. This results from the fact, that recombination within the mobile charge sheet is neglected. After integration well-known formula (8) [5] is obtained. It accounts for the channel length modulation in the same manner as in the saturation range for strong inversion conditions.

$$I_{c,f} = \frac{W}{L_{eff}} \cdot \mu_{c,f} \cdot V_t \cdot \left(Q_{c,f}(L_{eff}) - Q_{c,f}(0)\right) \tag{8}$$

2.4. A MODEL OF THE AC COMPONENT OF THE CHANNEL CURRENT

Extraction of the AC component of eq. (6a) leads to the set of equations (9).

$$\begin{cases} i^*_{c,f}(L_{eff}) - i^*_{c,f}(y) = W \cdot j\omega \cdot \int_y^{L_{eff}} q^*_{c,f}(y') dy' \\ \int_y^{L_{eff}} i^*_{c,f}(y') dy' = W \cdot \mu_{c,f} \cdot V_t \cdot \left(q^*_{c,f}(L_{eff}) - q^*_{c,f}(y)\right) \end{cases} \tag{9}$$

After joining them together one obtains the integral equation (10)

$$q^*_{c,f}(y) = q^*_{c,f}(L_{eff}) - \frac{i^*_{c,f}(L_{eff})}{W \cdot \mu_{c,f} \cdot V_t} \cdot (L_{eff} - y) + \frac{j\omega}{\mu_{c,f} \cdot V_t} \cdot \int_y^{L_{eff}} \int_{y'}^{L_{eff}} q^*_{c,f}(y'') dy'' dy' \tag{10}$$

which may be expressed in a very compact form (11).

$$q_{c,f}^{*}(y)=f(y)+j\omega\cdot F\left(q_{c,f}^{*}(y)\right) \tag{11}$$

Position-dependent function f(y) and linear integral operator F are defined as follows [6].

$$f(y)=q_{c,f}^{*}(L_{eff})-\frac{i_{c,f}^{*}(L_{eff})}{W\cdot\mu_{c,f}\cdot V_t}\cdot(L_{eff}-y) \tag{12}$$

$$F\left(q_{c,f}^{*}(y)\right)=\frac{1}{\mu_{c,f}\cdot V_t}\cdot\int_{y}^{L_{eff}}\int_{y'}^{L_{eff}} q_{c,f}^{*}(y'')\,dy''\,dy' \tag{13}$$

Solution of the integral equation (11) may be expressed like in [6] as a sum of an infinite series (14).

$$q_{c,f}^{*}(y)=\left(1+j\omega\cdot F+(j\omega)^2\cdot F^2+(j\omega)^3\cdot F^3+\ldots\right)f(y) \tag{14}$$

Ignoring higher order terms a first-order approximation of the solution may be obtained:

$$q_{c,f}^{*}(y)=\left(1+j\omega\cdot F\right)f(y) \tag{15}$$

Then after integration a spatial distribution of the phasor $q^{*}_{c,f}$ of the mobile charge density in the channel may be explicitly obtained. Finally, if we apply the resulting formula at the source end of the channel (y=0) the equation (16) expressing a phasor of the AC component of the current at the drain end of the channel may be obtained.

$$i_{c,f}^{*}(L_{eff})=\frac{W\cdot\mu_{c,f}\cdot V_t}{L_{eff}}\cdot\frac{-q_{c,f}^{*}(0)+q_{c,f}^{*}(L_{eff})\cdot\left(1+j\omega\cdot\frac{L_{eff}^{2}}{2\cdot\mu_{c,f}\cdot V_t}\right)}{1+j\omega\cdot\frac{L_{eff}^{2}}{6\cdot\mu_{c,f}\cdot V_t}} \tag{16}$$

The same method as above may be used to obtain a phasor of the AC component of the current at the source. This time however the appropriate AC components of the equations (6b) must be used. The resulting equation (17) is given below.

$$i_{c,f}^{*}(0)=\frac{W\cdot\mu_{c,f}\cdot V_t}{L_{eff}}\cdot\frac{-q_{c,f}^{*}(0)\cdot\left(1+j\omega\cdot\frac{L_{eff}^{2}}{2\cdot\mu_{c,f}\cdot V_t}\right)+q_{c,f}^{*}(L_{eff})}{1+j\omega\cdot\frac{L_{eff}^{2}}{6\cdot\mu_{c,f}\cdot V_t}} \tag{17}$$

Comparison of (16) and (17) clearly indicates, that there is a phase shift between AC components of the channel current at the source and drain end of the channel. In order to find the complete (DC and AC) model of the diffusion current the following variables must be determined:

L_{eff}	-	channel shortening,
$Q_{c,f}(0)$, $q^{*}_{c,f}(0)$	-	DC and AC components of the mobile charge at the source,
$Q_{c,f}(L_{eff})$, $q^{*}_{c,f}(L_{eff})$	-	DC and AC components of the charge at the drain.

2.5. CHANNEL LENGTH MODULATION

In the weak inversion range the electric field distribution near the drain is similar to the distribution in the strong inversion range under saturation conditions. Thus a well known simple formula [5] describing the channel shortening may be adopted. It results from the 1-D solution of Poisson equation in the "pinch-off" region, what imposes appropriate boundary conditions. This time however they may be simplified, while surface potential in the channel varies only slightly along the channel distance. Hence the effective channel length may be calculated from the formula (18)

$$L_{eff} = L - \sqrt{\frac{2 \cdot \varepsilon_{Si}}{q \cdot N_B} \cdot \left(V_D - \Psi_{s,f}\right)} \tag{18}$$

where $\Psi_{s,f}$ denotes surface potential at the front Si-SiO_2 interface.

2.6. COMPONENTS OF THE CHARGE DENSITY AT THE SOURCE $Q_{c,f}(0)$, $q^*_{c,f}(0)$

The mobile charge distribution in the channel under the weak inversion conditions is determined by the 1-D Poisson equation (19), where majority carriers (holes) term is neglected

$$\frac{d^2\psi}{dx^2} = \frac{q \cdot (N_B + n)}{\varepsilon_{Si}} \tag{19}$$

where the electrons term is given with the typical expression (20).

$$n = n_i \cdot \exp\left(\frac{\psi - \varphi_n}{V_t}\right) \tag{20}$$

Under the gradual-channel approximation (GCA) conditions the solution of (19) leads to the well known [5] formula (21)

$$q_{c,f}(y,t) \approx -\sqrt{2 \cdot \varepsilon_{Si} \cdot q \cdot N_B} \cdot \sqrt{\psi_{s,f} - \psi_B} \cdot \frac{V_t}{2 \cdot \left(\psi_{s,f} - \psi_B\right)} \cdot \left(\exp\left(\frac{\psi_{s,f} - \psi_B}{V_t}\right) - 1\right) \cdot \exp\left(\frac{-2 \cdot \phi_F + v_{BS}}{V_t}\right) \tag{21}$$

where all the variables have their usual meanings. It is worthwile to mention, that in (21) forward bias of the source-body junction is explicitly taken into account. It lowers the potential barrier fór electrons between the source and the channel. Thus for the given DC bias it increases the channel current as compared to the bulk MOSFETs case.

In order to find DC and AC components of the charge density distributions the sinusoidal steady-state analysis (S^3A) method is used. After extraction of both components of the equation (21) and after setting y=0 the charge at the source end of the channel may be expressed with formulae (22) and (23)

$$Q_{c,f}(0) \approx -\sqrt{2 \cdot \varepsilon_{Si} \cdot q \cdot N_B} \cdot \frac{\sqrt{V_t}}{2} \cdot \exp\left(\frac{-2 \cdot \phi_F + V_{BS}}{V_t}\right) \cdot \frac{\exp\left(\frac{\Psi_{s,f} - \Psi_B}{V_t}\right) - 1}{\sqrt{\frac{\Psi_{s,f} - \Psi_B}{V_t}}} \tag{22}$$

$$q^*_{c,f}(0) \approx$$

$$-\sqrt{2\cdot\varepsilon_{Si}\cdot q\cdot N_B}\cdot\frac{\sqrt{V_t}}{2}\cdot\exp\left(\frac{-2\cdot\phi_F+V_{BS}}{V_t}\right)\cdot\left[\frac{\exp\left(\frac{\Psi_{s,f}-\Psi_B}{V_t}\right)-1}{\sqrt{\frac{\Psi_{s,f}-\Psi_B}{V_t}}}\cdot\frac{v^*_{bs}}{V_t}+\right. \tag{23}$$

$$\left.+\frac{\exp\left(\frac{\Psi_{s,f}-\Psi_B}{V_t}\right)-\frac{V_t}{2\cdot(\Psi_{s,f}-\Psi_B)}\cdot\left(\exp\left(\frac{\Psi_{s,f}-\Psi_B}{V_t}\right)-1\right)}{\sqrt{\frac{\Psi_{s,f}-\Psi_B}{V_t}}}\cdot\frac{\psi^*_{s,f}-\psi^*_b}{V_t}\right]$$

$\Psi_{s,f}$ - Ψ_B , $\psi^*_{s,f}$ - ψ^*_b quantities, which appear in the above formulae may be found using the typical approach, which consists in use of the 1-D Gauss law under depletion approximation conditions. Thus the surface potential relative to body potential is given with the formula (24).

$$v_{Gf}(t)-V_{FB,f}-\psi_{s,f}(t)-\gamma_f\cdot\sqrt{\psi_{s,f}(t)-\psi_B(t)}=0 \tag{24}$$

Again using the S3A method DC and AC components of the required quantities may be easily determined.

$$\Psi_{s,f}-\Psi_B=\frac{\gamma_f^{\,2}}{4}\cdot\left[\sqrt{1+\frac{4}{\gamma_f^{\,2}}\cdot(V_{Gf}-V_{FB,f}-\Psi_B)}-1\right]^2 \tag{25}$$

$$\psi^*_{s,f}-\psi^*_b=\frac{2\cdot\sqrt{\Psi_{s,f}-\Psi_B}}{\gamma_f+2\cdot\sqrt{\Psi_{s,f}-\Psi_B}}\cdot\left(v^*_{gf}-\psi^*_b\right) \tag{26}$$

$$\psi^*_b=v^*_s+v^*_{bs} \tag{27}$$

The above mentioned formulae account explicitly for DC and AC components of the body-source forward bias.

2.7. MOBILE CHARGE DENSITY AT THE DRAIN $Q_{c,f}(L_{eff})$, $q^*_{c,f}(L_{eff})$

Close to the drain the split between quasi-Fermi levels of majority and minority carriers is not explicitly known as at the source end of channel. Thus it is not possible to calculate the charge density there analogously to (22). The results of accurate numerical simulations indicate, that similarly to the case of saturation range the carriers are pushed away from the surface by electric field. There they travel towards the drain with almost maximum available velocity υ_{max}. Thus a simple formula (28) may be used to describe this very fast transport.

$$i_{c,f}(L_{eff},t)=-W\cdot q_{c,f}(L_{eff},t)\cdot\upsilon_{max} \tag{28}$$

On the other hand at border between the channel and the "pinch-off" region the GCA conditions are still approximately valid. Hence neglecting υ_{max} variations it is possible to compare formulae (8) and (16) with the appropriate simple expressions associated with the DC and AC components of the current expression mentioned above. This approach allows to get analytical expressions (29) and (30) relating the searched variables to the appropriate data at the source end of channel.

$$Q_{c,f}(L_{eff}) = \frac{Q_{c,f}(0)}{1 + \frac{\upsilon_{max} \cdot L_{eff}}{\mu_{c,f} \cdot V_t}} \tag{29}$$

$$q^*_{c,f}(L_{eff}) = \frac{q^*_{c,f}(0)}{1 + \frac{\upsilon_{max} \cdot L_{eff}}{\mu_{c,f} \cdot V_t} + j\omega \cdot \frac{L_{eff}^2}{2 \cdot \mu_{c,f} \cdot V_t} \cdot \left(1 + \frac{\upsilon_{max} \cdot L_{eff}}{3 \cdot \mu_{c,f} \cdot V_t}\right)}. \tag{30}$$

3. MODEL CHARACTERISTICS

The proposed model was used to calculate several DC characteristics of the PD SOI MOSFET. The parameters of the device are as follows: $W=100\mu m$, $L=9.4\mu m$, $T_{Oxf}=32.8nm$, $T_{Oxb}=400nm$, $T_{Si}=150nm$, $N_B=9\times10^{16}cm^{-3}$, $V_{GbS}=0V$. The meanings of these data are typical.

A family of $I(V_{GfS})$ curves of the transistor is shown in Fig. 2. These curves exhibit significant decrease of the threshold voltage $V_{TH,f}$ for increasing V_{DS} voltage, significant increase of transconductance for increasing V_{DS} voltage and quite smooth transition between weak and strong inversion ranges.

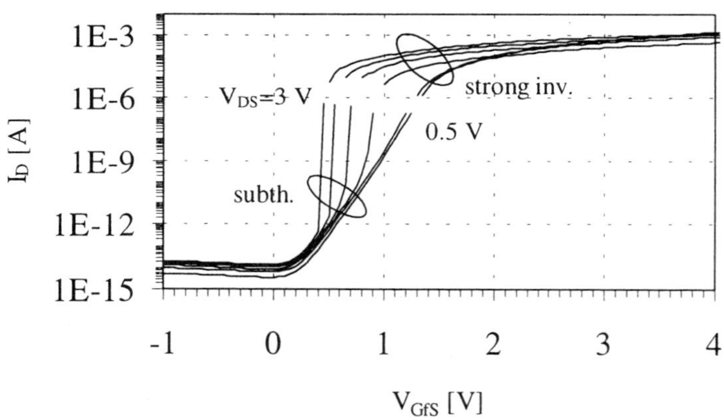

Fig. 2. $I(V_{GfS})$ characteristics of the PD SOI MOSFET (its parameters are given in the test) in the accumulation, weak inversion and strong inversion ranges.

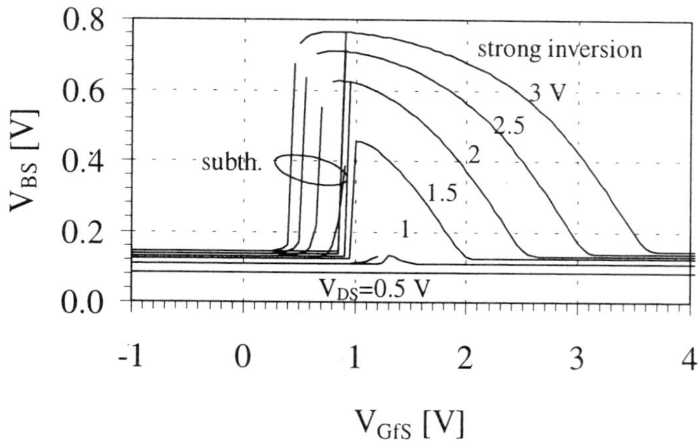

Fig. 3. $V_{BS}(V_{GfS})$ characteristics of the PD SOI MOSFETin the accumulation, weak inversion and strong inversion ranges.

A dependence of the body-source voltage V_{BS} on the front gate bias is shown in Fig. 3. There are two families of curves on this plot. One of them was calculated using only strong inversion model. The other was calculated according to the weak inversion model. The main conclusion resulting from this chart is that very strong increase of the body potential, which is responsible for the kink-effect, occurs not only in the saturation range but also in the subthreshold range. This explains strong dependence of the transconductance ant threshold viltage on the drain bias.

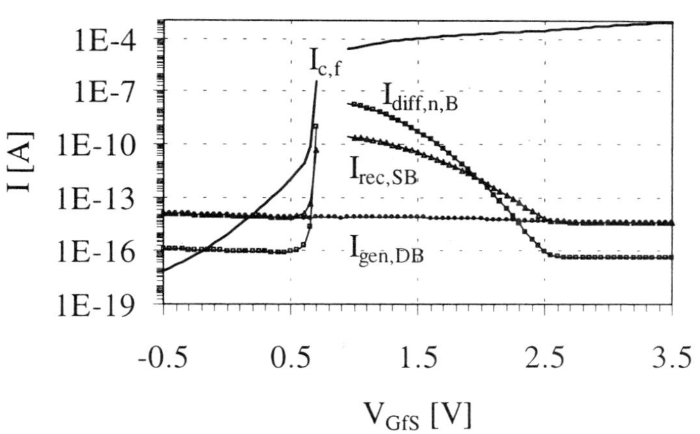

Fig. 4. Comparison of several current components in the PD SOI MOSFET.

Fig. 4 shows a comparison of several components of the total DC current of the device: diffusion current at the surface $I_{c,f}$ and in the bulk $I_{diff,n,B}$, source-body junction recombination current $I_{rec,SB}$ and drain-body junction generation current $I_{gen,DB}$. Again a very significant role of the subthreshold current model for the proper determination of the operation conditions of the MOSFET can be seen.

4. CONCLUSIONS

The analytical, NQS, small-signal model of the PD SOI MOSFETs operation in the weak inversion conditions was derived. It accounts for the main physical phenomena in important for this class of devices. Simulation results show that the proposed model exhibits significant dependence of the threshold voltage and transconductance on the drain bias. Moreover it shows quite smooth transition between weak and strong inversion. At present further work towards smooth connecting of the AC models in the weak inversion and strong inversion ranges is in progress.

Acknowledgement

The work was supported by the State Committee of Scientific Research under grant no. 8T11B 04 017

5. References

1. Flandre D. (1993) Analysis of Floating Substrate Effects on the Intrinsic Gate Capacitance of SOI MOSFET's Using Two-Dimensional Device Simulation *IEEE Trans. El. Dev.* **40**, p.1789
2. D.Tomaszewski, A.Jakubowski, J.Gibki (1999) A Small-Signal Non-Quasistatic Model of PD SOI MOSFETs *Electron Technology* **32**, pp.104-109
3. Colinge J.-P. (1991) Kluwer Academic Publishers, 1991
4. Laux S.E. (1985) Techniques for Small-Signal Analysis of Semiconductor Devices *IEEE Trans. on Electron Devices* **32**, pp.2028-2037
5. Tsividis Y.P. (1987) *Operation and Modeling of the MOS-Transistor*, Mc Graw-Hill, 1987
6. Hauser J.R. (1965) Small Signal Properties of Field Effect Devices *IEEE Trans. on Electron Devices* **12**, pp.605-618

ON SCALING THE THIN FILM Si THICKNESS OF SOI SUBSTRATES

A Perspective on Wafer Bonding for Thin Film Devices

K.D. HOBART, F.J. KUB, M.E. TWIGG and M. FATEMI
Naval Research Laboratory
Electronics Science and Technology Division
Washington, D.C 20375

1. Introduction

It is expected that as deep submicron CMOS technology fabricated on SOI substrates is further scaled, the thin film Si thickness will be scaled to the 10 nm regime. Presently, there are several SOI substrate fabrication concepts that have addressed the requirements of partially depleted (PD) CMOS circuits. The needs of fully depleted (FD) CMOS have begun to be addressed, but progress in scaling the thin film Si thickness below 50 nm has been limited. This paper reviews the current status of these technologies and examines the prospects of scaling down the thin film Si thickness to the 10 nm regime.

Fig. 1 shows the areas of SOI wafer applications where the emphasis here is in scaling for the FD or nanoelectronic regime. In this area, where the Si overlayer thickness is <100 nm, scaling challenges can be expected to be significant. The single largest task will be to improve the uniformity of ultra-thin SOI films. The performance of circuits based on FD MOSFETs depends critically on maintaining highly uniform SOI films since the threshold voltage, V_T, of FD MOSFETS is strongly dependent on the Si thickness, t_{Si} [1]. Innovative device structures such as the dual-gate (or double-gate) MOSFET, however, appear to have reduced dependencies of device characteristics on the silicon thickness uniformity [2] and fabrication of such devices represents one of the next significant challenges in the device scaling evolution.

Presently, there exist several competing SOI substrate manufacturing technologies based on the direct wafer bonding process as well as SIMOX (separation by implantation of oxygen). A comparison of the key technologies is shown in Table I. Successful technologies based on wafer bonding use precision thin film separation techniques such as the *Smart-Cut* process used in SOITEC's *UNIBOND* wafers [3], the water jet splitting process used in Canan's *ELTRAN* (Epitaxial Layer TRANsfer) substrates [4], and Silicon Genisis' *NanoCleave*, a process that uses high pressure nitrogen to split wafers at damaged regions formed by ion implantation [5]. Such layer transfer processes have eclipsed traditional Bond-and-Etch-back SOI (BESOI) fabrication techniques which are now used primarily for thick SOI applications such as MEMS and power electronics.

As shown in Tablel I, all current fabrication technologies suffer in uniformity when the Si overlayer thickness is scaled to less than 50 nm. All technologies using wafer

F. Balestra et al. (eds.), Progress in SOI Structures and Devices Operating at Extreme Conditions, 299–308.

bonding employ a precision thin film transfer technique as noted above. In each case further processing is required to produce a surface with sufficiently low surface roughness. In order to understand the subtle process details that lead to thin film Si nonuniformities, each process is described briefly. In addition, a new process developed at the Naval Research Laboratory and targeted at the fabrication of extremely thin Si films is described. A full description of the Ultra-Cut process is the subject of the sections 2 and 3.

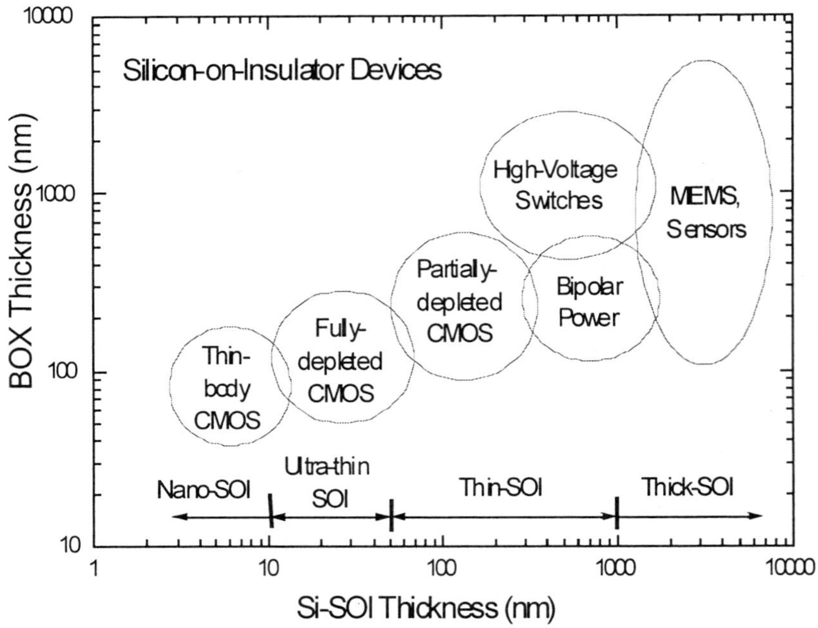

Fig. 1. Areas of SOI technology insertion.

1.1 UNIBOND

UNIBOND SOI wafers are prepared by first growing a thermal oxide film on the "donor" wafer to create the buried oxide (BOX). This wafer is then implanted with a high dose ($5x10^{16}$–$1x10^{17}$ cm^{-2}) of protons. The donor wafer is next joined to a "handle" wafer by direct wafer bonding. Next the Smart-Cut process is used to split the donor wafer near the projected range of the proton implant by heating the wafer pair to ~500°C. The *UNIBOND* process concludes with a "touch polish," or brief chemical mechanical polish (CMP), since the surface roughness following *Smart-Cut* separation is ~8 nm. This step produces sufficiently low surface roughness and, although extremely well developed, is essentially a bulk process that ultimately compromises the Si overlayer thickness uniformity.

Table I. Comparison of leading SOI technologies with ultra-thin Si films

	H^+ Implant?	Etch Stop/ Epitaxy	Separation/ Final Prep	Si Thick., t_{Si}
UNIBOND *SOITEC*	Yes	No	Thermal / CMP	50±5 nm
ELTRAN *Canon*	No	Porous Si/ Si Epitaxy	Water Jet Split / H_2 Anneal	10±1.75 nm
SOI Silicon *Genesis*	Yes	Possible	NanoCleave / Wafer Smoothing (HCL $+H_2$)	15±3 nm
Low-Dose SIMOX *IBIS*	Oxygen Implant	N/A	N/A Hi Temp Anneal	50±5 nm
Ultra-Cut *NRL*	Yes	SiGe / Si Epitaxy	Thermal / Selective Etch Back	5±0.5 nm

1.2 ELTRAN

Canon's *ELTRAN* SOI fabrication process uses a unique separation concept based on water jet penetration into a buried porous Si layer [6]. Initally, the porous Si layer is formed on the donor wafer by anodic oxidation. A Si film is next epitaxially grown on the porous Si layer followed by thermal oxidation to produce the BOX. The donor wafer is joined to the handle wafer by direct wafer bonding. Separation is performed by splitting the donor wafer from the handle wafer, BOX, and epitaxial Si film by injecting a high pressure water jet into the porous Si layer. Following separation, the porous Si layer is removed by a highly selective wet chemical etch. Low surface roughness is achieved by a high temperature H_2 anneal. The primary source of nonuniformity originates at the porous Si–epitaxial Si interface which rougher than typical epi/substrate interfaces. The H_2 anneal partially corrects the interface effect and reduces the overall nonuniformity to an impressive ±1.75nm for a 10.2 nm Si film while maintaining excellent surface characteristics.

1.3 SILICON GENESIS PROCESS

Silicon Genesis (SiGen) uses a process nearly identical to the UNIBOND process with the exception of the actual cleaving process. The Silicon Genesis cleavage process is purely mechanical (dubbed *NanoCleave*) and uses a high pressure gas (N_2) jet to separate the wafers at the damaged implant region. The process is performed at room temperature and a unique plasma bonding process is required to achieve sufficiently high bond strength such that debonding does not occur during the *NanoCleave* process. Following *NanoCleave* the surface roughness is lower than with Smart-Cut but post processing is required nonetheless. Low surface roughness and ultra-thin Si films are generated by a "Wafer Smoothing" process that consists of etching the top Si in an epitaxial reactor under H_2 and HCl ambient [7]. While low surface roughness is achieved, the Wafer Smoothing process results in fairly large Si film nonuniformity.

1.4 ULTRA-CUT PROCESS

The Ultra-Cut process combines the two SOI fabrication techniques *Smart-Cut* and BESOI. This new process combines the advantages of each process and eliminates their primary disadvantages. By including a SiGe etch stop layer [8,9] *within* the transferred Smart-Cut film, touch polishing is replaced by wet chemical etching. Requirements of the etch stop are also reduced in this process flow compared to traditional BESOI since the Si thickness that must now be removed is (1) significantly reduced ($<1\,\mu m$ versus full wafer thickness) and (2) is extremely uniform in thickness due to the ion implantation process. Additionally, with this approach the final thin Si film is determined by the thickness and uniformity of the epitaxial layer, a process that is scaleable down to 2 nm and for which the uniformity can be less than 1%. The epitaxial process also eliminates defects associated with bulk substrates such as crystalline oriented particles. As will be shown in the remainder of the paper, the Ultra-Cut process is capable of producing SOI substrates with thin Si films less than 5 nm and of excellent uniformity and surface roughness.

2. Experiment

The Ultra-Cut process begins with preparation of the "donor" wafers. The overall process flow is shown in Fig. 2. Beginning with (100) Si substrates, commercial low pressure epitaxy proceeds with the growth of the SiGe etch stop followed by the thin Si film. In the experiments discussed here, the $Si_{0.7}Ge_{0.3}$ etch stop is nominally 30 nm thick

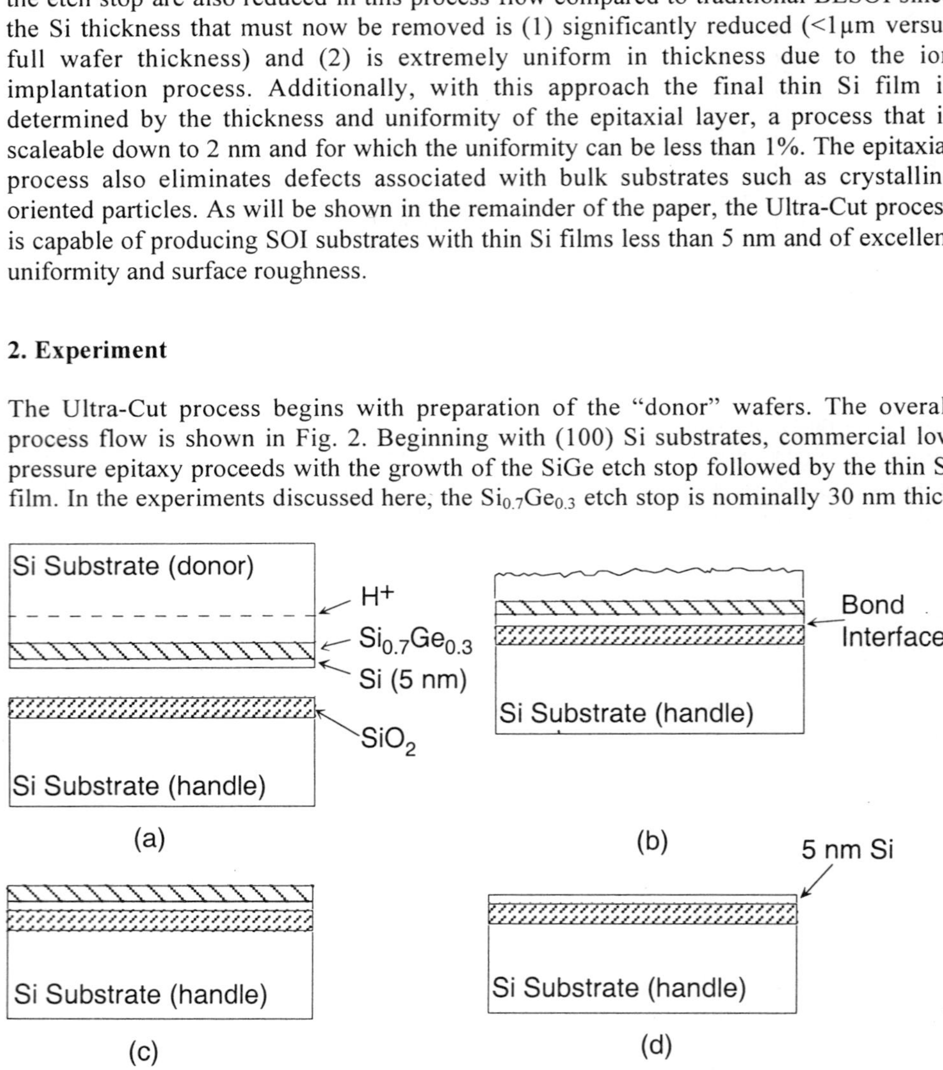

Fig. 2. Schematic of Ultra-Cut process sequence.

and the thin Si film is nominally 5 nm thick. The donor wafers are next implanted with H_2^+ at a dose and energy of $4.5x10^{16}$ cm^{-2} and 180 keV, respectively. The projected range of the protons is ~800 nm and thus the protons are implanted far below the SiGe etch stop. The "handle" wafer is prepared by growing 120 nm of dry thermal oxide. Both handle and donor substrates are cleaned first in a UV/ozone chamber for 10 minutes followed by wet chemical treatment in an SC-1 solution (NH_3OH:H_2O_2:H_2O::1:1:5). After spin-dry the wafers are bonded in atmosphere under IR observation. This process typically produces void free bonding with no degradation observed as a result of the epitaxial process. Following bonding the wafer pairs are subjected to a low temperature anneal (250 °C) to increase bond strength prior to further processing. *Smart-Cut* cleavage is achieved by annealing the wafer pair at 480°C for 10 minutes thereby separating the donor substrate from the Si/SiGe/Si/SiO_2 stack on the Si handle substrate (Fig. 2b). The remaining 800 nm of Si is selectively etched in 10% KOH (by wt.) at 35–40 °C (Fig. 2c). The etch selectivity of Si over $Si_{0.7}Ge_{0.3}$ in KOH is approximately 100. The SiGe layer is next selectively removed in an etch solution of CH_3COOH:H_2O_2:HF::3:2:1 developed by Carne's et al. [10] (Fig. 2d). This process has an etch selectivity of ~1000 and the etch rate of Si is less than 0.1 nm/min. It is this extreme etch selectivity that proves the utility of the SiGe etch stop and provides exceptional process latitude. Ultra-Cut substrates were characterized by AFM and TEM.

3. Results and Discussion

The Ultra-Cut wafers were found to be of very high quality with few observable surface imperfections. One type of void that was routinely observed in very low density appeared as a result of "bubbles" formed at the bond interface following Smart-Cut

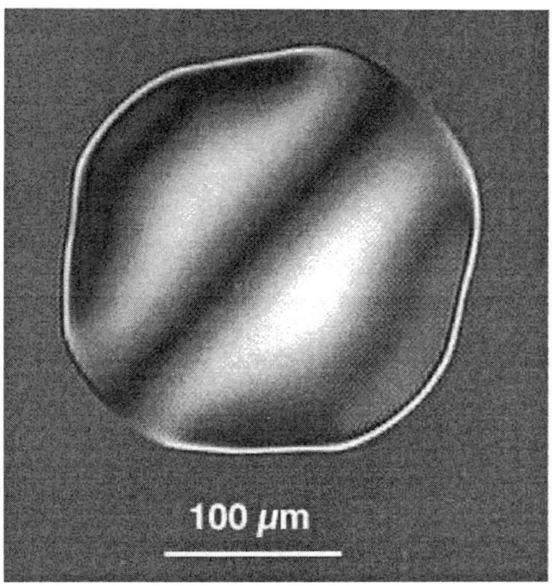

Fig. 3. "Bubble" defect observed in the as-cleaved film.

cleavage. The density varied from approximately 0.1–10 cm^{-2}. A micrograph taken with Nomarski Interference Contrast Microscopy is shown in Fig. 3. The micrograph shows one such bubble where the diameter is several hundred micrometers across. Stylus profilometry indicates that the bubble is 3–4 micrometers in height at the center. These bubbles are believed to form following the Smart-Cut cleavage process since they are not visible as microvoids prior to cleavage. Such defects may arise from hydrocarbon contamination or water trapped at the interface. It is supposed that following thin film cleavage, the trapped gas expands laterally to produce the observed bubbles. During subsequent etching steps the thin Si membrane that comprises the bubble is lifted off resulting in a void in the Si film of similar lateral dimensions.

The ultra-thin Si films were further characterized by TEM and AFM. Fig. 4 shows a high-resolution TEM cross-section of a nominally 5 nm Si film fabricated by Ultra-Cut. Careful examination shows that the Si film thickness is closer to 4 nm with a variation of approximately ±0.2 nm (±5%) over the sampled area. It is difficult to evaluate the Si thin film thickness with techniques other than TEM for such thin films. Surprisingly, visual examination of these thin films is possible due to modification of the optical path length through the oxide (the films are not visible if the oxide is less than 10–20 nm). Through visible examination, the films appear to be highly uniform, although it is clear that the films become slightly thinner within 1 cm of the wafer edge. Overall, the thin Si films fabricated by Ultra-Cut appear to be extremely uniform and currently represent the thinnest SOI film successfully produced with such low thickness variation.

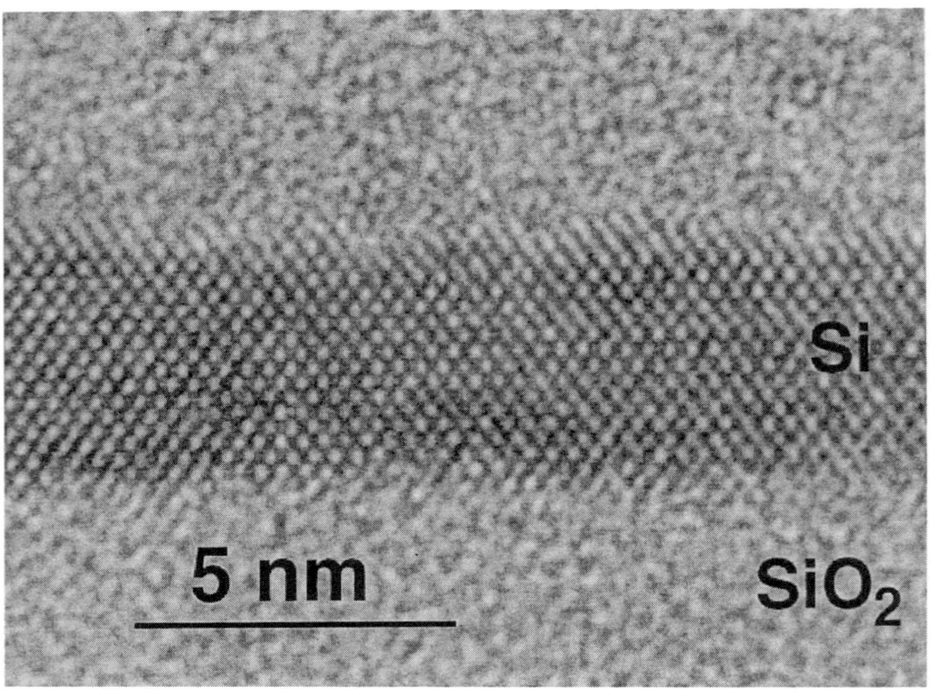

Fig. 4. Cross-sectional TEM micrograph of 4 nm Ultra-Cut SOI film.

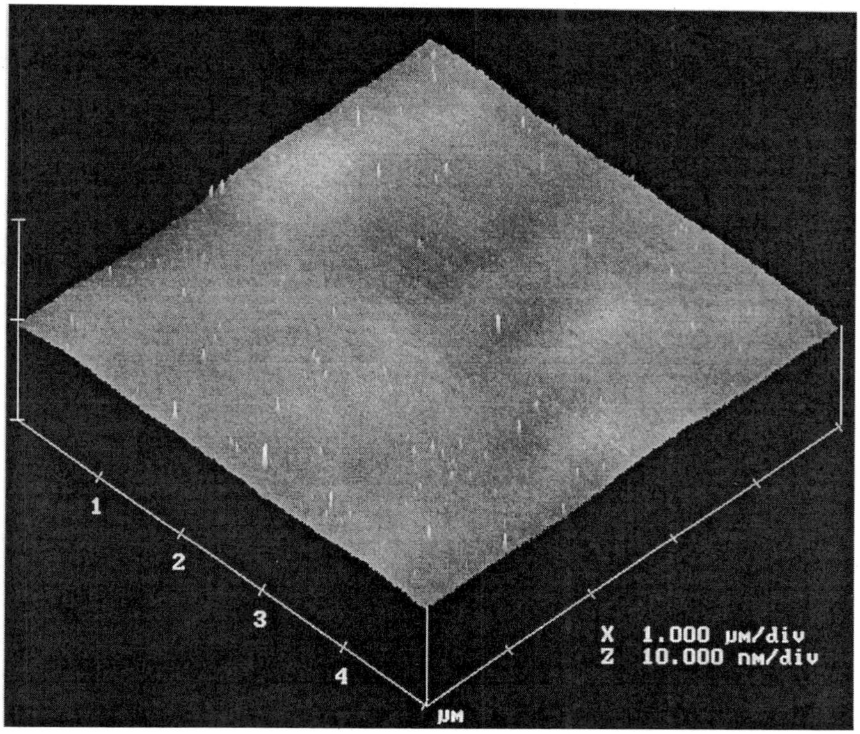

Fig. 5. AFM image of SiGe etch stop surface after KOH etching.

AFM results are shown in Figs. 5 and 6. The first image is of the SiGe surface following the KOH etch. The KOH etch process reduces the surface roughness of the as-cleaved surface from ~8 nm to 0.13 nm following etching. Regardless of the limited etch selectivity of the KOH etch, the surface roughness is nonetheless reduced to the atomic level as shown in Fig. 5. In general, the surface roughness is reduced at each selective etch step *and* by the ratio of etch rates or etch selectivity. It should be pointed out that other selective etch solutions are available without light ions such as tetramethyl ammonium hydroxide, TMAH [11]. With the bulk of the Si cleaved by Smart-Cut removed, the SiGe etch must remove the thin SiGe layer without degrading the surface roughness. The $CH_3COOH{:}H_2O_2{:}HF$ etch developed by Carnes et al. is extremely effective at removing the SiGe layer without etching the thin SOI Si film. The high selectivity and very low etch rate of Si is the key to the Ultra-Cut process. The high etch selectivity produces a highly reproducible and manufacturable process. Fig. 6 shows the AFM image of the as-etched Si surface following SiGe removal. The measured RMS roughness of this surface is 0.14 nm, which indicates that the surface quality of the etch stop is preserved.

One limitation of the Ultra-Cut process is the lack of a thermally generated Si/SiO_2 interface at the thin film Si–BOX interface. The limited thermal budget of the SiGe etch stop layer precludes oxidation of the donor wafer prior to wafer bonding as shown in

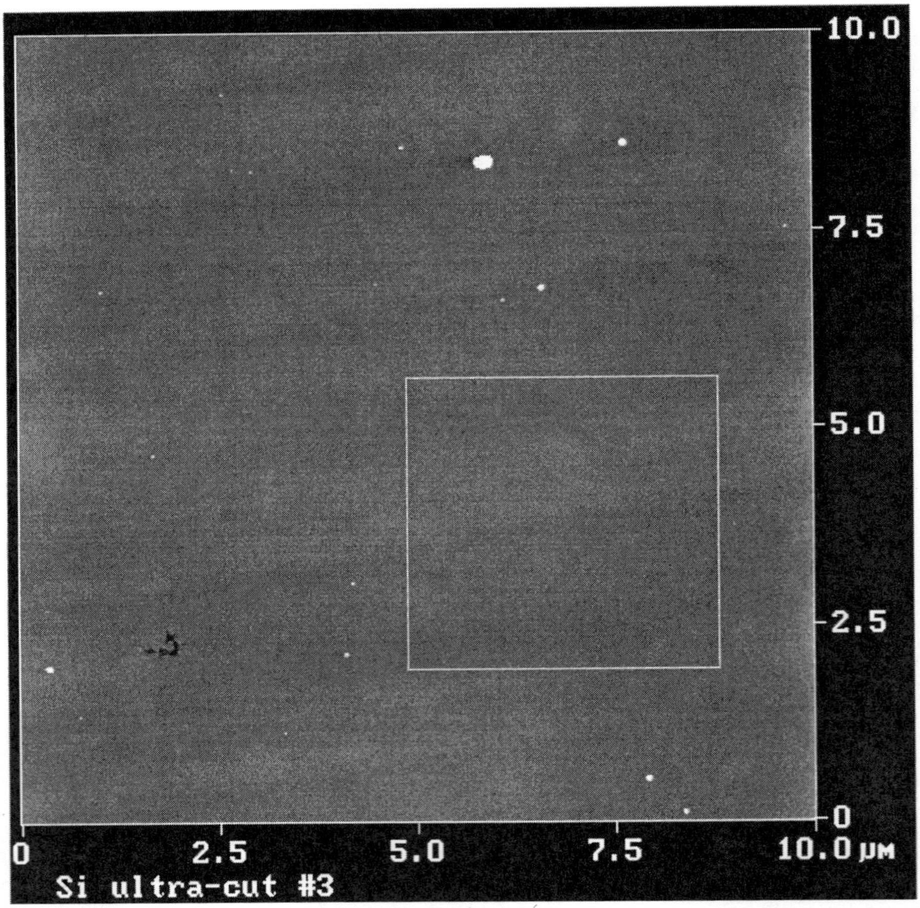

Fig. 6. Plan view AFM image of final, as-etched Ultra-Cut surface (periodic diagonal features are an artifact of the measurement).

the x-ray diffraction analysis of the SiGe etch stop of Fig. 7. At temperatures between 800°C and 900°C Ge begins to diffuse from the etch stop as evidenced by the shift in the SiGe peak toward the Si peak. Above 900°C the SiGe film begins to relax due to dislocation generation. Low temperature preparation of the buried oxide may be one solution. Recent work however shows that the bonded interface may be of sufficiently high quality that thermally grown thin film Si–BOX interface is not required. Anatonova et al. have quantified the interface state density by DLTS of bonded and thermally grown Si/SiO_2 structures [12]. Most significantly, the DLTS measurements determined that magnitude of interface states are no higher at the bonded interface than at the thermally-grown interface. Additionally, the interface states of the bonded structure are confined to a relatively narrow energy range compared to the more typical and distributed interface state density of thermal Si/SiO_2 interfaces.

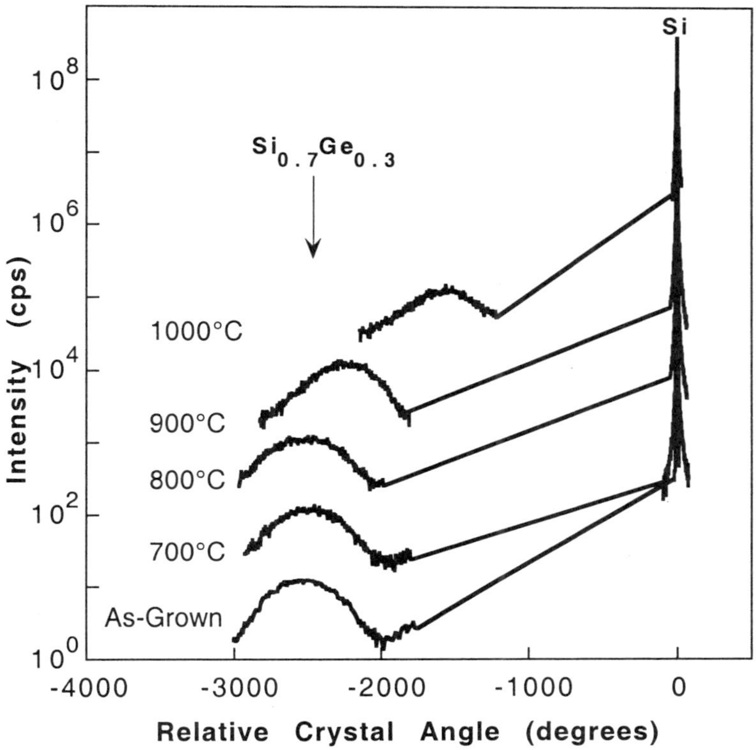

Fig. 7. X-ray diffraction spectra of Si/SiGe/Si(100) etch stop shown for a series of anneal temperatures. The $Si/Si_{0.7}Ge_{0.3}/Si(100)$ heterostructure was annealed at various temperatures for 90 min. in N_2. Shown are the 004 reflection spectra of Si and SiGe Bragg reflections.

4. Summary and Conclusions

SOI substrates for deeply scaled sub-0.1 micron CMOS will eventually encounter a technological barrier on which the survival of SOI depends: the SOI thin film Si thickness must be scaled down to below 50 nm and it is presently unclear which if any SOI fabrication technology can address this formidable task. In this paper a novel SOI fabrication process has been described. The new method, dubbed Ultra-Cut, combines the most attractive aspects of both Smart-Cut and BESOI to produce a process that exceeds the capabilities of presently available fabrication processes. Such a process may be required as Si MOSFET requirements are continually challenged. The process is suitable for the fabrication of dual-gate MOSFETs that require sub-10 nm films with very low ionized impurity concentration.

Additionally, the process demonstrates that strained films such as SiGe can be successfully transferred from the "seed" or donor substrate to a second handle substrate using Smart-Cut. Such a film can be used as a new or "virtual" substrate for the epitaxial growth of films with a lattice constant that dissimilar from the original seed substrate (see reference [13] for details). The application of Smart-Cut and other precision thin film separation techniques to the separation of arbitrary semiconductors and heterostructures demonstrates the truly unique capabilities these new processing techniques bring to the challenges of continued device scaling.

5. References

1. Lim, H.K. and Fossum, J.G., (1983) *IEEE Trans. on Electron Devices 30*, 1244.
2. Balestra, F. *these proceedings*.
3. Bruel, M. (1995) Silicon on insulator material technology. *Electron. Lett. 37*, 1201–2.
4. Ito, M., Yamagata, K., Miyabayashi, H., and Yonehara, T. (2000) Scalability potential in ELTRAN SOI-epi wafer. *Proceedings 2000 IEEE International SOI Conference*, October 2–5, 2000, Wakefield, Massachusetts, 10–11.
5. En, W.G., Malik, I.J, Bryan, M.A., Farrens, S., Henley, F.J., Cheung, N.W., and Chan, C. (1998) The Genesis Process: A new SOI fabrication method. *Proceedings 1998 IEEE International SOI Conference*, October 5–8, Stuart, Florida, 163–164.
6. Yonehara, T., Sakaguchi, K., and Sato, N. (1994) Epitaxial layer transfer by bond and tech back of porous Si. *Appl. Phys. Lett. 64*, 2108.
7. Thilderkist, A, Kang, S., Fuerfanger, and M., Malik, I.J. (2000) Surface finishing of cleaved SOI films using epi technologies. *Proceedings 2000 IEEE International SOI Conference*, October 2–5, 2000, Wakefield, Massachusetts, 12–13.
8. Godbey, D.J., Twigg, M.E., Hughes, H.L., Palkuti, L.J., Leonov, P., and Wang, J.J. (1990) Fabrication of bond and etch back silicon on insulator using a strained $Si_{0.7}Ge_{0.3}$ layer as an etch stop, *J. Electrochem. Soc. 137,* 3219–23.
9. U.S. Patent No. 5013681, Method of producing a thin silicon-on-insulator layer.
10. Carnes, T.K., Tanner, M.O, and Wang, K.L (1995) Chemical etching of $Si_{1-x}Ge_x$ in $HF:H_2O_2:CH_2COOH$. *J. Electrochem. Soc. 142*, 1260–6.
11. Tabata, O. , Asahi, R., Funabashi, H., Shimaoka, K. and Sugiyama, S. (1992) Anisotropic etching of silicon in TMAH solutions. *Sensors and Actuators A34*, 51-57.
12 Antonova, I.V., Naumova, O.V., Stano, J., Nikolaev, D.V., Popov, V.P., and Skuratov, V.A. (2001) Comparative charge deep-level transient spectroscopy studies of bonded Si/SiO_2 and Si/thermal SiO_2 interfaces in SOI structures, *these proceedings.*
13. K.D. Hobart, F.J. Kub, M.E. Twigg, M. Fatemi, and P.E. Thompson (2000) Compliant substrates: A comparative study of the relaxation mechanisms of strained films bonded to high and low viscosity substrates. *J. of Electronic Materials 29*, 897-900.

OXIDIZED POROUS SILICON BASED SOI: UNTAPPED RESOURCES

V. BONDARENKO, V. YAKOVTSEVA, L. DOLGYI, N. VOROZOV,
S. VOLCHEK, M. BALUCANI*, G. LAMEDICA*, A. FERRARI*
Belarusian State University of Informatics and Radioelectronics
Brovka 6, 220027 Minsk, Republic of Belarus
** INFM Unit E6 of Rome University "La Sapienza"*
Eudossiana 18, 00184 Rome, Italy

Abstract

This paper discusses new potentialities of oxidized porous silicon (PS) based SOI structures that have hitherto escaped attention of the researchers. Oxidized PS (OPS) has the property of controlling its characteristics depending on regimes of formation of PS layer and its further oxidation. The manner by which electrical and optical properties of the OPS layer may be varied with oxidation regimes of PS layers with various porosities is presented. It is of great importance that oxidized PS regions of different characteristics (porosity, thickness etc.) may be formed simultaneously within the same wafer.

The paper points out the meaning of oxidized PS. It is emphasized, by one way or another, that oxidized PS involves a whole series of dissimilar materials. PS may be oxidized either completely or partially and the formed OPS layer will have its properties depending on both the oxidizing regimes and the parameters of initial PS. The kinetics of PS oxidation and densification, the electrical and optical parameters of the material obtained and the feasibility to dope OPS with a desirable dopant are presented.

Dopants in OPS are considered on the example of rare-earth elements.

In addition, consideration is given to the ideas of on-chip integration PS-based SOI with different PS-based micro- and optoelectronic devices.

1. Porous silicon based SOI

Technology based on porous silicon (PS) figured prominently in the development of SOI structures. However, PS-based SOI has been in low esteem for the last few years. At the same time this technology has many untapped resources. For example, the possibility of PS to be used in optoelectronics deserves attention. We argue that PS-based technology is well worth to a spreader look.

F. Balestra et al. (eds.), Progress in SOI Structures and Devices Operating at Extreme Conditions, 309–327.

1.1. TIME HISTORY OF PS-BASED SOI

PS-based SOI technology was preceded by IPOS (Isolation by Porous Oxidized Silicon) in the late '60s. Fig. 1 depicts by convention the time history of PS-based SOI. Referring to Fig. 1, PS-based SOI technology takes its rise in IPOS. In 1969 Watanabe and Sakai from Nippon Telegraph and Telephone Public Co. (NTT) conceived the first idea to form PS-based SOI [1]. The idea was founded on the combination of porous anodisation, silicon epitaxy, oxidation of PS and precise polishing. Note that at the time, however, the common term "SOI" was not yet generally acknowledged.

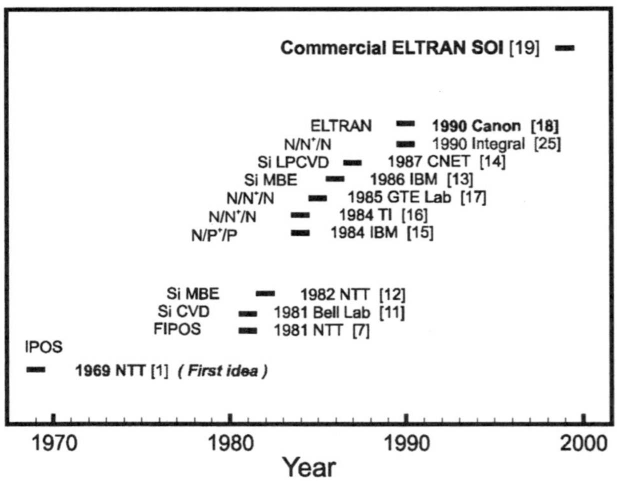

Figure 1. Time history of PS-based SOI.

From 1969 to 1981 several operational decisions for the PS-based SOI were advanced [2 - 6]. Even the idea of converting into PS the whole silicon substrate, except for a thin epitaxial layer, followed by oxidation of PS was proposed [5]. All these decisions were attractive but difficult if not impossible to realize, and they had no practical applications. These are the archetype of SOI.

The situation radically altered in 1981 when Imai from NTT originated the method called FIPOS (Full Isolation by Porous Oxidized Silicon) [7]. The method is based on the temporary change of silicon conductivity from *p* to *n* type in the selective regions by proton implantation to provide selective porous anodisation of *p*-type silicon regions around the *n*-type. SOI structures fabricated by FIPOS consisted of 0.1μm silicon layer on top of 8μm layer of buried oxidized PS [7, 8]. High speed 16K and 64K CMOS SRAM with 2 and 1.5 μm features and access time of 30 and 20 ns respectively were fabricated using the FIPOS process [9, 10]. In 1984 these were the fastest devices of this class. Unfortunately, FIPOS did not find acceptance because this method imposes stringent constrains on the size of silicon islands and buried oxidized PS takes up a disproportionate share of the room as compared to the thickness of the isolated silicon islands.

Nevertheless, FIPOS method renewed interest in PS-based SOI structures. A lot of scientific centers like Bell Lab. (USA), Sandia National Lab. (USA), CNET (France), RSRE (now DERA, UK), etc. were involved in this research. Moreover, companies of eminence like NTT (Japan), IBM (USA), Texas Instruments (USA), etc. invested in R&D activities to develop a proprietary PS-based SOI technique. As for the countries of the former USSR, Belarusian State University of Informatics & Radioelectronics has been deeply involved in IPOS and PS-SOI research for Integral Inc. (Republic of Belarus).

Among the new processes which have appeared, SOI based on Si epitaxy on PS followed by oxidation of buried PS layer should be pointed out [11, 12]. A serious effort was made to develop a low-temperature epitaxial method to prevent PS sintering. Molecular beam epitaxy, CVD epitaxy, to name but a few, worked for this aim [11-14]. At the same time, technologies based on a preferential anodisation of p^+-layer within a sandwich $n/p^+/p$ structure or n^+-layer within a $n/n^+/n$ structure were developed [15-17]. And in 1990 ELTRAN (Epitaxial Layer Transfer) technology was developed by Yonehara from Canon (Japan) [18]. Canon's ELTRAN process is made in principle by bonding a first silicon substrate with a PS layer on which an epitaxial silicon layer is grown and a second thermally oxidized wafer. Next, a water-jet on the PS layer splits the two bonded wafers, and the remaining exposed PS is subsequently etched to reach the epitaxial silicon, and a hydrogen annealing finishes the surface [19]. While ELTRAN utilizes PS only as a sacrificial layer along which splitting of bonded wafers is performed, the research into PS (e.g. PS formation, epitaxial growth on PS, thermal behavior of PS, etc.) is the basis for this method.

1.2. CLASSIFICATION OF PS-BASED SOI

A variety of PS-based SOI methods can be classified into three main groups. The first is a formation of PS layers selectively underneath the silicon islands. This group exploits the strong dependence of PS formation rate on both type and concentration of dopant. FIPOS and techniques of selective anodisation of p^+-layer within $n/p^+/p$ structure and n^+-layer within $n/n^+/n$ structure may be cited as typical representatives of this group. The second is a direct epitaxial silicon growth onto the PS layer previously made on the surface of a silicon substrate. In these groups, the SOI structures have the buried oxide (BOX) layer made from oxidized PS. To convert PS into oxidized PS, high temperature thermal oxidation is performed [20].

In the third group, PS layer is used as a sacrificial layer that is responsible for wafer splitting. ELTRAN is a representative of the third group. With ELTRAN, BOX of SOI consists of thermal silica and not of oxidized PS[19].

1.3. THE PROGRESS OF PS-BASED SOI

In the past years much effort was expended to produce PS-based SOI. Considerable progress has been made towards that goal. The essential results obtained by various research groups are collected in Table 1. The data in Table 1 are predominantly referred to the first SOI group based on the selective anodisation of silicon [8 - 10, 21-29]. The second group dealt with epitaxy on the PS layer is represented by the most up-to-date results from the technology developed by Romanov et al from the Institute of Semiconductor Physics of Siberian Division of Russian Academy of Sciences (ISP RAS) [30]. With this method, silicon is grown epitaxially onto the surface of the PS layer specially oxidized. This technology is the further development of the method proposed by Bomchil et al. from CNET in 1987 [14]. The method allows the formation of the continuous SOI, as differentiated from the methods of the first group.

TABLE 1. Parameters of PS-based SOI.

Parameter	**FIPOS**	**$N/N^+/N$**					**Si MBE**	**ELTRAN**
	NTT (1984)	**TI** (1986)	**CNET** (1987)	**RSRE** (1989)	**Integral** (1992)	**Authors** (1998)	**ISP RAS** (1998)	**Canon** (1999)
T_{Si}, μm	0.05-0.4	0.45	0.35	0.1-0.14	0.3	0.15-0.3	0.1-0.5	0.1-0.2
T_{BOX}, μm	8-10	~2	1.2	0.8	1.8	0.8-2.5	1.0	0.1-0.2
Warpage, μm	< 20	< 20	-	-	< 20	< 5	-	-
W_{max}, μm	10	20	-	-	40	100	-	No limit
Junction leakage, A/μm	10^{-12}	10^{-13}	10^{-13}	10^{-13}	10^{-13}	10^{-13}	-	-
μ_e, cm²/V·sec	750	570	680	520	550	550	-	-
μ_h, cm²/V·sec	200	150	-	235	200	200	135	-
Subthreshold slope, mV/dec	110	90	100	80	70-80	70-80	-	-
Maturity	**R***	R	R	R	**R***	**R***	R	I
References	[8-10]	[22]	[23]	[24]	[25, 26]	[27-29]	[30]	[19]

R – research, R* – demo of IC, I – industry

All the methods collected in Table 1 were studied at the pilot-line level. Only NTT [9, 10] and Integral Inc. [25] have reported SOI-based VLSIs. The approach favored by Integral Inc. was based on a preferential anodisation of n^+-layer within a sandwich $n/n^+/n$ structure. Performance of Integral's SOI shown in Table 1 are date to 1992. Logic SOI/CMOS ICs of AC 74 series were produced. Devices were subjected to the prolonged climatic and radiation testing and successfully withstood the tests. Low price and simplicity of fabrication of PS-based SOI explain Integral's interest. However, the USSR disintegration interfered with Integral's plans and in 1992 SOI technology stopped.

In 1995 our group (Belarusian State University of Informatics & Radioelectronics and INFM E6 of Rome University "La Sapienza") continued the advancement of SOI technology based on anodisation of n^+-layer within $n/n^+/n$- structure. The corresponding column of the Table 1 represents performance of SOI structures resulted from the modificated technology. Referring to Table 1, improvement of the SOI parameters was gained. The thickness of the silicon film was reduced from 0.3 to 0.1 μm, the range of the BOX thickness was expanded from 1.5 - 1.8 μm to 0.8 - 3.0 μm. Warpage of the wafers was reduced from 20 to 5 μm. The use of low-temperature epitaxy and optimization of both the processes of formation of the buried n^+-layer and its further anodisation made possible above-mentioned improvements. Test matrix chips were realized using Integral facilities.

As for ELTRAN, nowadays this method is brought up to the industrial production. Referring to Canon's report [19], ELTRAN permits a fabrication of SOI with a wide range of the thickness of both a Si layer and BOX. However, no data for electronic devices fabricated in ELTRAN SOI are available.

2. PS-based technology potentialities of value

Over the years SOI based on selective anodisation fell from favor. This is certainly attributable to the fact that the prime interest of wafer manufacturers is SOI technologies suitable for the production of saleable wafers. To date, two basic SOI technologies, separation by implanted oxygen (SIMOX) and bonded wafers, came to the forefront. It is clear that SOI based on selective anodisation cannot defy competition with SIMOX or bonded wafers. Both SIMOX and wafer bonding techniques yield wafers with a continuous silicon layer isolated from the semiconductor substrate with a continuous silica layer, while all except ELTRAN PS-based technologies offer island SOI. That is the reason why largest silicon wafer manufacturers are interested in the development of precisely SIMOX and wafer bonding technologies, which therefore are rather well established. Moreover, the supply, the material quality and the cost exert the main influence on the entry of SOI wafers into mainstream production. On this evidence, the grand total of these factors has been holding companies back from embracing PS-based SOI for high-volume applications.

Thus, current SOI technologies are pointed toward a direction of producing continuous SOI. It is not our intention to change radically the course of SOI development. We only would like to attract attention to the little path, which branches of the mainstream. This trend may be of considerable interest in future technical applications that deal with on-chip integration of electronic SOI ICs. Porous silicon is just the material, which provides great scope for such integration.

And what is more, we believe that the main disadvantage of the SOI based on selective anodisation - discontinuity of the SOI structures - can metamorphose into its main advantage. Really, PS technology provides the following benefits:

- It is possible to form local PS regions, that is, only specified wafer regions may be selectively transformed into porous state.
- Porous silicon may be oxidized either partially or completely and the oxidation process is controllable.
- Physical properties of both PS and oxidized PS regions depend on the layer thickness, porosity, dimensions of pores and silicon matrix crystallites, composition,

etc. and are controlled by regimes of PS formation and oxidation.

- Introducing of desired dopants may readily change physical properties of both PS and oxidized PS.
- PS and oxidized PS regions of different characteristics (porosity, thickness, uniformity, etc.) may be created simultaneously within the same wafer.

Fig. 2 illustrates some of the SOI structures, which are available with PS technology. Fabrication of these structures and their possible applications are discussed below.

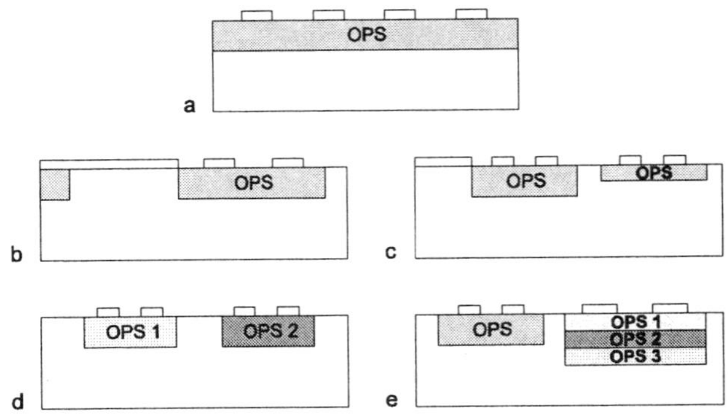

Figure 2. SOI structures available with PS technology.

Fig. 2a demonstrates a conventional SOI structure with the continuous BOX layer made from oxidized PS on the entire surface of the silicon substrate. A lot of silicon islands of the desired configuration for circuit components to be formed are disposed onto the BOX layer. Unlike the SOI realized by well known BESOI, UNIBOND or ELTRAN, the distinctive property of the SOI structure with the continuous BOX layer from this technique is the island structure of the isolated silicon regions. Such structures offer no gain in comparison with above-mentioned methods. However, sometimes both SOI and bulk silicon regions must be created within the same wafer. With the continuous BOX layer in the SOI wafers, BOX should be removed from the substrate in the definite regions. For thin SOI, BOX removal should present no problems. As for thick SOI, BOX removal is a more complicated challenge and the surface planarity will be impaired. With PS, this problem is simple to solve by selective anodisation. Fig. 2b shows such an integrated structure consisted of SOI and bulk silicon regions.

Let us consider in more detail some plausible application of the semiconductor structures shown in Fig. 2b. Fig. 2b represents a case when SOI regions are made within a chip while bulk silicon regions are left between the chips. Along these bulk silicon regions the separation of wafer into chips can be made. Bulk silicon regions also may be left within the chip. Of course, such structures essentially enhance the capabilities of device designers. To cite one example, contacts from the chip surface to the semiconductor substrate may be formed throughout these regions. Moreover, on-chip integration of SOI and bulk silicon regions provide a useful integration of, for example, low-voltage CMOS devices made in the SOI regions with high-power devices made in the bulk silicon regions. The small bulk silicon regions inserted into the SOI regions may serve as heat sink reducing device heating in the operation. This structure offers the gain in the temperature decrease of about 100°C for MOS transistors operating in the pulse regime [31].

Nowadays gain in functionality and cost reduction is of a prime importance for integrated circuits. On-chip integration of digital and/or analog devices, and/or low, medium and high power devices would provide a greater flexibility, increased reliability, and lower cost of electronic systems. The structure shown in Fig. 2c offers the unique possibility to modulate the thickness of the BOX layer. The structure may be realized as follows. First, a step of forming a buried n^+-type layer to be converted into PS is performed using photolithography and ion implantation of Sb or As into the exposed regions. Then high temperature annealing activates Sb (As). To provide the substrate with thinner n^+-type regions, second photolithography and ion implantation of Sb (As) into the exposed regions is performed. In this case implantation energy is much lower than the first one. Second high temperature annealing activates Sb (As) in these regions. To provide a SOI-structure with buried isolation region of ~0.3 - 0.5 μm, shallow n^+-type regions should be formed. These regions are made possible by multiple doping steps and rapid thermal annealing of implanted dopants.

Using PS, even more complex structures may be readily formed. For example, extra BOX layer of another thickness, in addition to just mentioned one, is made feasible by the PS-based technology, as shown in Fig. 2c. In principle, it is possible to make any desired number of SOI-structures heterogeneous in BOX thickness [27-29].

The ability of PS to be oxidized in variable degree may be used to form any oxidized PS regions with changeable properties. Fig. 2d shows an example of the complex structure consisted of the oxidized PS regions of the same thickness but different in porosity and oxidation level. These oxidized PS regions refer to either partially or completely oxidized PS. Such structures are useful for on-chip integrating of microelectronic and optoelectronic devices as to be discussed below.

PS technology makes feasible PS (and correspondingly oxidized PS) layers heterogeneous in porosity with the layer thickness. For oxidized PS, heterogeneity in porosity implies heterogeneity in oxidation level and physical properties. A combination of such the structure with PS-based SOI is shown in Fig. 2e.

Structures shown in Figs. 2d and 2e combines regions consisted of PS oxidized to a variable extent. These are only particular cases of the possible combinations of different regions made of PS and exposed to different treatments. Noteworthy is a wide variety of new conceivable materials with unique properties resulting from PS treated by one means or another. Let us use the oxidation of PS to illustrate the type of evidence supporting this conclusion.

3. Formation and properties of oxidized PS

The first paper relating to the oxidation of PS was published in 1975 by Yamanaka et al from NTT [32]. Behavior of IR spectra of PS layers oxidized at high temperatures in dry oxygen was discussed. In 1976 Imai and Yashiro from NTT reported C-V characteristics of oxidized PS [33]. They were the first to establish that PS may be oxidized either fully or partially depending on oxidation temperature. The present views of oxidation processes for PS are based on research made by the experts of NTT [32-35], IBM [36], CNET [37, 38], and DERA [24,39]. According to these views, oxidation of PS is controlled by a surface reaction at both the surface of PS and the wall of pores in PS. It was determined that oxidation kinetics and characteristics of oxidized PS depend on both parameters of initial PS and oxidation regimes.

The discussion that follows points out the meaning of oxidized PS. It should be emphasized that oxidized PS involves a whole series of dissimilar materials. We will look more closely at the kinetics of PS oxidation and densification, electrical and optical parameters of the material obtained, and feasibility to dope oxidized PS with desirable rare-earth dopants.

3.1. PS OXIDATION AND DENSIFICATION

The oxidation kinetics of PS can be characterized by measuring the increase of PS weight during oxidation and calculating the oxidized fraction (p_{ox}), defined as the percentage of silicon oxide in the porous material [35, 37]. Fig. 3 shows the variations in the oxidized portion p_{ox} with time in dry oxygen and steam for the PS sample of 50% porosity. It is seen from the Fig. 3 that after a rapid growth during the first few minutes of PS oxidation in dry oxygen, p_{ox} changes very slowly: only a few percent in more than 5 h remaining below 15 %. It means that oxidation of PS at low temperature in dry oxygen is far from complete. Partially oxidized PS consists of three fractions: silicon nanocrystallites, silicon oxide and voids. Characteristics of partially oxidized PS depend on the percentage composition of these fractions. Much higher values of oxidized portion can be reached in steam atmosphere and by increased temperature. It was found that complete oxidation of PS could be achieved in rather short times provided that the temperature is high enough. Moreover, a complete PS oxidation can be achieved if the initial porosity of PS is higher than 46%. Below this critical porosity value, the PS cannot be fully oxidized even at high temperature [35, 37].

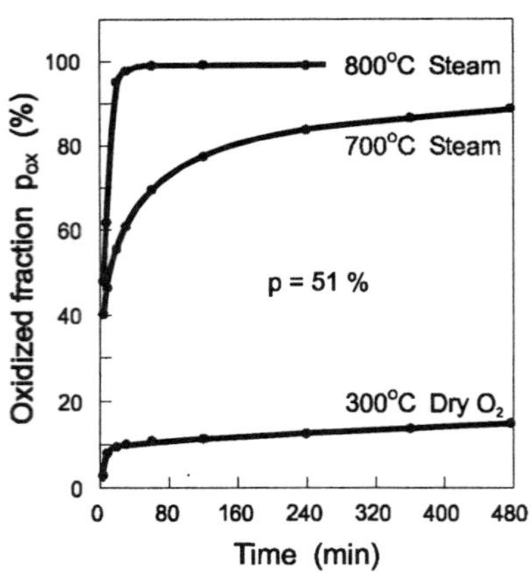

Figure 3. Variation of the oxidized fraction with time in dry oxygen and steam [37].

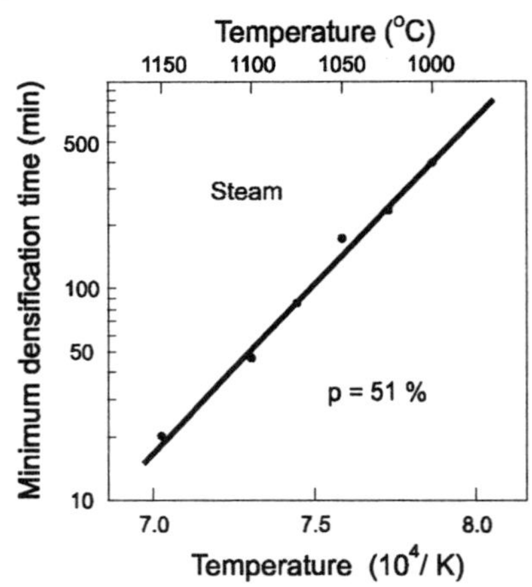

Figure 4. Temperature dependence of minimum time necessary for the densification of porous oxide in steam ambient [37].

One of the main features of the oxidized PS obtained in the temperature range of 300 - 900°C is that it remains porous even after complete oxidation. In this case the oxidized material is characterized by open porosity. The porous material resulted from above oxidation is much different from the porous material provided by high-temperature oxidation of PS at 1100 - 1150°C. The latter creates close pores inside the oxidized layer isolated from the outside of the sample.

To densify the porous oxides obtained from PS, an additional high-temperature annealing/oxidation should be performed. Fig.4 shows the minimum time necessary for the densification of porous oxide in steam ambient at different temperatures. The variations of the minimum densification time with temperature are well described by an Arrhenius-type law, with activation energy of about 3 eV. The densification refers to the viscous flow of the oxidized fraction in oxidized PS. That is why the densification temperature should be as high as 1150 - 1200°C to provide softening and viscous flow of the oxide resulting in pore healing [36,37].

3.2. DIELECTRIC PROPERTIES OF OXIDIZED PS

The dielectric constant ε is a principal characteristic of dielectric. It can be derived from the capacitance measurements. Fig.5 shows dielectric constant ε vs. the temperature T for the PS oxidation in wet and dry oxidizing atmospheres [35, 37]. A dashed line at ε = 3.8 refers to quartz and high-quality thermal silica.

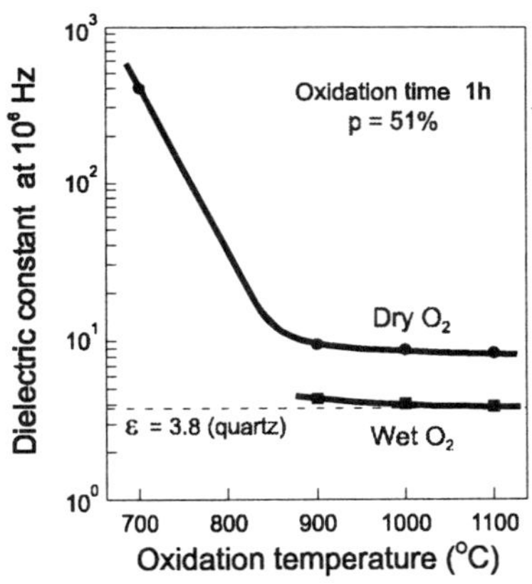

Figure 5. Temperature dependence of dielectric constant for the PS oxidation in wet and dry oxidizing atmospheres [35].

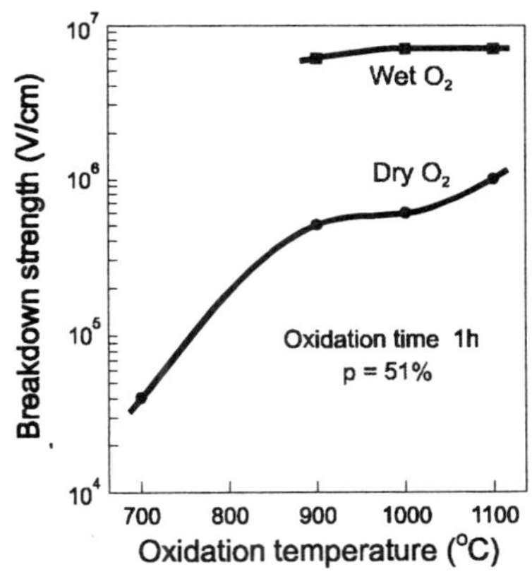

Figure 6. Variation of breakdown strength of oxidized PS with temperature for the PS oxidation in wet and dry oxygen [35].

For PS oxidation in wet oxygen, the dependence of ε on the oxidation temperature is weak. PS oxidation at temperatures above 1000°C brings material to have the same properties of silica.

As for the oxidation in dry oxygen, ε strongly depends on the oxidation temperature.

When oxidized at temperatures above 900°C, obtained material shows ε of 8 - 10. This is a good evidence that material includes the remainder of silicon matrix along with oxide fraction. It is particularly remarkable that for oxidation at temperature below 900°C values of ε are very high ranging up to 500. This value is considerably in excess of ε of single-crystal silicon, which is equal to 12. It is reasonable to suppose that strong polarization effects are responsible for such high values of ε (in excess of ε of all conceivable components involved). However, the question calls for further investigations.

Fig. 6 shows the variation of breakdown strength of oxidized PS (E_b) with temperature for the PS oxidation in wet and dry oxygen. Like ε, breakdown strength of oxidized PS strongly depends on the oxidation regimes. When densified at high temperature, fully oxidized PS displays breakdown strength $E_b = (6 - 7)\cdot 10^6 \text{ V}\cdot\text{cm}^{-1}$ to be very close to E_b of thermal silica. For partially oxidized PS E_b is about one order of magnitude less [35, 37].

It should be noted that dielectric parameters of non-densified oxidized PS are very sensitive to environmental conditions. This is due to the porous structure of as-oxidized PS resulting in advanced adsorption from environment. This effect is useful in sensor devices, e.g. gas sensors, chemical microsensors, biosensors etc [40 - 43].

3.3. REFRACTIVE INDEX OF OXIDIZED PS

Studies of the optical properties of oxidized PS are far from complete. It seems likely that in-depth research into this field will open up radically new fields to the use of oxidized PS in optoelectronics. Some surprises are in store in using oxidized PS with controlled variable parameters. We only have touched some optical characteristics of waveguide interest, the central of which is the refractive index (n).

The refractive index of oxidized PS is very sensitive to structural properties, composition, and density of the material. The possibility to change the parameters of PS and oxidized PS allows periodic structures with different porosity and thickness to be easily fabricated in a silicon wafer.

Recall values of n for some materials used for waveguides. For single-crystal silicon n is equal to 3.5. For natural quartz n is in the range of 1.458 – 1.460. Silicon oxides obtained by thermal oxidation have refractive indices varying from 1.452 to 1.466, depending on the oxidation atmosphere and temperature. In the absence of impurities, one can say with a fair degree of confidence that when any silicon oxide has n lower than natural quartz, the oxide should contain voids (pores). With impurities, the pattern is more complex. Nevertheless, increased n is unambiguously indicative of impurity presence in silicon oxide.

Fig. 7 shows n vs. porosity for unoxidized PS and PS oxidized at 970°C in steam for 15 min [36]. Referring to Fig. 7, for unoxidized PS an increase in porosity from 45 to 85% causes a decrease in n from 2.7 to 1.7. A thermal treatment at 970°C in steam leaves behind oxidized PS that shows a decrease in n to 1.22 – 1.44. Such low values of n (below the lower limits for natural quartz and compact silica) strongly suggest that oxidized PS resulted from above treatment is porous everywhere over the above porosity range.

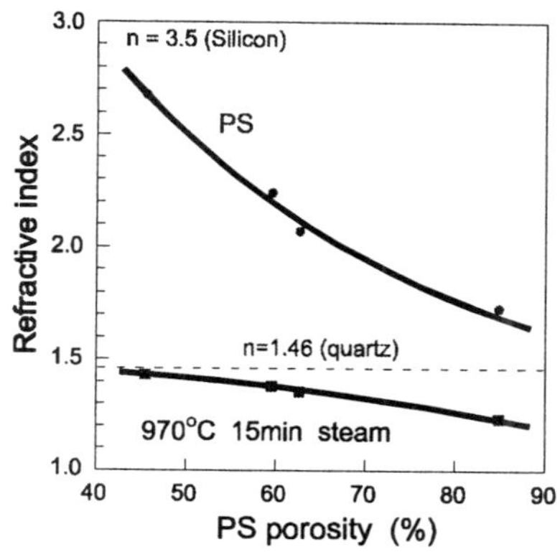

Figure 7. Effect of porosity on the refractive index of unoxidized PS and PS oxidized at 970°C for 15 min in steam [36].

Fig. 8 gives n as a function of porosity for oxidized PS realized in more drastic conditions at 1200°C in steam for 15 min [36]. As is seen from Fig. 8, the graph may be divided into three regions depending on the position of the refractive index of natural quartz (~1.46). For the region lettered by A $n > 1.46$, suggesting that PS is oxidized partially. The region lettered by B (n ~ 1.46) corresponds to compact fully oxidized PS comparable with high-quality thermal silica. In the region lettered by C $n < 1.46$, indicating the porous structure of oxidized PS. So, for this treatment, porosity of initial PS is responsible for the structure and composition of resulting material.

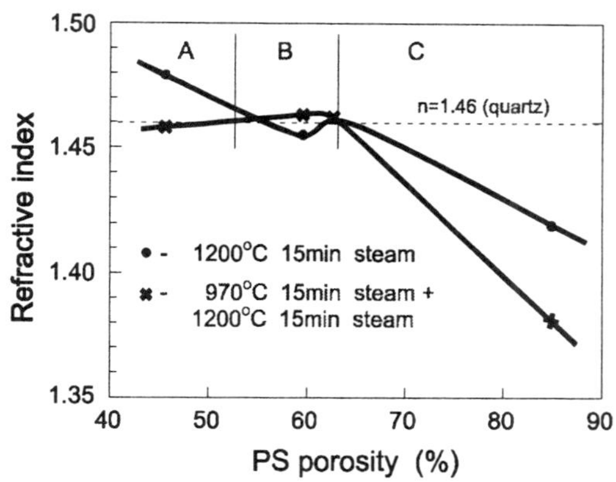

Figure 8. Effect of porosity on the refractive index of oxidized PS realized at 1200°C in steam for 15 min and 970°C for 15 min in steam + 1200°C for 15 min in steam [36].

Fig. 8 also illustrates the combined two-step PS oxidation: (i) 970°C for 15 min in steam + (ii) 1200°C for 15 min in steam. Referring to Fig. 8, such treatment is able to provide two types of fully oxidized PS: compact when the porosity of the initial PS layer is less than 62%, and porous when it is more than 62%.

Thus, it is possible to design oxidized PS layers whose optical properties can be controlled quite accurately. Up to now optical superlattices (e.g. Bragg filters with two distinct porosity values) as well as systems with continuous porosity modulation (and hence refractive index modulation) have been realized [44]. Integrated optical waveguides based on oxidized PS have been developed demonstrating in the visible range optical losses less than 0.5 dB/cm [45, 46]. Periodic structures made up of layers of varying porosity, with appropriate refractive index and thickness, can be used to produce optical filters suited for different visible red and infrared regions of the electromagnetic spectrum.

3.4. DOPANTS IN OXIDIZED PS

It is worthwhile to mention another unique possibility that PS-based technology offers, that is to change the characteristics of formed layers by incorporation of various dopants. It is common practice to introduce dopants into as-formed PS followed by conversion of PS to the oxidizing state. So, hereinafter it should be remembered that when it comes to the doping of oxidized PS, there is no alternative to the above procedure.

As example, let us consider the doping of oxidized PS with rare-earth elements. Introducing rare-earth elements (in particular, Er) into oxidized PS provides a way of producing active devices (e.g. amplifiers and lasers) from the passive ones such as PS-based waveguides. The characteristic, atomic like luminescence of rare-earth elements results from the incompletely filled inner 4f shell, which permits intra-shell optical transitions with long lifetimes and hence very narrow line width. The outer shell screens the influence of a host crystal very efficiently and makes the emission wavelength host and temperature independent.

Er may be readily introduced into PS by a cathodic electrochemical treatment in the solutions of Er salts [47]. Fig. 9 shows the photoluminescence spectra of Er-doped PS oxidized in various regimes [48]. As Fig. 9 depicts, depending on the type of oxidized

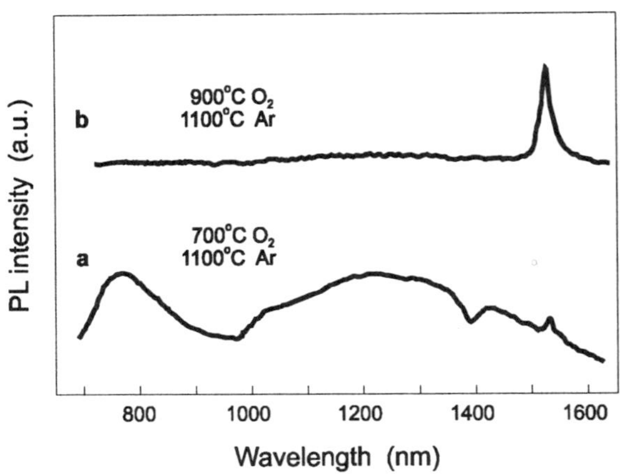

Figure 9. Photoluminescence spectra of (a) partially and (b) fully oxidized PS doped with Er [48].

PS by our classification, bands of photoluminescence at different wavelengths are observed. For partially oxidized PS photoluminescence bands from PS matrix and from Er^{3+} ions are revealed while fully oxidized PS demonstrates only sharp photoluminescence bands due to inner 4f shell transitions within Er^{3+} ions.

The excitation mechanism of Er in OPS was investigated by measuring the photoluminescence intensity at 1.53μm as a function of excitation wavelength [48]. Fig. 10 shows the photoluminescence excitation (PLE) spectra recorded at room temperature for partially and fully oxidized PS:Er. They consist of broad band (dashed line) and sharp peaks located at 381, 523, 654, and 980 nm. The broad band PLE spanning from 300 to 600 nm is attributed to Er ions excitation processes involving silicon nano-crystallites in the partially oxidized PS host. As the oxidation temperature increases the broad excitation band disappears due to the oxidation of nano-crystallites.

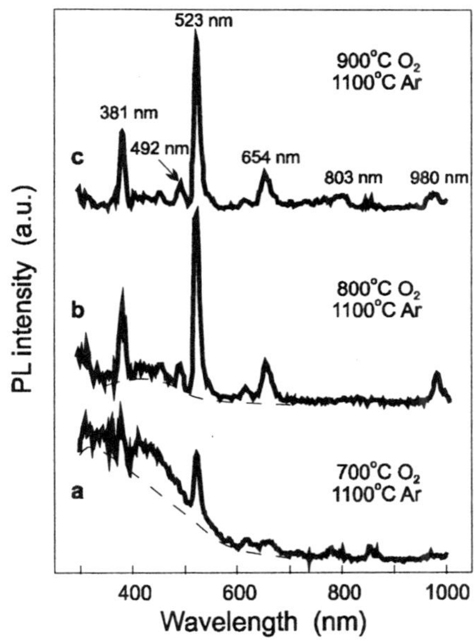

Figure 10. Room temperature PLE spectra of 1.53-μm Er-related emission from (a,b) partially and (c) fully oxidized PS:Er [48].

The PLE spectrum of fully oxidized PS:Er is seen to have a pronounced banded structure. Five peaks are exhibited at wavelengths of 381, 492, 523, 654, and 980 nm, reflecting the structure of the optical absorption bands of Er^{3+} ions. These results indicate that Er ions in fully oxidized PS are excited directly by exciting radiation.

So, it is possible to access the luminescent levels of rare-earth ions in partially oxidized PS via the broad-band absorption of the silicon matrix. 1.54 μm emission from Er-doped partially oxidized PS is largely independent of pump wavelength. This is in contrast to the Er-doped stoichiometric fully oxidized PS in which emission follows the Er absorption spectrum. Reported in [48] Er concentration in the oxidized PS is large enough to attain the amplification effect. This opens up possibilities for developing optoelectronic devices (e.g. lasers, amplifiers etc.) based on oxidized PS doped with rare-earth elements.

4. Possible application fields of oxidized PS

The data on the properties of various layers of oxidized PS reported thus far and its possible application fields are categorized in Table 2. As seen from Table 2, both electrical and optical properties of the oxidized PS layer may be varied with oxidation regimes of PS layers with different porosities.

TABLE 2. Properties of oxidized porous silicon and its application.

Parameter	Partially Oxidized PS (POPS)		Fully Oxidized PS (FOPS)	
	POPS	Densified (DPOPS)	FOPS	Densified (DFOPS)
PS porosity, %	20 – 45	20 – 45	45 – 85	45 – 60
Heat treatment	200-750°C, dry O_2	+ 1100-1200°C, dry Ar	750-1100°C, dry O_2, wet O_2, steam	+ 1100-1200°C, dry N_2, dry Ar, steam, wet O_2
ε (at 1 MHz)	10 – 30	10 – 50	1.5 – 3.8	~ 3.8
E, V/cm	$< 10^3$	$< 10^5$	$(1-6)\cdot10^6$	$(6-7)\cdot10^6$
n	1.5 – 2.5	1.5 – 2.5	1.2 – 1.45	1.45 – 1.46
Structure of OPS	Open porosity, crystallites	Nonporous, crystallites	Porous, no crystallites	Nonporous, no crystallites
Field of application	BS, GS, PD, LED	PD, WG, WG-D	WG, WG-D	SOI WG, WG-D, HS

BS – biosensor, GS – gas sensor, HS – heat sensor, LED – light emitting diode, PD – photodetector, WG – waveguide, WG-D – waveguide-based devices

PS may be oxidized either completely or partially with properties of the formed oxidized PS layer dependent on both oxidizing regimes and parameters of initial PS. The oxidized PS layers provided by variations in the regimes of its formation may be classified into four types according to the oxidation extent.

Fig. 11 illustrates schematically a generalized pattern of basic types of oxidized PS that can be put to practical use in combination with PS-based SOI.

A thermal treatment at 200 - 750°C in dry oxygen oxidizes PS partially covering pore sidewalls with a thin silicon dioxide (SiO_2) film. Partially oxidized PS consists of three fractions: silicon nano-crystallites of the host matrix, SiO_2 and voids. Characteristics of partially oxidized PS depend on the percentage composition of these fractions. The thickness of SiO_2 on the pore sidewalls and dimensions of Si nano-crystallites have a determining effect on the carrier transport, optical characteristics of the material and its temporal stability. The fields of application of this material are gas sensors, light-emitting devices, photodetectors etc. [40-44, 50].

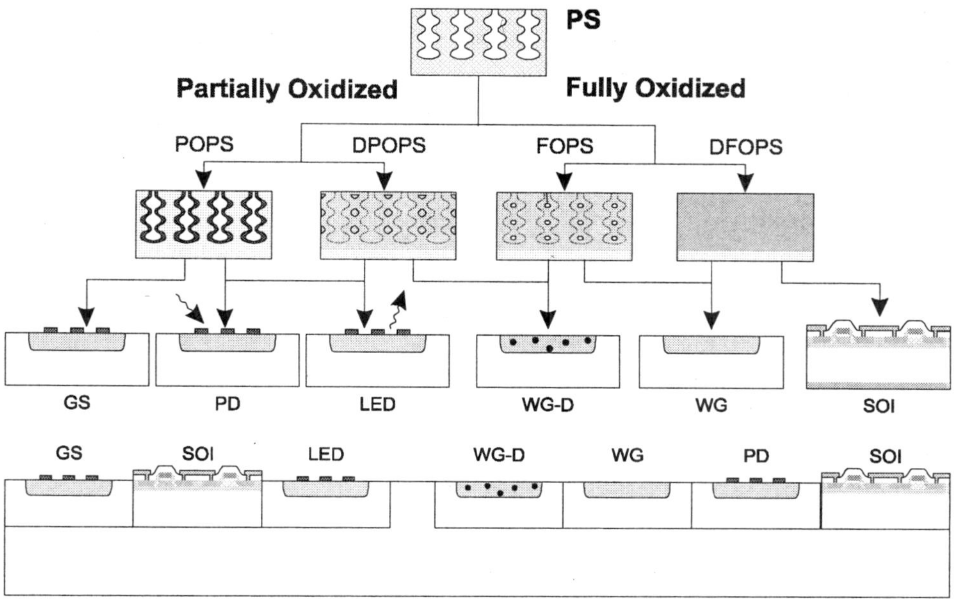

Figure 11. A generalized pattern of basic types of oxidized PS that can be put to practical use in combination with PS-based SOI.

An additional thermal treatment of partially oxidized PS at the temperature above 1000°C in inert atmosphere causes voids "healing". The formed material is SiO_2 containing silicon crystallites. Refractive index of the material may by varied from 1.5 (refractive index of SiO_2) to 2.5. This material is a representative of low-dimensional materials with associated application fields [44, 49, 50].

A thermal treatment at 750 - 1100°C in an oxidizing atmosphere causes complete oxidizing of PS. Such a material represents SiO_2 that does not contain silicon crystallites. However voids can occur inside this material. Dielectric characteristics of the material are worse than of dense thermal SiO_2. Refractive index of the material is changed from 1 to 1.5. The material may be used in different sensors and as cladding layers for in integrated optical waveguides and SOI structures [40, 45, 46].

An additional thermal treatment at the temperature 1100 - 1200°C in inert atmosphere clears above material of voids resulting in non-porous completely oxidized PS. Characteristics of such a material are much like those of a conventional thermal SiO_2. This material may be utilized as a core region in integrated optical waveguides [45, 46, 49] and as an insulator layer in SOI [50-52].

5. Conclusion

Oxidized PS as storehouse of untapped resources was analyzed. PS may be oxidized either completely or partially with properties of the formed OPS layer dependent on both oxidizing regimes and parameters of the initial PS. It is of great importance that PS and oxidized PS regions of different characteristics (porosity, thickness etc.) may be created simultaneously within the same wafer.

The kinetics of PS oxidation and densification, electrical and optical parameters of the material obtained, and feasibility to dope oxidized PS with a desirable dopant were discussed.

Finally, we pointed out a few more ideas relevant to the on-chip integration of PS-based SOI with different PS-based micro- and optoelectronic devices.

It should be emphasized that the discussed material is only the top of iceberg. PS offers the way for researchers to create an abundance of quite new materials with unique properties. Who knows what the new matter will result from the studies of PS treated either in nitrogen, carbon, magnetic or other agents.

6. References

1. Watanabe, Y., Sakai, T. (1972) Semiconductor Device and method of producing the same. *Jap. Patent* 49-19030 (*US Patent* 3640806).
2. Hirosy, S., Watanabe, Y., Arita, M., Asaoka, K., (1978) Semiconductor devices. *Jap. Patent Appl.* 53-45675.
3. Sakurdi, T., Ukami, T. (1981) Method of manufacturing the semiconductor devices. *Jap. Patent Appl.* 52-84968.
4. Pogge, H.B. (1975) Method of fabricating semiconductor device embodying dielectric isolation. *US Patent* 3919060.
5. Magdo, I. E., Magdo, S., Nestork, W.J. (1976) Process for fabricating devices having dielectric isolation utilizing anodic treatment and selective oxidation. *US Patent* 3954523.
6. Pogge, H.B. (1978) Total dielectric isolation utilizing a combination of reactive ion etching, anodic etching, and thermal oxidation. *US Patent* 4104090.
7. Imai, K. (1983) Semiconductor devices and method of manufacturing the same. *US Patent* 4393577.
8. Imai, K., Unno, H. (1984) FIPOS technology and its application to LSI's. *IEEE Trans.Electron.Dev.* **31**, 297-302.
9. Mano, T., Baba, T., Sawada, H., Imai, K. (1982) FIPOS CMOS 16K-bit static RAM. *Technical Digest Symp. on VLSI Technology*, 12.
10. Ehara, K., Unno, H., Muramoto, S. (1985) 1,5 μm FIPOS/CMOS VLSI process with low wafer warpage and Si deposit-defect-free. *Electrochem.Soc.Ext.Abs.* **85**, 457.
11. Frye, R.C. (1983) Method of forming dielectrically isolated silicon semiconductor

materials utilizing porous silicon formation. *US Patent* 4380865.

12. Konaka, S., Tabe, M., Sakai, T. (1982) A new silicon-on-insulator structure using a silicon molecular beam epitaxial growth on porous silicon. *Appl.Phys.Lett.* **41**, 86-88.
13. Lin, T.L., Chen, S.C., Wang, K.L., Iyar, S. (1986) Si-MBE SOI. *Mat.Res.Soc.Proc.* **53**, 193-197.
14. Bomchil, G., Halimaoui, A., Herino, R. (1989) Procede de fabrication d'une structure de silicium sur isolant. *Patent of France* 2620571.
15. Kinney, W.I., Lasky, J.B., Nesbit, L.A. (1985) Method of manufacturing semiconductor structures having an oxidized porous silicon isolation layer. *US Patent* 4532700.
16. Zorinsky, E.J., Spratt, D.B. (1986) Method for obtaining full oxide isolation of epitaxial islands in silicon utilizing selective oxidation of porous silicon. *US Patent* 4628591.
17. Holmstrom, R.P., Chi, W.J-Y. (1986) Method of forming an isolated semiconductor structure. *US Patent* 4627883.
18. Yonehara, T. (1994) Semiconductor member and process for preparing semiconductor member. *US Patent* 5371037.
19. http://www.canon.co.jp/ELTRAN
20. Tsao, S.S. (1987) Porous Silicon Techniques for SOI Structures. *IEEE Circuit & Device Magazine* **3**, 3-7.
21. Matloubian, M., Zorinsky, E.J., Spratt, D. (1986) Total dose radiation characteristics of SOI mosfets fabricated using islands technology. *IEEE Transactions on Nuclear Science* **35**, 1650-1652.
22. Zorinsky, E.J., Spratt, D.B., Virkus R.L. (1986) The "islands" method – a manufacturable porous silicon SOI technology. *Technical Digest of IEDM*, 431-434.
23. Barla, K., Bomchil, G., Herino, R., Monroy, A. (1987) SOI Technology Using Buried Layers of Oxidized Porous Si. *IEEE Circuit & Device Magю* **3**, 11-14.
24. Thomas, N.J., Davis, J.R., Keen, J.M., Castledine, J.G., Brumhead, D., Goulding, M., Alderman, J., Farr, J.P.G., Earwaker, L.G., L'Ecuyer, J., Stirland, I.M., Cole, J.M. (1989) High-Performance Thin-Film Silicon-on-Insulator CMOS Transistors in Porous Silicon. *IEEE Elect.Dev.Let.* **10**, 129-131.
25. Bondarenko, V.P., Yakovtseva, V.A., Dolgyi, L.N Dorofeev, A., Vorozov, N., Troyanova, G. (1994) SOI – structures based on oxidized porous silicon. *Microelektronika* **23**, 61-68 (in Russian).
26. Bondarenko, V.P., Bogatirev, Y.V., Dolgyi, L.N., Dorofeev, A.M., Panfilenko, A.K., Shvedov, S.V., Troyanova, G.N., Vorozov, N.N., Yakovtseva, V.A. (1995) 1.2 μm CMOS/SOI on porous silicon. in J.P.Colinge, A.N.Nazarov (eds.). *Physical and Technical Problems of SOI Structures and Devices*, Kluwer Academic

Publishers, Dordrecht, pp.275-280.

27. Balucani, M., Bondarenko, V., Dolgyi, L.N., Ferrari, A., Lamedica, G., Yakovtseva, V.A. (2000) Process for forming structures with different conductivity showing a hyperfine transition region, for forming porous silicon. *International Patent*: PCT/IT00/00329
28. Balucani, M., Bondarenko, V., Dolgyi, L.N., Ferrari, A., Lamedica, G., Yakovtseva, V.A. (2000) Process for the two-step selective anodization of a semiconductor layer for forming porous silicon. *International Patent*: PCT/IT00/00330
29. Balucani, M., Bondarenko, V., Ferrari, A., Lamedica, G., Panfilenko, A.K., Yakovtseva, V.A. (2000) Process for the forming of isolation layers of a predetermined thickness in semiconductors wafers for the manufacturing of integrated circuits. *International Patent*: PCT/IT00/00331
30. Romanow, S.I., Dvurechenskii, A.V., Kirienko, V.V, Grötschel, R., Gutakovskii, A., Sokolov, L.V., Lamin, M.A. (2000) Homoepitaxy on porous silicon with a buried oxide layer: full-wafer scale SOI. in H.L.F.Hemment, V.S.Lysenko, A.N.Nazarov (eds). *Perspectives, Science and Technologies for Novel Silicon on Insulator Devices*, Kluwer Academic Publishers, Dordrecht, pp.29-46.
31. Balucani, M., Bondarenko, V., Lamedica, G., Ferrari, A. (2000) Opto-electronic Silicon On Insulator integrated circuits by porous silicon technology. *Electronic Group Meeting*, Parma 9/06/2000.
32. Yamanaka, H., Sakamoto, M. (1975) Oxygen sensitivity of porous silicon formed by anodic reaction. *Jap.J.Appl.Phys.* **14**, 303-304.
33. Imai, K., Yashiro, T. (1976) Oxidized Porous Silicon-Silicon Interface Investigation by C-V Method. *Jap.J.Appl.Phys.* **15**, 553-554.
34. Arita, Y., Kuranari, K., Sunohara, Y. (1976) Thermal behavior of porous silicon. *Jap.J.Appl.Phys.* **15**, 1655-1664.
35. Unagami, T. (1980) Oxidation of Porous Silicon and Properties of Its Oxide Film. *Jap.J.Appl.Phys.* **19**, 231-241.
36. Aboaf, J.A., Broadie, R.W., Pliskin, W.A. (1977) Integrated circuit isolation structure and method for producing the isolation structure. *US Patent* 4016017.
37. Yon, J.J., Barla, K., Herino, R., Bomchil, G. (1987) The kinetics and mechanism of oxide layer formation from porous silicon formed on p-Si substrates. *J.Appl.Phys.* **62**, 1042-1048.
38. Barla, K., Herino, R., Bomchil, G. (1986) Stress in oxidized porous silicon layers. *J.Appl.Phys.* **59**, 439.
39. Pickering, C., Beale, M.I., Robbins,D.J., Pearson, P.J., Greef, R. (1985) Optical properties of porous silicon films. *Thin Solid Films* **125,** 157-163.
40. Burkhardt, P.J., Robert, M. (1977) Porous silicon dioxide moisture sensor and method for manufacture of a moisture sensor. *US Patent* 4057823.

41. Schechter, I., Ben-Chorin, M., Kux, A. (1995) Gas Sensing Properties of Porous Silicon. *Anal.Chem.* **67**, 3727-3732.
42. Balucani, M., Bondarenko, V., Dolgyi, L. La Monica, S., Masini, G., Yakovtseva, V., Ferrari, A. (1997) Humidity Sensor on Partially Oxidized Porous Silicon. *Solid State Phenomena* **54**,75-83.
43. Sailor, M.J. (1997) Sensor applications of porous silicon, in L.Canham (eds.). *Properties of porous silicon. EMIS Datareviews Series No 18*, INSPEC, The Institution of Electrical Engineers, London, pp. 364-370.
44. Theiss, W., Hilbrich, S. (1997) Refractive index of porous silicon, in L.Canham (eds.). *Properties of porous silicon. EMIS Datareviews Series No 18*, INSPEC, The Institution of Electrical Engineers, London, pp. 223-228.
45. Yakovtseva, V., Dolgyi, L., Vorozov, N., Kazuchits, N., Bondarenko, V., Balucani, M., Lamedica, G., Franchina, L., Ferrari, A. (2000) Oxidized Porous Silicon: From Dielectric Isolation to Integrated Optical Waveguides. *J.of Porous Materials* **7**, 215-222.
46. Benson, T.M., Arrand, H.F., Sewell, P., Niemeyer, D., Loni, A., Bozeat, R.J., Krüger, M., Arens-Ffischer, R., Thönissen, Lüth, H. (1999) Progress towards achieving integrated circuit functionality using porous silicon optoelectronic components. *Materials Science and Engineering* **B69-70**, 92-99.
47. Bondarenko, V., Dolgyi, L., Dorofeev, A., Kazuchits, N., Leshok, A., Troyanova, G., Vorozov, N., Maiello, G., Masini, G., La Monica, S., Ferrari, A. (1997) Porous silicon as low-dimensional host material for erbium-doped structures. *Thin Solid Films* **297**, 48-52.
48. Bondarenko, V.P., Vorozov, N., Dolgyi, L., Yakovtseva, V., Petrovich, V., Volchek, S., Kazuchits, N., Grom, G., Lopez, N.A., Tsybeskov, L., Fauchet, P.M. (1999) Formation and luminescent properties of oxidized porous silicon doped with erbium by electrochemical procedure. *Mat.Res.Soc.Symp.Proc.* **536**, 69-74.
49. Bondarenko, V.P., Yakovtseva, V.A. (1997) Optoelectronic applications of porous silicon, in L.Canham (eds.). *Properties of porous silicon. EMIS Datareviews Series No 18*, INSPEC, The Institution of Electrical Engineers, London, pp. 356-363.
50. Miller, R.O. (1990) Method of making a silicon integrated circuit waveguide. *US Patent* 4927781.
51. Bondarenko, V.P., Dorofeev, A.M. (1995) Why porous silicon for SOI? in J.P.Colinge (eds.) *Physical and Technical Problems of SOI Structures and Devices*, Kluwer Academic Publishers, Netherlands, pp.15-26.
52. Bondarenko, V.P., Yakovtseva, V.A. (1997) Microelectronic applications of porous silicon, in L.Canham (eds.). *Properties of porous silicon. EMIS Datareviews Series No 18*, INSPEC, The Institution of Electrical Engineers, London, pp. 343-348.

ELECTRON-HOLE PAIR REVERSED DRIFT IN SOI STRUCTURE

V.N. DOBROVOLSKY, L.V. ISHCHUK, and G. K. NINIDZE
Department of Radiophysics, Kiev University
2 Glushkov Av., build. 5, 03127 Kiev, Ukraine

1. Introduction

It is known that in the n-semiconductor electron-hole pairs generally drift along the electrical field. However the electron and hole field heating can reverse the drift direction [1]. This non-obvious phenomenon is named reversed drift. Till now it was observed only in germanium [1] and indium antimonide [2]. Here we present a calculation of electron-hole pair drift velocity and the reversed drift observation in silicon. We used the SOI structures. Their properties ensured the experiment simplicity.

2. Theory

We will consider a current flowing in a sample containing electron-hole plasma. There electron and hole concentration can be written as $n = n_0 + \Delta n$, $p = p_0 + \Delta p$, where n_0, p_0 and $\Delta n, \Delta p$ are their equilibrium values and deviations from these values, respectively, $\Delta n = \Delta p$. The electron and hole conservation equations are $\frac{\partial \Delta n}{\partial t} = (g_n - r_n) + \frac{1}{e}\frac{\partial j_n}{\partial x}$ and $\frac{\partial \Delta p}{\partial t} = (g_p - r_p) - \frac{1}{e}\frac{\partial j_p}{\partial x}$, where $g_{n,p}$ and $r_{n,p}$ are theirs generation and recombination rates, $g_n - r_n = g_p - r_p$, $j_n = ev_n(E)n + eD_n\,\partial\Delta n/\partial x$ and $j_p = ev_p(E)p - eD_p\,\partial\Delta p/\partial x$ are current densities. In last equations $v_{n,p}$ and $D_{n,p}$ are the electron and hole drift velocities, and diffusion coefficients, E is field strength. From above expressions after the apparent transformation [3] we obtain

$$\frac{\partial \Delta n}{\partial t} = (g_n - r_n) - V_{\text{amb}}\frac{\partial \Delta n}{\partial x} + D_{\text{amb}}\frac{\partial^2 \Delta n}{\partial x^2},$$

where $V_{\text{amb}} = \frac{v_p(\partial v_n/\partial E) - v_n(\partial v_p/\partial E)(p/n)}{(\partial v_n/\partial E) + (\partial v_p/\partial E)(p/n)}$ and $D_{\text{amb}} = \frac{D_n D_p (n+p)}{D_n n + D_p p}$ are the ambipolar drift velocity and diffusion coefficient. We have neglected the influence of the electron and hole field heating on the diffusion.

F. Balestra et al. (eds.), Progress in SOI Structures and Devices Operating at Extreme Conditions, 329–332.

According to the article [4] in silicon at 300 K $v_n = \frac{1.44 \cdot 10^3 E}{\left(1+5.46 \cdot 10^{-5} E^{1.11}\right)^{0.90}}$,

$v_p = \frac{4.65 \cdot 10^2 E}{\left(1+6.9 \cdot 10^{-6} E^{1.21}\right)^{0.82}}$, where v_n, and v_p are measured in [cm/s], and E in [V/cm]. Using these formulae we calculated velocity V_{amb}. Curves 1-4 in Figure 1 present the relationships V_{amb} vs E at the different p/n ratio.

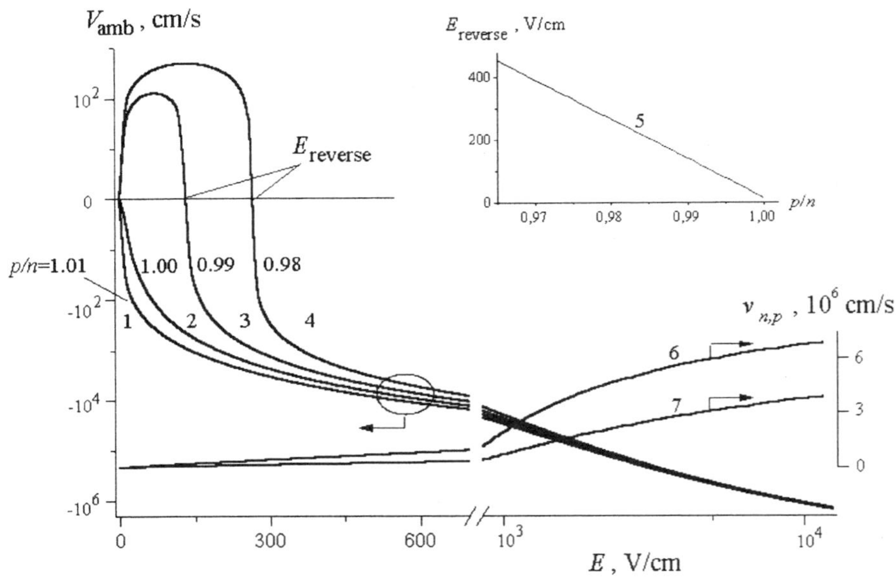

Figure 1. Ambipolar drift velocity vs electric field curves for the different p/n values: 1 – 1.01, 2 – 1.00, 3 – 0.99, 4 – 0.98. Curve 5 - the field of drift reversing E_{reverse} vs p/n. Curve 6,7 – the electron and hole drift velocities vs electric field.

At field $E < E_{\text{reverse}}$ velocity $V_{\text{amb}} > 0$, and the ordinary drift takes place. But at $E > E_{\text{reverse}}$ velocity $V_{\text{amb}} < 0$, and the reversed drift exists. The p/n rising leads to the E_{reverse} decreasing (curve 5). Curves 6 and 7 present relationships $v_{n,p}$ vs E.

3. Experimental structure

The *Si*-film of the SOI structure investigated is shown in Figure 2. Its thickness d is 0.4 μm. The buried oxide and substrate thicknesses are 1 μm and 400 μm. Donor

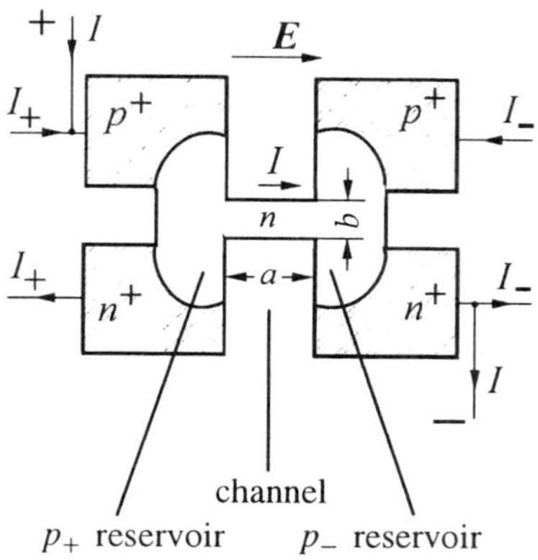

Figure.2. Experimental SOI structure.

concentration N_d and electron-hole pair diffusion length L_{dif} in the Si-film are $5 \cdot 10^{15}$ cm^{-3} and 2 μm. The narrow channel connects two wide parts of the film. Its length a and width b are 4 μm and 2 μm, respectively. Currents I_+ and I_- flow through the $p^+ - n$ and $n^+ - n$ junctions. They create the electron-hole pair reservoirs at the channel ends. We will designate the pair concentrations in these reservoirs as p_+ and p_-. The I current flows through the channel. It causes voltage U drop between the $p^+ - n$, and $n^+ - n$ junctions.

For $U/a >> E_{\text{dif}} = kT/eL_{\text{dif}}$, when the pair drift predominates over diffusion, our structure design allows to determine the drift type. Really, the pair drift into the channel from one of the reservoirs and increase the I current through the channel. Under the ordinary drift the pairs drift from the p_+ reservoir. But under the reversed drift they move from the p_- reservoir. The I_+ current rising increases the p_+ concentration. Therefore, when the ordinary drift is realized the I_+ current rise leads to current I increasing. However, in the case of reversed drift the current I_- rise leads to current I increasing.

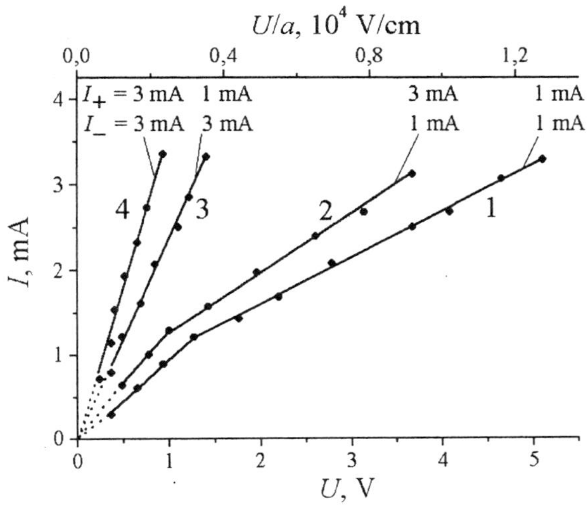

Figure 3. I-U curves for different ratio of the I_+ and I_- currents.

4. Results and discussion

The I vs U relationships are shown in Figure 3 for the several I_+ and I_- values. These values are written above corresponding curves. These relationships are same for pulsed (pulse duration is 1 μs) and stationary current measurements. This proves that a channel heating is small.

One could see that the I_- current rise increases current I several times (compare curves 1 and 3 (2 and 4)). This is the proof that just the reversed drift takes place in our experiments.

The I current rise by the I_+ current increasing is small (compare curves 1 and 2 (3 and 4)). The p_+ reservoir resistance decreasing by the I_+ current explains it.
Drift increases the pair concentration in the channel only till maximum value p_-. After this the channel current is $I = ebd \cdot \left[v_n(N_d + p_-) + v_p p_-\right]$. If the field strength is small, then $v_{n,p} = \mu_{n,p} E$ (curves 6 and 7 in Figure 1), and $I \propto U$. Here $\mu_{n,p}$ is the electron (hole) mobility. The p_- concentration reducing by $I > I_-$ and decreasing of the $\upsilon_{n,p}$ velocities by high field strength can explain the breaks of the 1 and 2 lines in Figure 3.
The condition $E_{\text{dif}} << U/a < E_{\text{reverse}}$ is needed for the ordinary drift observation. We used above expression for the channel current at the small field strength, and from the curves in Figure 3 found that p/n = 0.97 -0.99 are realized in our experiments. According to the curve 5 in Figure 1 values E_{reverse} = (390 -130) V/cm are corresponding to these p/n values. The E_{dif} value is 125 V/cm. This value and the E_{reverse} values are closely spaced, and above condition is not fulfilled. Therefore we did not observe the ordinary drift.

5. Conclusion

The calculation of electron-hole pair drift velocity was performed in n-silicon for different field strength values. It was shown that the velocity changes its sign at field greater than some critical value, and the reversed ambipolar drift takes place. This field value is the less the closer electron and hole concentration ratio to one.
The results of calculation were verified by experiments. They were performed on the SOI structure. This structure allows us to create the high electric field in the silicon film without any sufficient film heating.

References

1. Prior, A.C. (1960) A reversed carrier transport effect in germanium. *Proc. Phys. Soc.* **76**, 465-480.
2. Tosima, S. and Ando, K. (1967) Hot carrier double injection in n-InSb. *J. Phys. Soc. Japan* **23**, 812-819.
3. Lampert, M.A. and Mark, P. (1970) *Current Injection in Solids*, Academic Press, New York and London.
4. Canali, C., Majni, G., Minder, R., and Ottaviani, G. (1975) Electron and hole drift velocity measurements in silicon and their empirical relation to electric field and temperature. *IEEE Trans. Electron Devices* **ED-22**, 1045-1047.

A NOVEL DEPLETED SEMI-INSULATING SILICON MATERIAL FOR HIGH FREQUENCY APPLICATIONS

MIKAEL JOHANSSON AND STEFAN BENGTSSON
Microtechnology Centre at Chalmers and
Solid State Electronics Laboratory,
Department of Microelectronics
Chalmers University of Technology
SE-412 96 Göteborg, Sweden

Abstract

Wafer bonding has been used to manufacture a silicon material intended as substrate for high frequency applications. The space charge regions surrounding the bonded silicon/silicon interface deplete the silicon thereby causing semi–insulating behaviour at high frequencies. The material has been characterized electrically for frequencies up to 40 GHz using metal transmission lines and crosstalk structures on its surface. Measurements on the depleted silicon/silicon structures has been compared to similar measurements on bulk silicon wafers of different resistivities, SIMOX wafers and quartz. The results show that the material has potential to be used at high frequencies.

Introduction

Cost is a major issue for wireless communication products at 1-10 GHz. As a consequence increased interest is focused on silicon materials and devices. At higher frequencies the influence of the substrate and the interaction of substrate and devices need to be taken into account. In industry, scaling has enabled integration of both analog and digital circuits at the same chip. Signal switching in the digital part introduce noise in the more sensitive analog part, which can reduce the performance of the circuit. As a consequence the influence of the substrate on circuit performance is becoming increasingly important when mixed signal solutions are coming in wider use.

At high frequencies the electromagnetic field of a signal transmitted along a line extends outside the line. If free carriers are available in the substrate the transmitted signal loses energy to the carriers. At even higher frequencies the carriers are no longer in phase with the signal and carrier caused losses decrease. At high frequencies a semiconductor therefore exhibits dielectric behavior with losses determined by polarization [1]. Figure 1 shows the approximate onset of dielectric behavior for silicon

F. Balestra et al. (eds.), Progress in SOI Structures and Devices Operating at Extreme Conditions, 333–341.

of different resistivities. The straight line marking the division between semiconducting and dielectric behavior is the dielectric relaxation frequency, f_{dr}. It can be determined from [2]

$$f_{dr} = \frac{\sigma_0}{2\pi\varepsilon_r\varepsilon_0} \tag{1}$$

where σ_0 is the DC conductivity of the silicon material, ε_0 is the dielectric constant of vacuum and ε_r is the relative dielectric constant of silicon. From Fig. 1 it can be concluded that insulators or materials depleted of free carriers can be expected to be good substrates for high frequency applications.

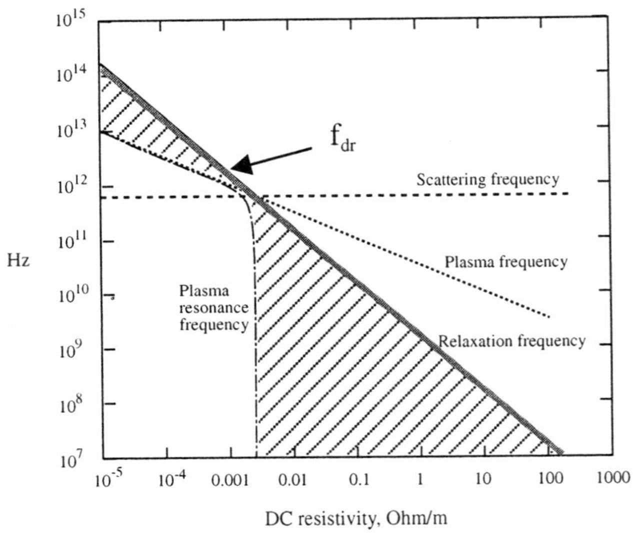

Figure 1. Resistivity-frequency chart. Above the solid line, f_{dr}, the semiconductor behaves as a dielectric. In the plasma resonance frequency region, the semiconductor behaves as a lossy metal.

In III-V technology the influence of the substrate is minimized at high frequencies by use of semi-insulating gallium arsenide substrates. In silicon technology no semi-insulating substrate is available. Silicon-On-Insulator (SOI) materials, especially SOS [3] and MICROX (highly resistive SIMOX) [4], offer advantages over bulk silicon at high frequencies. Intrinsic silicon or silicon materials depleted from free charge carriers are expected to be useful as RF substrates. In this paper we report a novel silicon material made by wafer bonding.

Material fabrication

To create an alternative to conventional SOI materials wafer bonding was used to manufacture a depleted silicon material. The novel substrate was made by bonding of medium doped "device layer" wafers to "substrate" wafers of different type and resistivity. Following the bonding procedure, the device wafer was thinned to an approximate thickness of 0.3–0.5 μm. In our case a SIMOX wafer served as the source of the thin silicon film, where the buried oxide was used as an etch stop. The manufacturing procedure is shown in Fig. 2.

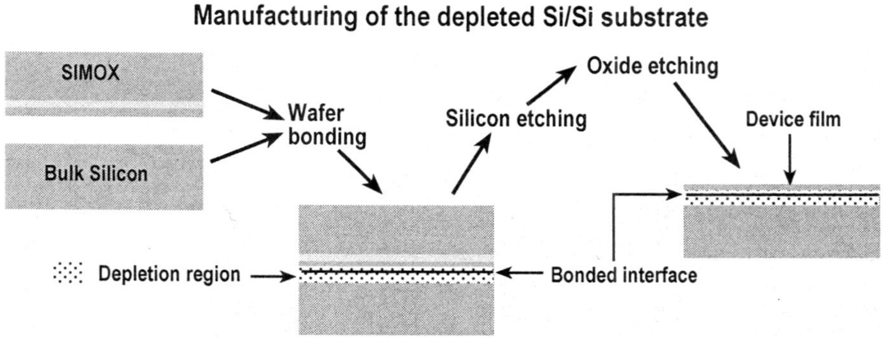

Figure 2. The manufacturing procedure of the semi-insulating depleted silicon/silicon material.

The structure and the band diagram of the material are shown in Fig. 3. The space charge region surrounding the bonded silicon/silicon interface depletes the silicon thereby causing a semi–insulating behavior at high frequencies.

The interface charge of bonded silicon/silicon interfaces and thereby the widths of the surrounding depletion regions are determined by the cleaning procedure used prior to wafer bonding [5]. If the wafers are of different resistivities the depletion region will extend to different depths on the two sides of the bonded junction.

To maximize the width of the depletion region the silicon surfaces were exposed to hot nitric acid before bonding. To compensate for the interface charges a depletion of free charges will occur at the interface.

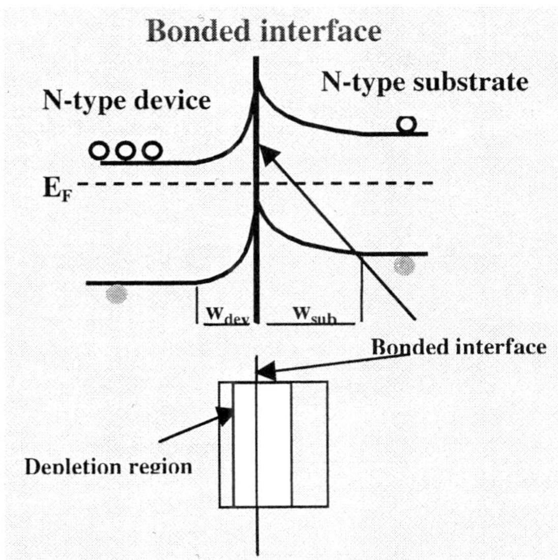

Figure 3. Structure and band diagram of the bonded silicon/silicon material.

Measurements details

The formed material was characterised using measurements on co-planar aluminium transmission lines, shown in Fig. 4 and on crosstalk structures shown in Fig. 5. The metal structure were formed on the surfaces of the materials by deposition and photolithography.

The results were compared to similar measurements on bulk silicon wafers, SIMOX and quartz. Some samples were oxidised before metal deposition. Data on the different samples are given in Table I.

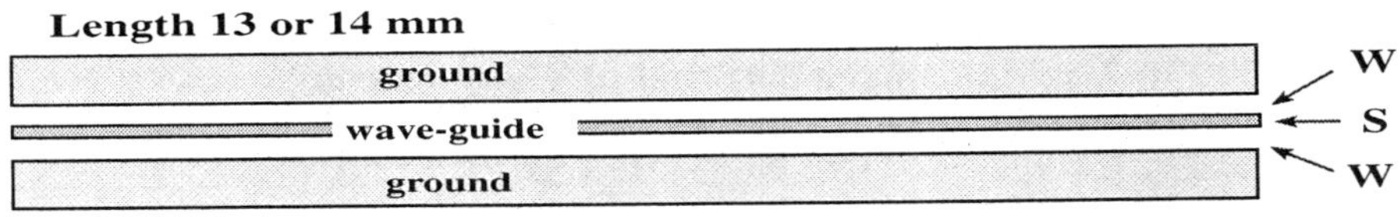

Four different dimensions (in μm)

	Type **a**	Type **b**	Type **c**	Type **d**
S	60	80	40	95
W	40	55	15	22

Figure 4. *Co-planar transmission lines and dimensions. S is the width of the wave-guide and W is the distance between the wave-guide and ground-plane.*

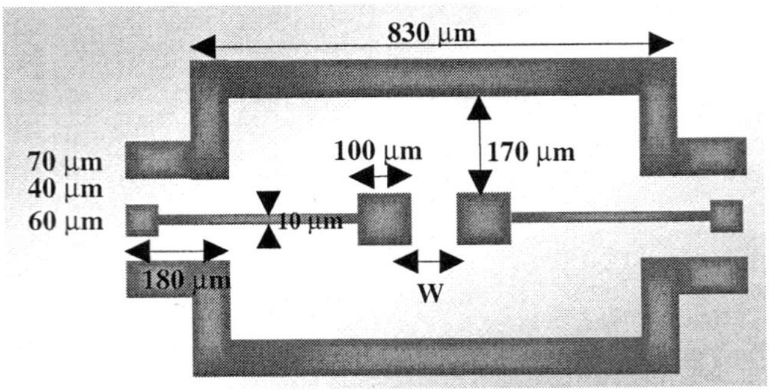

Figure 5. *Co-planar metal structures used to measure crosstalk in the substrates of Table I. The pad separation distance W was 50, 100 or 150 μm.*

TABLE I. Description of samples.

SUBSTRATE	DESCRIPTION
• Medium • Medium + oxide	Medium resistive n-doped substrate 10 Ωcm, without and with 1 μm oxide layer
• High • High + oxide	High resistive n-doped substrate 10 kΩcm, without and with 1 μm oxide layer
• Si/Si (medium) + oxide	Depleted Si/Si substrate, medium resistive n-doped bulk 10 Ωcm, with medium resistive n-type device film 20 Ωcm, with 0.3 μm oxide layer
• Si/Si (high) • Si/Si (high) + oxide	Depleted Si/Si substrate, high resistive n-doped bulk 10 kΩcm, with medium resistive n-type device film 20 Ωcm, without and with 0.3 μm oxide layer
• Si/Si (pn) • Si/Si (pn) + oxide	Depleted Si/Si substrate, medium resistive p-doped bulk 10 Ωcm, and n-type device film 20 Ωcm , without and with 0.3 μm oxide layer
• SIMOX • SIMOX + oxide	Industry manufactured n-doped SIMOX, medium resistive, 20 Ωcm, n-type bulk and device film, without and with 0.3 μm oxide layer
• Quartz	Quartz wafer

The S-parameters of the transmission lines were measured in the frequency range 45 MHz-40 GHz using a HP8510C Vector Network Analyzer with a test signal of about 1 mW power. The Loss-power, L_P, is extracted from the S-parameters by:

$$L_P = 1 - \frac{|S_{21}|^2}{\left(1 - |S_{11}|^2\right)} \tag{2}$$

Where S_{11} and S_{21} is the reflection and transmission coefficient. The total attenuation, α_{tot} is given by:

$$\alpha_{tot} = 10 \cdot \log\left(\frac{1}{1 - L_P}\right) \tag{3}$$

To further investigate the high frequency properties of the novel materials cross-talk structures shown in Fig. 5 were formed on top of the materials of Table I.
The crosstalk is given by the parameter S_{21} which was measured using a HP8510C Vector Network Analyzer. The crosstalk, CT, is simply given by the magnitude of S_{21} (in dB):

$$\mathrm{CT} = 10 \cdot \log\left(|S_{21}|^2\right) \tag{4}$$

To distinguish crosstalk from losses a through-line structure with the same dimensions as the crosstalk structure in Fig. 5 was formed. This structure can be seen in Fig. 6. To compensate for losses in the crosstalk structure, the attenuation in the through-line was evaluated in the same way as for the transmission lines. The compensated crosstalk, CT_C, is thus given by (in dB):

$$\mathrm{CT_C} = 10 \cdot \log\left(\frac{|S_{21}|^2}{1 - L_{PTL}}\right) \tag{5}$$

where S_{21} is the transmission coefficient of the crosstalk structure while L_{PTL} is the loss-power of the through-line structure on the same substrate.

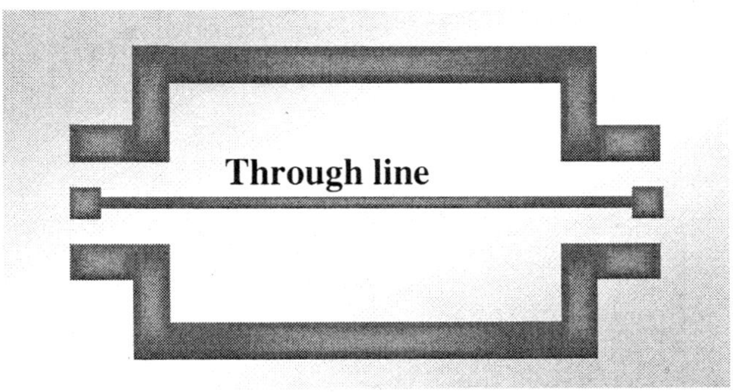

Figure 6. *Short-circuit structures with the same dimensions as the crosstalk structure in Fig. 5 but the pads are dismissed and the wave-guides connected.*

Results and Discussion

In Fig. 7 attenuation coefficients for the transmission lines as a function of frequency are shown for the samples of Table I. The influence of the doping level of the reference samples is clearly seen and as expected the loss decreases with increasing resistivity.

The bonded structures show losses corresponding to what was achieved on SIMOX materials. It can also be seen that the bonded structures consisting of a medium doped device layer on top of a low doped substrate caused small losses in the signal transmission in spite of the medium doped device layer.

It can also be seen that the presence of a SiO_2 layer between the transmission line and the silicon further decreases the losses.

In Fig. 8 crosstalk results are shown for the samples of Table I. For all results in Fig. 8 the distance between the pads was 100 μm.

From Fig. 8 it can be clearly seen that for the lower part of the frequency range (<10 GHz) the crosstalk is larger in medium doped silicon and SIMOX as compared to highly resistive silicon and the bonded sample type Si/Si (pn) + oxide. At higher frequencies the crosstalk increases for all samples and slightly converging. The magnitude of S_{21} obtained for medium doped samples show a

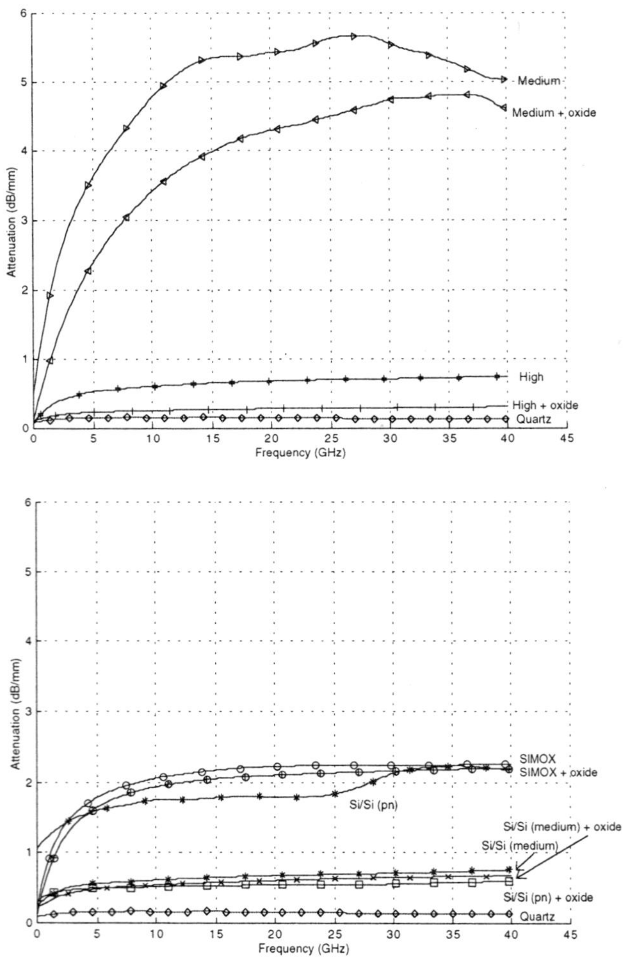

Figure 7. Attenuation in co-planar transmission lines made on the substrates of Table I.

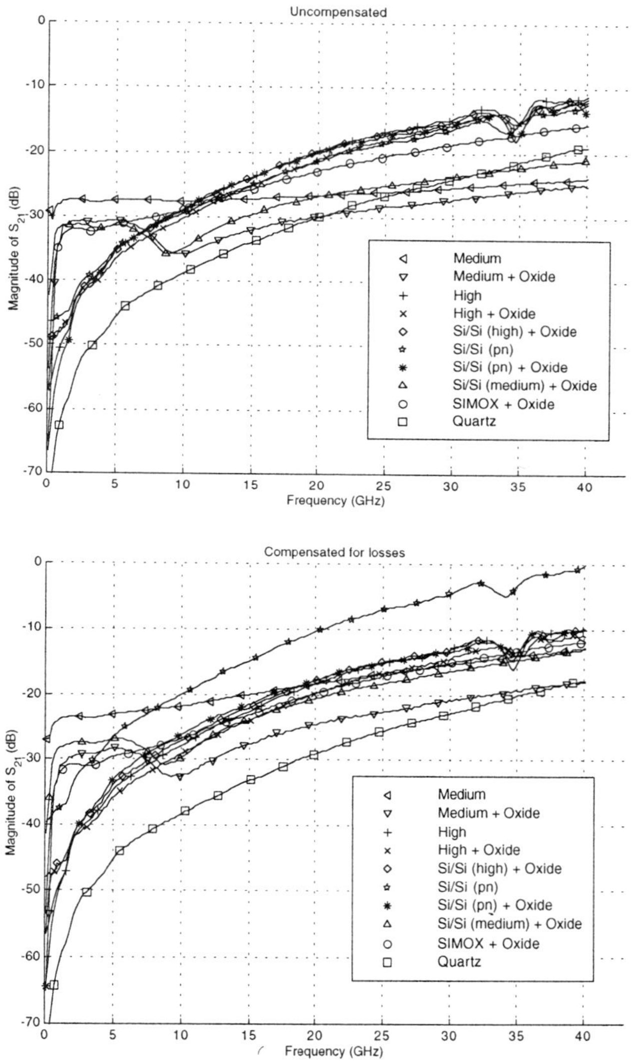

Figure 8. *The magnitude of S_{21} measured for the test structures shown in Fig 5 on the substrates of Table I. The spacing between the metal pads was 100 µm. Figure 8 a shows magnitude of S_{21} while Fig. 8 b shows the crosstalk compensated for losses.*

relatively flat frequency response. This has been explained to be due to a crosstalk dominated by a resistive coupling (in contrast to capacitive coupling) [6].

As can be seen from Fig. 7 the losses varies a lot for the different investigated substrates. An attempt was therefore made to compensate for the losses due to the substrate connecting the crosstalk pads to the pads where the probes were connected (see Fig. 5). To estimate the losses structures of the type shown in Fig. 6 were used. The attenuation for the short-circuit structures on the different substrates can be seen in Fig. 9. The magnitude of S_{21} compensated for the losses in the lines are shown in Fig. 8b. The same behaviour is seen, but relative values of different samples is changed. The crosstalk for the quartz substrate is changed marginally while the value for medium resistive substrate is largely changed. In Fig. 8b quartz show superior behaviour over the entire frequency range, which is expected.

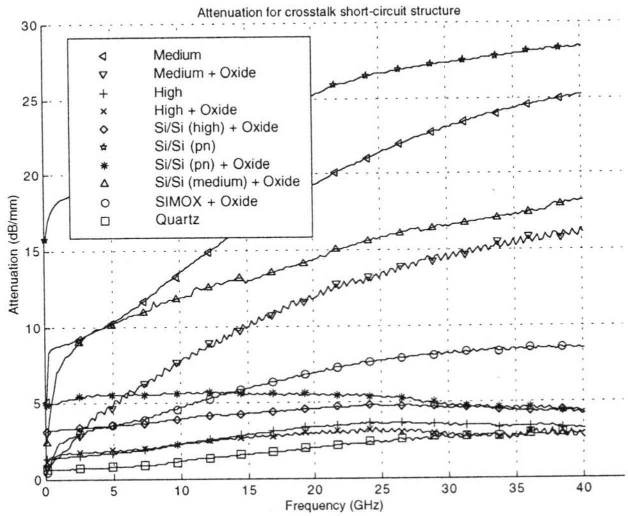

Figure 9. Attenuation for short-circuit crosstalk lines. The attenuation is plotted as dB/mm.

In Fig. 10 the dependencies of the width of the gap on the magnitude of S_{21} (dotted lines) and the compensated crosstalk (solid lines) is shown for two samples. Three different pad separations (W in Fig. 5) have been used, 50, 100 and 150 µm. As can be seen the crosstalk decreases with increasing width as expected. The increased spacing seam to be give a constant downward shift over the hole frequency region, which is in agreement with simulations [6].

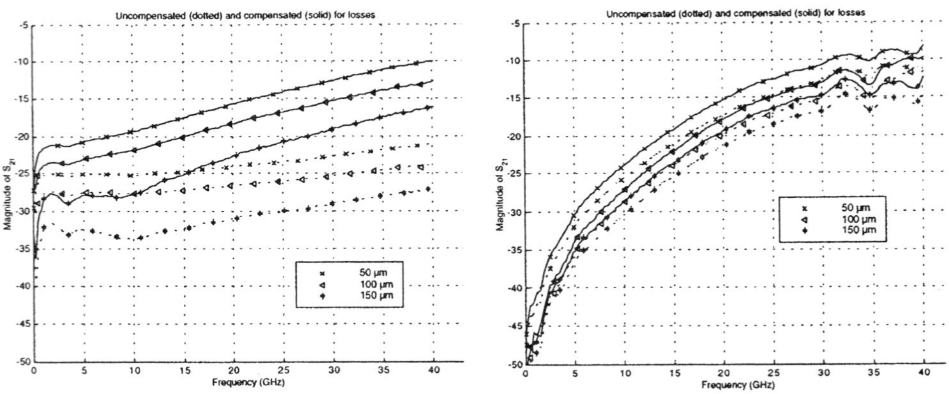

Figure 10. The magnitude of S_{21} for different pad separations 50, 100 and 150 µm. Data is given for the samples "Medium resistive" and "Si/Si (high) + oxide".

Conclusions

The formation of a depleted silicon substrate by wafer bonding and etch-back has been demonstrated. The substrate shows a behaviour that is normally attributed to semi-insulating materials. The wafer bonding procedure is tuned to maximize the interface charge at the bonded interface thereby depleting the surrounding silicon. The use of differently doped wafers in the bonding step allows for different depletion widths on either side of the bonded junction. Electrical characterisation of transmission lines and crosstalk structures made of aluminium on top of these materials shows that they have the potential to be a silicon material for high frequency applications. The novel material behave in its main characteristics as can be expected for a material where the charge carrier concentration has been reduced from its normal value given by the resistivity. Alternative designs of the structure may include a thermally grown or deposited film of silicon dioxide or silicide forming metal particles at the bonded interface.

References

1. S. Gevorgian, Int. J. RF and Microwave *Computer Aided Eng.* **8**, 433 (1998)
2. See for instance: N.W. Ashcroft and N.D. Mermin in *Solid State Physics*, Chapter 1 (Holt-Saunders International Editions, Philadelphia, 1976)
3. R.A. Johnson, P.R. de la Houssaye, C.E. Chang, P.-F. Chen, M.E. Wood, G.A. Garcia, I. Lagnado and P.M. Asbeck, *IEEE Trans. Electron. Dev.* **45**, 1047 (1998)
4. M.H. Hanes, A.K. Agarwal, T.W. O'Keeffe, H.M. Hobgood, J.R. Szedon, T.J. Smith, R.R. Siergiej, P.G. McMullin, H.C. Nathanson, M.C. Driver and R.N. Thomas, *IEEE Electron. Dev. Lett.* **14**, 219 (1993)
5. S. Bengtsson and O. Engström, *J. Appl. Phys.* **66**, 1231 (1989)
6. J-P. Raskin, A. Viviani, D. Flandre and J-P. Colinge, *IEEE Trans. Electron. Dev.* **44**, No 12, December (1997)

SELF-ORGANIZING GROWTH OF SILICON DOT- AND WIRE-LIKE MICROCRYSTALS ON ISOLATED SUBSTRATES

A. I. KLIMOVSKAYA, I. P. OSTROVSKII*, A. A. EFREMOV,
A. V. SARIKOV, S. A. KOSTYUKEVICH
Institute of Semiconductor Physics NAS, Kyiv, Ukraine
**Lviv Polytechnical University, Lviv, Ukraine*

1. Introduction

Practical needs in Si-based superlattices and bulk circuits with high integration have initiated the intensive search for the new processes and technological approaches in SOI technologies (SIMOX, SIMNI, BISOI, including bounded–and–backside–etched and smart–cut technologies, etc.). Various kinds of etching techniques for bulk materials and/or films remain the key step in modern technologies. Etching generates a number of surface defects and gives rise to instabilities of the superstructures. In recent years an alternative resolving this problem has attracted much attention. Alternative methods are based on the ability of matter to self-organise. They involve creation of seeds for crystal growth and growing dot– or wire–like microcrystals, called whiskers. Like in the smart–cut technology, these methods allow growing any type of microcrystals on any substrate. Although self-organising microcrystal growth on substrate (SOGMOS) has been used for the first time many years ago [1-2] and suggested for application in the devices of vacuum [3] and quantum microelectronics [3-4], there are still many unresolved problems in the technology and basic knowledge of these microcrystals.

The present work gives an overview on our recent research in SOGMOS–technology and results on microcrystals characterisation (morphology, crystalline structure, optical properties, surface state).

2. Technology of Superstructure Preparation

The main processes of the used technology are as follows:

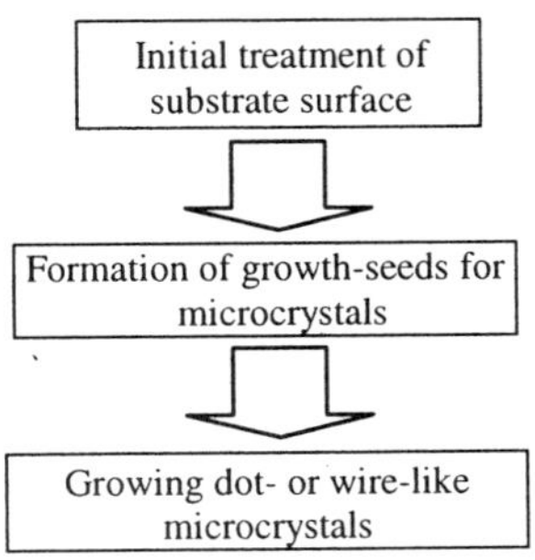

F. Balestra et al. (eds.), Progress in SOI Structures and Devices Operating at Extreme Conditions, 343–348.

The aim of the first process is to remove initial contamination from the surface. If a conductive (silicon) substrate is used, the deposition of amorphous silicon dioxide is applied. The second process consists in fabrication of metal droplets on the substrate surface. For this purpose a number of metals with low melting temperature, e. g. Au, Pt, Cu, Ni, is used. The size and size distribution of formed droplets mainly influence the parameters of grown microcrystals. Therefore, second process is one of the key processes in the technology of superstructure preparation.

There are several methods being used for droplets formation:

- deposition of thin metal film on the substrate surface with subsequent heat treatment or electron/ultraviolet lithography;
- use of the gas–phase reactions similar to those during crystal growth [1-2].

All these methods enable producing droplets with sizes in the range of 20 to 5000 nm. To form smaller droplets the additional etching is needed. The methods involving lithography yield a controllable site and size droplets distribution. Therefore, although expensive, these methods nonetheless are very important in a production of microelectronic devices integrated in Si–based IC.

The most attractive procedure is based on the ability of matter to self–organize, i. e. on the above–mentioned methods but without lithography.

The next process of the technology is growth of microcrystals by vapour-transport reactions in sealed tube containing bromine (or iodine) and the initial silicon:

$$2Si + 2Br_2 \Leftrightarrow 2SiBr_2 \Leftrightarrow Si + SiBr_4. \tag{1}$$

To provide these reactions with precipitation of pure silicon on the substrate, the temperature gradient is to be created in the reaction tube. In the hot part of the tube the reaction with output of volatile compound $SiBr_2$ takes place. Due to the temperature and concentration gradient this compound diffuses towards the colder part where it decomposes with an output of pure silicon and new volatile compound $SiBr_4$. Due to the drastic difference in adhesion coefficients to metal and dielectric, the rejected silicon atoms precipitate on metal droplets and the growth of microcrystal occurs. The new compound $SiBr_4$ diffuses towards the hot part of the tube where it transforms back into $SiBr_2$. This process repeats resulting in the delivery of silicon atom to the growing crystals. The investigations of influence of growth conditions on the size distribution, morphology, and composition [5-6] have shown that controllable distribution can be achieved under definite conditions by means of self-organisation processes.

3. Dependence of the size distribution function on growth conditions

The comprehensive analysis [7] of the dependence on growth conditions of the size distribution function was performed. In simulation of growth process two approaches were used: continuous and discrete.

In the first approach, the evolution of the distribution function F(d, t) is described by the Focker-Plank equation:

$$\frac{\partial F}{\partial t} = -\frac{\partial}{\partial d}\{\upsilon(d) \cdot F(d,t)\} + \frac{\partial}{\partial d}\left\{D(d) \cdot \frac{\partial F}{\partial d}\right\} \tag{2}$$

Here $v(d) \propto d$ is a phenomenological coefficient similar to velocity in a space of sizes. It describes the irreversible atoms capture (due to supersaturation) by growing microcrystal. D(d) is a diffusion coefficient corresponding to the capture with a following re-evaporation. Both coefficients were calculated in microscopic model taking into account the parameters of the growth process.

The second approach is based on the solution of a system of equations for a N_k (t) clusters with definite number of particles. The clusters interact with a gas medium by capture (release) of particles with the probability w_k which depend on the cluster size:

$$\frac{dN_k(t)}{dt} = -w_k N_k + w_{k-1} N_{k-1} + w_{k+1} N_{k+1}, \qquad k = 1, 2, 3, \ldots$$

The numerical estimations on both equation sets have given good agreement with the experimental results.

In Fig. 1 the size distribution functions of microcrystals grown at different conditions with Au growth-seeds created by the method of gas-phase reactions during the crystal growth are shown.

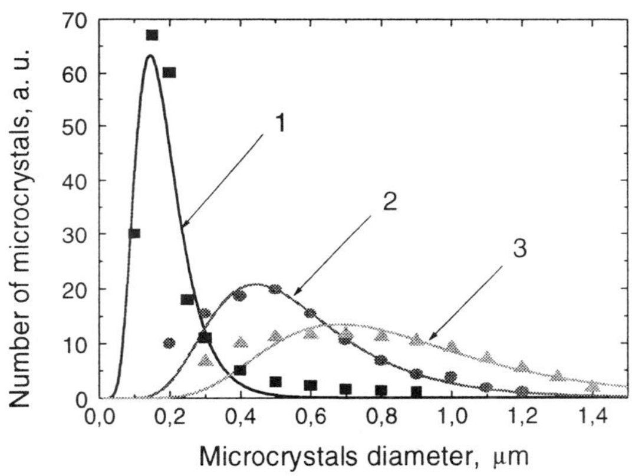

Figure 1. Time evolution of distribution function of microcrystals grown during different times (1, 2, 3) under similar growth conditions. Dots are experimental data, solid lines represent theoretical dependences.

Good agreement between theoretical and experimental distributions suggests that the process of self-organisation may be controlled and predicted results may be obtained. The smallest size distribution is obtained at smallest growth times. But the latter depends on the growth conditions created in the tube, therefore, it can be controlled. Self–organisation process appears to have a considerable potential in a production of silicon superstructures.

4. Morphology of the whiskers grown on substrate

By controlling the growth of whiskers one may not only change the size distribution, but also the crystals shape and structure. Typical shapes of whiskers grown under different conditions are shown in Figs 2 to 5.

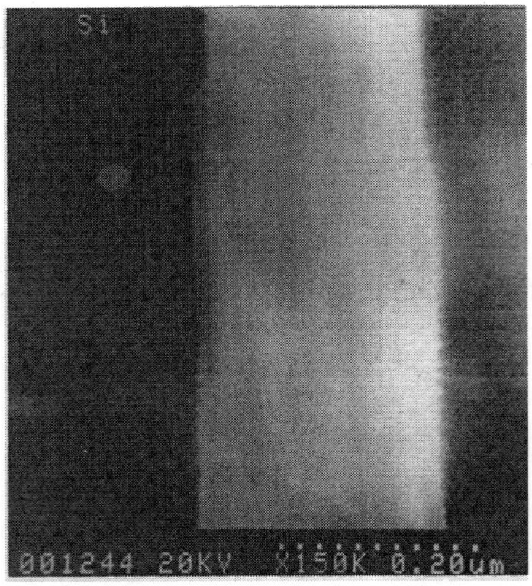

Figure 2. Whisker with regular prismatic shape.

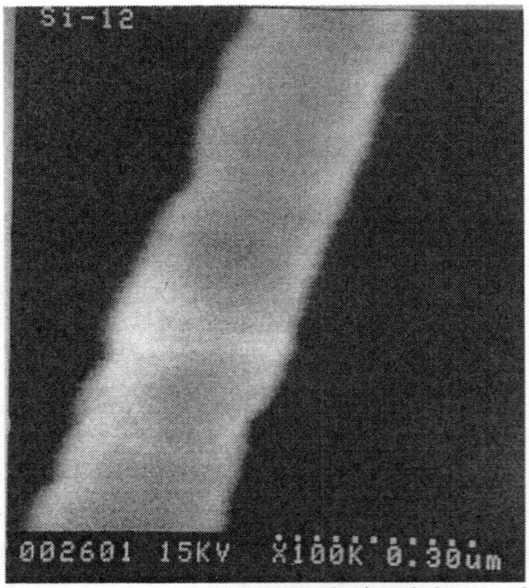

Figure 3. Whisker with quasi-cylindrical shape and longitudinal periodicity.

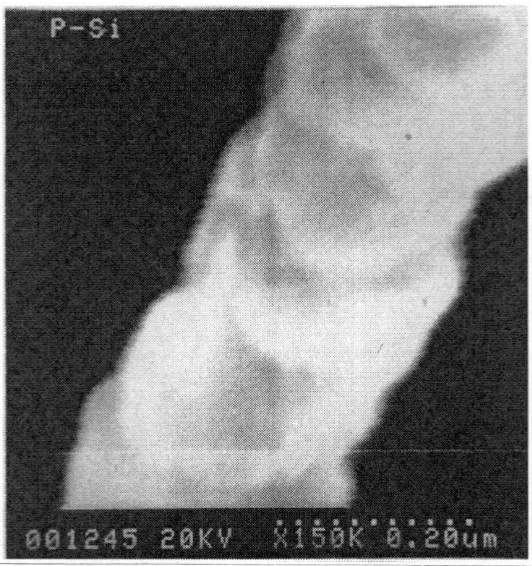

Figure 4. Whisker with semi-spherical nuclei arranged periodically on external surface.

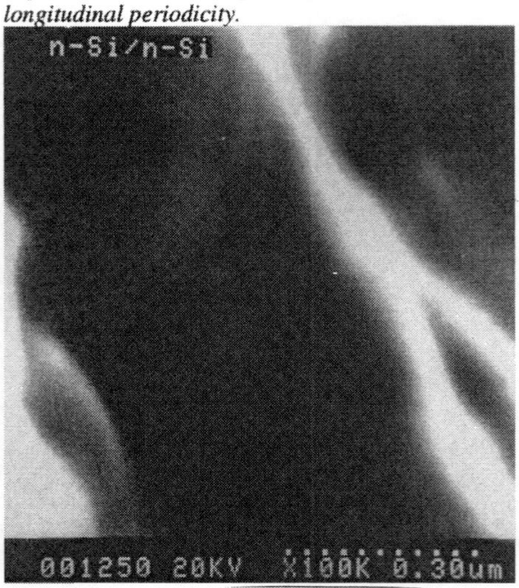

Figure 5. Whisker with irregular bends on the surface.

Crystals presented in Figs 2 to 4 were grown at low density of the growth-seeds on the substrate while crystal shown in Fig. 5 was grown under quite the opposite condition. Both cases of the crystal growth are presented in Figs 6 and 7, accordingly. In the latter case the distance between whiskers is smaller than their radius of a depletion layer induced by Gibbs-Thomson effect during the growth of crystals in vapour. This invokes the competition between neighbouring crystals to accumulate atoms from the vapour. If the distance along the big crystal axis varies and is comparable with the depletion length, the shape of grown crystals is similar to that shown in Figs 5 to 7.

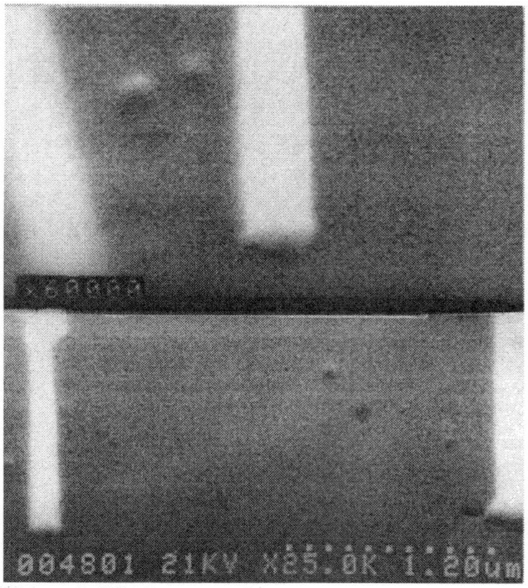

Figure 6. Whiskers on substrate.

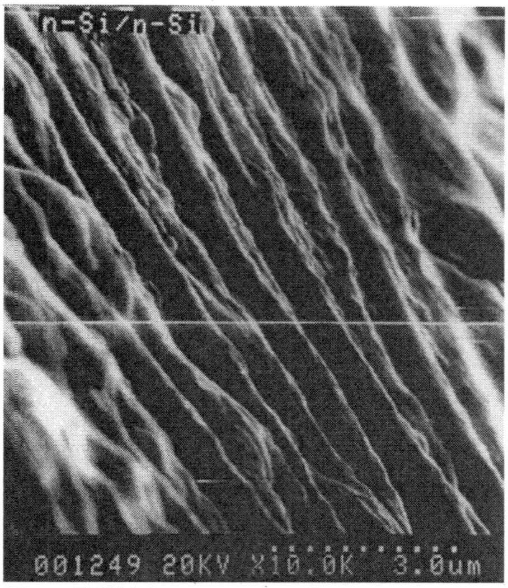

Figure 7. Whiskers grown on substrate with high density of growth-seeds.

5. Crystalline structure of whiskers

The crystalline structure of whiskers has been studied by X-ray double crystal diffraction, electronography and transmission electron microscopy. It was found that the crystalline structure depends on the thickness of whiskers, their length, and growth conditions.

Central parts of the whisker always present diamond-like strained lattice [8,9]. If the whisker length is much greater than its diameter, the whisker represents free–standing wire, where acting Laplace force $F_s = 2\sigma / d$ (with σ being the surface tension energy, d – whisker diameter). As a result, the bulk part of the crystal is usually compressed. On the other hand, in short whiskers on substrate the influence of the substrate is very large. Since a state of the substructure with a smaller surface is preferred to the state with a smaller free energy, substructure tends to minimise a surface area. Therefore, short whiskers on substrate are tensile.

Upper layers of whiskers are expected to form new reconstructed lattice, similarly to the reconstruction of plane surface [10]. The photoluminescence studies of the whiskers [11] have shown that the excitonic spectra of whiskers transform from those of bulk silicon to completely new spectra as the whisker diameter decreases. The critical diameter for such a transformation was found to be smaller than 100 nm. Therefore, photoluminescence studies suggest that the outer reconstructed layer has the same value. But such an explanation could call questions due to the large value of the layer thickness. According to the conventional viewpoint on surface reconstruction in a semi–infinite crystal approximation, the structure changes decay sharply as the distance from the surface increases and are very small already by one third to fourth the subsurface layer. On the other hand, in microcrystals of a quasi-cylindrical shape the role of the surface must be enhanced because of the greater ratio of the number of surface atoms to the number of atoms in the bulk part of the crystal. Moreover, in the crystals with a curved surface the Laplace forces act. All this should lead to an increase of the number of reconstructed atomic layers, as well as of their influence on inner part of microcrystal which should be stronger than that in crystals with a plane surface.

Recent study of the whisker structure by high–resolution transmission electron microscopy has strengthened the above-discussed assumption on a complex structure of the whiskers. They do consist of central bulk part of diamond-like silicon and extended reconstructed layer. The thickness of this reconstructed layer ranges from 50 to 300 nm depending on whiskers growth conditions.

6. References

1. Sandulova, A.V., Bogoyavlenskaya, P.S., Dronyuk, M.I., (1964) Patent USSR No. 160829, Prior. 06.07.1962. *Bull. Izobretenii I Otkritii* **5**.
2. Wagner, R.S., Ellis, W.C., Arnold, S.M., Jackson, K.A. (1964) Study of the filamentary silicon crystals grown from the vapor, *J. Applied Physics* **35,** 2993-3000.
3. Katsuyama, T., Morgan, G.P., Ogawa, K., Himura, K., Kakikayashi, H. (1990), in *Proceedings of 17th International Quantum Electronics Conference,* IEEE, New York.
4. Givargizov, E.I. (1975), *J. Crystal Growth* **31**, 20.
5. Klimovskaya, A.I., Ostrovskii, I.P., Ostrovskaya, A.S. (1996) Influence of Growth Conditions on Morphology, Composition, and Electrical Properties of n-Si Wires, *Phys. Stat. Sol.(a)* **153,** 465-472.
6. Klimovskaya, A., Ostrovskii, I., Gule E., Prokopenko, I. (1999) Self-organized growth of Si/Si and $Ge_xSi_{(1-x)}$/Si superlattices, in *Proceed. 17th Intern. Quant. Electronics Conf.(MIEL'2000),* Nis, Yugoslavia.
7. Klimovskaya, A.I., Ostrovskii, I.P., Efremov, A.A., Sarikov, A. (2001) Controlled Self-Organising Growth of Dot- and Wire-Like Microcrystals, *in print.*
8. Klimovskaya, A.I., Ostrovskii, I.P., Baitsar, R.I, Ostrovskaya, A.S. (1995) X-ray study of free-standing filament crystals of $Ge_x Si_{(1-x)}$ solid solution, *J. Phys.: Condens. Matter* **7**, 1229-1234.
9. Klimovskaya, A.I., Ostrovskii, I.P., Fadeyev, S.V., Fruginskii, M.S., Rudyi, I.O., (1998) Lattice parameter change in submicron Si whiskers, *Thin Solid Films* **318,** 84-86.
10. Bechstadt, F., Enderlein, R. *Semicoductor Surfaces and Interfaces. Their Atomic and Electronic Structures* (Akademie-Verlag, Berlin, 1988).
11. Gule, E.G., Rudko, G.Yu., Klimovskaya A.I.,Valakh, M.Ya., Ostrovskii, I.P. (1997) Visible light emission from free-standing filament crystals of silicon, *Phys. Stat. Sol (b)* **161**, 565.

AUTHOR INDEX